T0252187

Fly by Night Physics

How Physicists Use the Backs of Envelopes

A. Zee

PRINCETON UNIVERSITY PRESS   PRINCETON AND OXFORD

Requests for permission to reproduce material from this work
should be sent to permissions@press.princeton.edu

Published by Princeton University Press
41 William Street, Princeton, New Jersey 08540
6 Oxford Street, Woodstock, Oxfordshire OX20 1TR

press.princeton.edu

Library of Congress Control Number: 2020942704

ISBN 978-0-691-18254-4
ISBN (e-book) 978-0-691-20773-5

British Library Cataloging-in-Publication Data is available

Editorial: Ingrid Gnerlich and Arthur Werneck
Production Editorial: Karen Carter
Text Design: Lorraine Doneker
Jacket/Cover Design: Karl Spurzem
Production: Jacqueline Poirier
Publicity: Matthew Taylor and Amy Stewart
Copyeditor: Cyd Westmoreland

This book has been composed in Sabon with ITC Stone Sans for display

Printed on acid-free paper. ∞

Printed in the United States of America

10 9 8 7 6 5 4 3 2 1

To the memory of my late colleague* Joseph Polchinski

---

*I feel sure that Joe would have liked this kind of physics.

Contents

Preface

Handwaving, horseback, and back of the envelope physics

Never never calculate unless you already know the answer!
[John Wheeler to a group of freshmen[1] he handpicked]

My first week in college, before my first physics homework assignment* ever,[2] I heard those shocking words, shocking to a naive freshman who thought that theoretical physicists were supposed to calculate till smoke came out of their ears. That was in an experimental (in the sense of pedagogy) course[3] John Wheeler was never allowed[4] to teach again (or so I heard).

The kind of physics Wheeler wanted to instill in these hapless freshmen is also known variously as handwaving physics, horseback[5] physics (the term that Wheeler used), and back of the envelope[6] physics. Wheeler's point was of course that before madly calculating, one should cultivate the habit of thinking through to the answer, or failing that, guessing the answer.

Since few physicists these days are in the habit of jumping on a horse and galloping off, I choose to call this kind of physics "fly by night physics." Instead of equine transport, a long bus ride or a sleepless night imprisoned in a jet, without pen or paper, could work just as nicely.[7] The absence of writing material sharpens the brain, and often, with no possibility of worrying about the 2s and the πs, allows one to get to the heart of the problem, to see the forest rather than the trees.

---

*Turned out it was standing in silence for five minutes in front of the house Einstein (by then long dead) had lived in.

For whom this book is intended

I address this book to advanced undergraduates, those wondrous persons with some exposure[8] to the four core subjects of physics: classical mechanics, quantum mechanics, electromagnetism, and statistical physics. But I hope that others will also find useful stuff here. Certainly, it is up to you to define an advanced undergraduate. You don't even have to be an actual undergraduate. From the emails I have received about my three previous textbooks, I have learned that many of my readers are autodidacts long out of school, and a few are autodidacts long before they even get to college. If you feel your mastery of physics is at the level of an advanced undergraduate, then you are ready for this book.

Genesis of this book

I had been thinking of writing this book for quite a long time. Several years ago, I proposed a special topics course on fly by night physics to my physics department. My colleague in charge of the undergraduate curriculum, Harry Nelson, was enthusiastically supportive. When I taught the course for a second time, with a better sense of what undergraduates at my university were capable of and interested in, I decided to start writing. By the time this book is published, I would have taught the course a fourth (!) time. Thus, input by flesh and blood undergraduates[9] has played an important role. By now, I have a much better idea of what advanced undergraduates are capable of.

I hired an undergraduate assistant to help me with this book. Ashley, in her words: "Undergraduate physics students crave direction, but have been directed to think of physics as a set of equations to memorize and manipulate. But this book reveals to them the importance of understanding the equations so that they could make an educated guess, flying by night to arrive at an answer."

The role of precise calculations

> I understand what an equation means if I have a way of figuring out the characteristics of its solution without actually solving it. [Paul Dirac as quoted by Richard Feynman]

I hope that no reader thinks that I do not like fly by day physics. Of course I do. I revel in it. And I have even written three textbooks about it. Indeed, I acknowledge the failure of fly by night physics when it happens, such as in chapter VI.3.

Surely you do not think that I advocate not doing exact calculations.[10] Of course you should if you could! But even if you could, do a back of the envelope calculation first and get the essential physics right. That will guide you in your exact deliberations and provide a reality check of your answer. At the end of a precise calculation, see if you could recall where every 2 and π came in. That would eventually endow you with a sixth sense, enabling you to throw in some 2s and πs even while flying by night.

So, this is not just a book about getting order of magnitude estimates. Actually, one of my purposes is to show the reader that a bit of fly by night physics can help you understand some fairly profound physics, such as Hawking radiation from black holes or the two-neutrino hypothesis. I include these more "cutting edge" topics[11] because I know well that many undergrads are eager to go beyond electromagnetism and the Bohr atom. I also want to impart another lesson. Silly to do a precise calculation when the theory is not yet fully formulated! Conversely, nobody is impressed by your ability to calculate precisely when the theory is already in textbooks.[12]

Undergraduates tend to think of physics as logically and precisely derived. By necessity and for good reason, the undergraduate curriculum presents physics as tightly reasoned. But still, too much calculating, too little understanding! I realize, of course, that understanding often can only come from doing detailed calculations repeatedly. All too often, however, homework exercises merely train undergrads, even the better ones, to look through the textbook chapter just covered, search for a likely looking equation, plug in, and obtain a precise answer, which almost any dodo can do.

Homework versus research

But homework is not the same as research.[13]

Much of physics research involves applying known laws to novel situations. I am not denigrating this wing of theoretical physics by any means. Applying known physics often requires much ingenuity and insight. But suppose you don't know the laws. Particle theory in the 1950s and 1960s offers some spectacular examples of successful grasping at straws. As another example, in part V, I tried to imagine, after the Planck distribution was proposed, how one might have guessed the Fermi-Dirac and Bose-Einstein distributions. (The correct derivation is of course given in any standard textbook on statistical physics.)

The processing space freed up by not calculating exactly may allow you to make an important discovery. Let Planck be your inspiration! He didn't derive his eponymous distribution for black body radiation. How could he? He did it way before quantum mechanics, let alone quantum statistics.

In some areas of physics, for example astrophysics, even though the underlying laws are largely known, it is impossible to nail down all the input variables. Then fly by night physics is not only a good idea, but a necessity.

Even if you are not working at the edge of the unknown, a rough guess, followed perhaps by an order of magnitude estimate, suffices in many situations, or if not sufficient, is better than nothing.

Nature of this book

Perhaps I could best define this book by what it is not. Not a textbook, certainly. Instead, it tries to convey to you some of the skills needed to forge a path in the dark through unknown territory, rather than to guide you through a well-lit city with street signs. This sort of skill is particularly essential on the forefront of research, when one is literally in the dark.

When I mentioned that I was working on a book about fly by night physics, many colleagues immediately assumed that it would be about Fermi problems, perhaps the best-known example of which is to estimate the number of piano tuners in Chicago.* The secret to success in tackling such problems is to have a large number of factors, so that an overestimate for one factor might be cancelled by an underestimate for another. Another example might be to estimate the number of potato chips eaten per year around the world. The general public is often amazed that physicists are able to come up with rough estimates, within the ballpark at least. But typically, little or no physics is involved. Because of this, I do not cover this type of problems as much. Besides, fine books already exist on the subject.[14]

I have of course looked at a few, but emphatically not all, books that purport to do back of the envelope physics. Some I like,[15] some I don't. Let me give you an example from the latter. One book asks the reader how long it takes a tsunami originating in Asia (say, Japan) to reach the west coast of North America. The book then says, here is the formula for the speed of water wave, here is how wide the Pacific Ocean is, and plugging in these numbers, here is the answer.

I would not have liked it as a student.[16] I would have wanted to know where the formula came from.

I have to assume that advanced undergrads know how to plug in numbers. Indeed, when I taught the course that this book is based on, students in my class did not want me to spend valuable time doing numbers.[17] Some even felt insulted[18] when asked if they could plug in numbers.[19]

I am more interested in a fly by night derivation or guess of the relevant formula than in plugging numbers into a given formula. No doubt some people would have liked this book more if it had more numbers in it.

To summarize, this book in its emphasis and style differs from existing books in significant ways. It does not teach you physics already taught in standard textbooks, does not help you solve standard physics homework problems, does not glorify plugging in numbers for the sake of obtaining a number, and does not subscribe to the common notion that the ability to manipulate loads of formalism amounts to understanding.

In any case, I believe that this book is unique in its breadth, ranging from the pendulum to the charm quark, from water striders to the atomic bomb.

---

*One would have to multiply numerous factors together (such as the number of households in Chicago, the fraction of households with a piano, how often does a piano need to be tuned, and so on) and give reasonable estimates for each factor.

Joy and fun

In an authoritative review of my quantum field theory textbook in *Physics Today*, Zvi Bern, a distinguished quantum field theorist at UCLA, opined that the thought foremost in my mind was how to make quantum fields "as much fun as possible." What a mind reader Zvi was! Yes, physics should be fun.

Indeed, in writing my three previous textbooks,* I tried to recapture the joy I felt in learning quantum field theory, Einstein gravity, and group theory. So, what am I recapturing in this book? The joy of learning that physics is capable of describing the real world! The emphasis in this book is on fun, on the joy of understanding, not on technical mastery, and certainly not on the ability to multiply and divide.

So, sorry, this book is not for the students shaky in their mastery of the basics and hanging on by the skin of their teeth. You could hardly have fun riding if you are petrified that the horse will slam you to the ground at any moment.

From talking to advanced undergraduates and from my own student experience, I believe that many physics students, by the time they reach their junior or senior year, feel ground down by Bessel functions, magnetostatics, and the like, and are rearing at the bit to learn more exciting stuff, from quantum field theory to black holes. I hope that this book, besides deepening their understanding, also serves to fan their enthusiasm. Thus, I debated for a long time whether to include part IX, consisting of a whiff of particle physics. Many students did ask for it, and so in the end I did include it, partly for the reason given here. Indeed, some readers might want to read the prologue to part IX now.[†]

Put simply, this is the book I wished I had when I was an undergraduate.

How I chose and ordered the topics

For a textbook, the choice and order of material is fairly fixed. In writing *QFT Nut*, for instance, I knew that some topics need to precede others. In contrast, the nature of this book is such that I could choose and pick from a wealth of topics, and order them however I like. The freedom of choice is a challenge as well as an opportunity. Ultimately, I am guided mostly by asking what I wanted to learn when I was an advanced undergraduate, and by my contact with undergraduates. It is probably accurate to say that most physics undergrads are more interested in black holes and Hawking radiation than in, say, viscous flow.

So my choice of topics is totally idiosyncratic. I can already hear readers wondering, "Why did he talk about this, but not that?" No particular reason, I just feel like it. I also include[20] bits and pieces that totally do not belong, such

---

Quantum Field Theory in a Nutshell, *Einstein Gravity in a Nutshell*, and *Group Theory in a Nutshell for Physicists*, all published by Princeton University Press. Throughout I will refer to these three books as *QFT Nut*, *GNut*, and *Group Nut*, respectively.

[†]Indeed, in fall 2020, I had available the table of contents of this book to distribute to the students in my course. While I was finishing up part I, a few of them asked if we could skip to part IX right away. I said no.

as Feynman's identity, because I really liked being shown stuff like that when I was an undergrad. Thus, the three math medleys were thrown in this book for fun. Students in my course liked these. Still, this is a book about physics, not math tricks.

In teaching this course, I felt obliged to review quickly some basic materials from classical mechanics, quantum mechanics, electromagnetism, and statistical physics that the students allegedly knew. These are included in appendices at the end of the book, which also serve to establish notation and convention.

How much you get out of this book evidently depends on how much you know. Again, by the nature of this book, you don't have to read it from cover to cover. So it is perfectly okay to pick and choose what interests you.

A word about the exercises

Some readers might be inclined to criticize this book for its relative lack of exercises. Of course I could have generated an arbitrary number of exercises requiring merely the plugging in of numbers. But as was already mentioned above, not only I, but also the students in my course, feel that such exercises teach relatively little. (The reader is of course invited, indeed encouraged, to plug numbers into any and all of the fly by night results we obtained to "get a feel" for the orders of magnitude involved.) However, my experience indicates that only a handful of exceptional students are capable of deriving a fly by night result from scratch without my providing ample background material and guidance. But by the time I encumber an exercise with a detailed explanation and various nudges and hints, I might as well include the example in the text.

Consider the two exercises in chapter II.3, one asking for a numerical estimate of the Thomson cross section for the scattering of an electromagnetic wave off an electron, the other soliciting a guess of how this cross section would behave when quantum effects kick in. The first, besides plugging in numbers, also requires the student to recognize what value to use for e^2, and so is educational to some extent. But the second exercise actually prods the student to develop some physical intuition, which, after all, is one of the goals of this book.

A word to the rigorous

Friends tell me that the prefaces to my three previous textbooks are a bit combative. Darn well they are if I am to keep at bay the rigor mortis tribes[21] that patrol the Amazon! Finally, I get my chance to talk back. For this book, ha ha! The lack of rigor is the name of the game.* The discussions are sure as eggs not rigorous: they are not even exact!

---

*Of course, rigorous theorems have an important role to play in physics. I even invoke a few, for example, in chapter III.1 and appendix N.

A word to the wise

So, the bottom line that concerns some readers of this book: How does a prospective physicist develop intuition?

Probably the truly honest answer is that you are born with it. Like Mozart. But a competing honest answer is that you practice, practice, and then practice some more. Like Hendrix. And this book is an attempt to provide a practice guide.

So, let the physics begin!

Acknowledgments

I am pleased to thank Nima Arkani-Hamed, Yoni BenTov, Mark Bowick, Brandon Brown, Matteo Cantiello, Emily Jane Davis, Eric DeGiuli, Joshua Feinberg, Matthew Fisher, Sheldon Glashow, David Gross, Dan Holz, Greg Huber, Shamit Kachru, Pavel Kovtun, Paul Krapivsky, Ian Low, Harry Nelson, Ashley Ong, Sandip Pakvasa, Rafael Porto, Vikar Qahar, Srinivas Raghu, Eva Silverstein, Douglas Stanford, Arkady Vainshtein, David Wang, and Tzu-Chiang Yuan for commenting on various chapters, offering encouragement, and generally being of good cheer. Feedback from students has been particularly helpful. Computer help by Ricardo Escobar, Alina Gutierrez, and Craig Kunimoto proved to be indispensable. Ingrid Gnerlich, who has worked on all my Princeton University Press books, has been enthusiastically supportive since the inception of this project. Once again, she has entrusted the copyediting to my long-time collaborator Cyd Westmoreland, who, in spite of our ongoing minor disagreements over punctuations and such minutiae, has been a tremendous help. I also thank Karen Carter for overseeing the production process. Finally, I have benefited from the support and presence of Janice and Max.

Notes

[1] Yours truly among them.

[2] Since I didn't have the opportunity to take physics in high school.

[3] I learned later that Wheeler, having heard that at Caltech Feynman was giving an experimental course (which eventually turned into the big red books), wanted to do the same at Princeton.

[4] Too many bongs of the bell and egg crates (a joke for those who know what I am referring to).

[5] Although I did not learn to ride a horse till years later, when I attended the Boulder summer school on physics run by the University of Colorado, I understood immediately that the bumpy ride would make it difficult to calculate with pen and paper.

[6] In the old days, neither good nor bad, physics professors received an enormous amount of physical mail, and hence there were always a few envelopes at hand to scribble on. Next to the mailboxes in the Princeton physics department, as I recall, there was a table for discussion and a gigantic trash can to which one could jettison one's mail without opening it, or after opening it and saving the precious envelopes.

[7] Other experiences during which I have enjoyed doing fly by night calculations in my head include interminable faculty meetings when some people fell in love with the sound of their own voices, dull children's basketball games, and bouts of insomnia. Surely the reader could come up with some others.

[8] But who are not yet fossilized, if I may say so.

[9] I discourage the half-dead by announcing that class participation counts toward the final grades.

[10] I often teach the group theory course, sometimes back to back with the fly by night course. Some students take both, and so have to switch brains over a period of a few months. In many situations, nothing short of an "exact" answer is required. If an experimentalist wants to know the ratio of the production cross sections for $p + p \to d + \pi^+$ and for $p + n \to d + \pi^0$, you can't just tell him or her that the ratio is order 1 by dimensional analysis. (See *Group Nut* page 306 for the answer.) There are certainly limits to flying by night.

[11] It is rather sobering to realize that these topics are already either close to or more than a half century old!

[12] See chapter IX.3. The curious reader might want to read endnote 4 there now.

[13] Consider the phenomenon of people who do fabulously well in courses but fail at producing original research.

[14] For example, L. Weinstein and J. Adam, *Guesstimation: Solving the World's Problems on the Back of a Cocktail Napkin*, Princeton University Press, 2008.

[15] A few decades ago, Ed Purcell, whom I admired enormously, had a column in the *American Journal of Physics* called "Back of the envelope physics." For example, one problem is whether a helium balloon filled with the helium gas from a tank of liquid helium can lift that tank. See chapter II.3 in this book for an example of back of the envelope physics at its best. It does not involve plugging in numbers!

[16] I am paraphrasing some of the blurbs about my textbook on quantum field theory, *QFT Nut*.

[17] As an example, asking you to derive the attenuation rate of infrared sound (see exercise (4) in chapter VI.2) surely exercises your brain more than asking you to plug various frequencies into a formula given to you for the attenuation rate.

[18] People underestimate what undergrads are capable of; I know that I do. In my classes at the University of California, Santa Barbara, those who have just transferred from community colleges, and hence the most eager beavers, often have surprised me.

[19] Whether everybody in the class is capable of even doing this is of course also open to doubt, but those who can't soon drop out.

[20] But then lots of material ended up on the cutting room floor, alas. However, I felt much less regret after my editor Ingrid stated that a sequel is certainly possible.

[21] I would have some of these ignorant people know that I actually came out of the rigorous tradition, having written my senior thesis on axiomatic field theory with Arthur Wightman. While the jungle complains of the lack of rigor in my books, leading mathematicians and mathematical physicists have reviewed them favorably.

Dimensions and fundamental constants

As explained in chapter I.1, I use the square bracket $[X]$ to denote the dimension of a physical quantity X in terms of M, L, and T.

Temperature is denoted by T; in most cases, confusion with T, the unit for time, seems unlikely. Nevertheless, I often alert the reader.

Often, instead of M, it is more convenient to use E, the unit for energy (not to be confused with the electric field $\vec{E}$ of course).

Dimensions

A couple of examples:

Boyle's law $P = nT$: $[P] = E/L^3 = [n][T] = (1/L^3)E$.

Newton's law of gravity $E = -GM_1M_2/r$: $[E] = [G][M_1M_2]/[r] = M(L/T)^2$. Hence $[G] = L^3/MT^2$.

$[E] = M(L/T)^2 = ML^2/T^2$

$[g] = (L/T)/T = L/T^2$

$[P] = (ML/T^2)/L^2 = M/(LT^2) = (ML^2/T^2)/L^3 = E/L^3$

$[T] = E$ temperature

$[D] = L^2/T$ diffusion constant

$[\mu] = FT/L = ET/L^2$ coefficient of friction

$[e^2] = EL = ML^3/T^2$

$[e] = M^{\frac{1}{2}}L^{\frac{3}{2}}/T = E^{\frac{1}{2}}L^{\frac{1}{2}}$.

$[\vec{E}] = [e/r^2] = (M/L)^{\frac{1}{2}}/T$

$[\vec{B}] = [e/r^2] = (M/L)^{\frac{1}{2}}/T$

$[eE] = ML/T^2$

$[E^2] = [B^2] = M/LT^2 = (ML^2/T^2)/L^3 = E/L^3$

$[\vec{A}] = M^{\frac{1}{2}}L^{\frac{1}{2}}/T$

$[e\vec{A}/c] = ML/T$

$[G] = L^3/MT^2$

$[\omega] = 1/T$

$[\vec{k}] = 1/L$

$[\hbar] = ML^2/T = ET$

$[c] = L/T$

$[\hbar c] = ML^3/T^2 = EL$

$[\nu] = L^2/T$ kinematic viscosity

$[\mu] = M/LT$ dynamical viscosity

$[\gamma] = E/L^2$ surface tension

$[L] = E/T$ luminosity

Natural or Planckian units

$[\varepsilon] = [\rho] = M^4$ energy density or mass density

$[P] = M^4$ pressure

$[S] = 1$ entropy

$[G] = 1/M^2$

$[GM] = L$ black hole radius

Particle physics units

$\alpha \equiv \frac{e^2}{4\pi} \simeq \frac{1}{137}$ (Heaviside-Lorentz units)

$[G_F] = 1/M^2$

$[\tau] = 1/M$ lifetime of an unstable particle

Notation

I will not bother to state precisely what I mean by $\sim$, except to say that $\simeq$ is more equal than $\sim$, and $=$ is more equal than $\simeq$.

The nature of this book is such that overall signs do not matter, except when they do matter.

Notation: $p^2/(2m)$ is written as $p^2/2m$; parenthesis included only when there is an ambiguity.

Table of fundamental constants and some useful numerical values

Quantity	Symbol	Value
speed of light in a vacuum	c	2.998×10^{10} cm sec^{-1}
gravitational constant	G	6.67×10^{-8} g^{-1} cm^3 sec^{-2}
reduced Planck's constant	$\hbar = h/2\pi$	1.055×10^{-27} g cm^2 sec^{-1}
	$\hbar c$	3.16×10^{-17} g cm^3 sec^{-2} = 197 MeV fm
Planck's constant	h	6.625×10^{-27} g cm^2 sec^{-1}
	hc	1.99×10^{-16} g cm^3 sec^{-2}
Planck mass	M_P	2.18×10^{-5} g $\sim 1.3 \times 10^{19}$ m_p
Planck length	l_P	1.62×10^{-33} cm
Planck time	t_P	5.39×10^{-44} sec
In Heaviside-Lorentz units		
fine structure constant	$e^2/4\pi\hbar c$	1/137.036
electromagnetic coupling (with $\hbar c = 1$)	e	0.303
electromagnetic coupling (without $\hbar c = 1$)	e	1.70×10^{-9} g$^{\frac{1}{2}}$ cm$^{\frac{3}{2}}$ sec^{-1}
In Gaussian units		
fine structure constant	$e^2/\hbar c$	1/137.036
electron mass	m_e	0.911×10^{-27} g
proton mass	m_p	1.67×10^{-24} g
electron rest energy	$m_e c^2$	0.511 MeV
proton rest energy	$m_p c^2$	938.2 MeV
Fermi weak interaction constant	G_F	1.17×10^{-5} GeV^{-2}
Bohr radius	a	5.29×10^{-9} cm
Rydberg energy	$Ry = hcR_\infty$	13.6 eV
classical electron radius (Gaussian units)	$e^2/m_e c^2$	2.82×10^{-13} cm
Thomson cross section	σ_T	6.65×10^{-25} cm^2
solar mass	$M_\odot$	1.99×10^{33} g
solar radius	$R_\odot$	6.96×10^{10} cm
solar luminosity	$L_\odot$	3.83×10^{33} erg sec^{-1}
solar surface temperature	$T_\odot$	5.78×10^3 K
earth mass	$M_\oplus$	5.97×10^{27} g
earth radius	$R_\oplus$	6.38×10^8 cm
gravitational acceleration (on earth)	g	9.81 m sec^{-2}
atmospheric pressure	p_0	1.01×10^6 dyne cm^{-2}
density of water at 25° C	ρ	0.997 g cm^{-3} $\sim$ 1 g cm^{-3}
density of air at 20° C	ρ_0	1.225 kg m^{-3} $\sim$ 1 kg m^{-3}
surface tension of water-air at 20° C	γ	72.8 dyne cm^{-1}

Quantity	Symbol	Value
dynamic viscosity of water at 20° C	η	1.01×10^{-2} g cm^{-1} sec^{-1}
kinematic viscosity of water at 20° C	ν	1.01×10^{-2} cm^2 sec^{-1}
dynamic viscosity of air at 20° C	η	1.82×10^{-4} g cm^{-1} sec^{-1}
kinematic viscosity of air at 20° C	ν	1.51×10^{-1} cm^2 sec^{-1}
Julian year	yr	3.156×10^7 sec
light year	ly	9.46×10^{17} cm
astronomical unit	AU	1.496×10^{13} cm
parsec	pc	3.086×10^{18} cm $= 3.26$ ly
unit of energy	erg	1 cm^2 g sec$^{-2} = 6.24 \times 10^{11}$ eV
electron volt	eV	1.16×10^4 K $= 1.602 \times 10^{-12}$ erg
fermi	fm	10^{-15} m $= 10^{-13}$ cm
Angstrom	Å	10^{-8} cm
micron	μm	10^{-6} m
Avogadro's number	N_A	6.02×10^{23} mol^{-1}
atomic mass unit	amu	1.66×10^{-30} g
Stefan-Boltzmann	σ	5.67×10^{-5} erg cm^{-2} sec^{-1} K^{-4}
Boltzmann constant	k	1.38×10^{-16} erg K^{-1}
1 kiloton TNT		4.18×10^{19} ergs

A few remarks

In the course on which this book is based, all numerical results in homework and exams are quoted to one significant figure only, in accordance with the fly by night spirit. Hence, the contents of this table were originally also given to one significant figure. However, my colleague Greg Huber objected vociferously, insisting that tables of numbers in physics books should always be given to three (or more) significant figures, so as to allow the person using the table the luxury of rounding off as he or she sees fit. And so I yield to conventional wisdom. The Huberian rule also states that if the fourth significant figure is 5, then the 5 should be kept. Of course, the reader must not plug these constants into formulas obtained by fly by night methods and then claim the result to more accuracy than is warranted.

We avoid units unfamiliar to most people, such as cSt for centiStokes, and omit stuff that some people might think is important but is not mentioned in this book.

A fermi is a femtometer (10^{-15} m). Mnemonic: both start with "fe"! Another mnemonic: "femten" in Danish and Norwegian is cognate with "fifteen" in English.

For the peculiar unit that e is expressed in (without $\hbar c$ set equal to 1), see chapter II.1.

Part I

Dimensional analysis: from a not so secret
to an allegedly secret weapon

Dimensional analysis

a not so secret weapon

An apparently innocuous but surprisingly powerful remark

The secret weapon of the fly by night physicist is dimensional analysis.

Many small American towns boast a wooden sign on the road leading into the hamlet, stating some basic numerical facts about the place, for example,[1] "Welcome to Dum Dum Town, home of the Angry Ducks,[2] founded 1869, population 12, elevation 233 feet, total 2114."

You laugh. Good, it means that you understand one of the foundational principles of physics: We can add various quantities together only if they are expressed in the same unit. The sign for Dum Dum Town should have given the population as 24 feet.[3] Furthermore, meaningful numerical values should be convention independent, which 1869 is not. It follows that, in physics, we should not refer temperature to Fahrenheit's armpits.[4]

Even if various quantities are in the same unit, they still may not be added together. To underline this point, allow me to mention that my family and I were once at a bus stop on a desert road in Israel near the Dead Sea. A high-tech board displayed the destinations of various buses and their estimated times of arrival thus: Jerusalem 35 minutes, Beersheba 14 minutes, Eilat 29 minutes. My son Max, 6 at the time and having recently mastered addition, announced that the sum was 78. (What was meaningful about the sum, I pointed out to Max, was that it decreased by a multiple of 3 every time they updated the board. This prediction[5] was shortly verified observationally.)

We are allowed to add, or subtract, various quantities together only if they are expressed in the same unit. Thus, the left hand and right hand sides of an equation in physics must have the same physical dimension.[6] The reader surely knows that this apparently innocuous statement is surprisingly powerful in physics, but people outside physics are often amazed.

The harmonic pendulum

We begin with a canonical example from the early days of physics. The story is that one day in church, Galileo watched the various hanging incense burners swinging back and forth and silently measured their periods against his pulse.

So, consider a pendulum of length l subject to the acceleration of gravity g. What is its period for small oscillations?

Convention used throughout this book and a word of caution regarding the paucity of letters

To do physics, we need units for mass, length, and time. Three and only three dimensions, which we denote by M, L, and T, respectively. More about this in chapter IV.1. Why only three? Nobody knows. Could there be more? Very unlikely, at least until the day we discover that we need more.

No, Patrick, degree is not a fundamental unit of physics. Temperature is an energy, and should be measured in units appropriate for energy. Again, more in chapter IV.1.

For the time being, all I need to say is that I adhere to the convention of using square brackets $[X]$ to denote the dimension of a physical quantity X. For example, recalling that kinetic energy is $E = \frac{1}{2}mv^2$, we see that energy has dimension $[E] = M(L/T)^2 = ML^2/T^2$. A table of dimensions is given in the front of this book.

Acceleration is time rate change of velocity: thus $[g] = (L/T)/T = L/T^2$. Indeed, when we first started studying physics, it was drummed into us that numerically $g = 32$ feet per second squared, or 9.8 meters per second squared, depending on where you were.

Henceforth, I will use the three letters, M, L, and T, generically to denote dimensions of mass, length, and time, respectively. In many cases (and one will come up presently), I will have to use one single letter to denote more than one physical quantity. Please do not be confused; the alphabet contains only so many letters.

Also, occasionally, it turns out to be more convenient to use energy E rather than mass M, as we will see.

Period of a pendulum

So, what is the period T of a pendulum? Well, we are to form an expression with the dimension of time out of l and g, with dimension $[l] = L$ and $[g] = L/T^2$, respectively. To get rid of L, we see that the ratio l/g is the only possibility, with dimension $[l/g] = L/(L/T^2) = T^2$. (You were forewarned: T here stands for two different concepts.) Thus, the period has to be

$$T \sim \sqrt{l/g} \qquad (1)$$

Remarkably, we have determined the dependence of T on l, seemingly with no effort. For example, if we double the length of a pendulum, we know that the period T should increase by a factor of $\sqrt{2} \simeq 1.4$. This square root dependence of the period on the length is an experimentally verifiable prediction.

Lord Rayleigh was allegedly the first to systematically exploit dimensional analysis, which he called the "law of similitude," in the late 19th century. I am surprised that it occurred so late, but surely the requirement that in physics dimensions must match was known to the founding fathers, almost by definition a bunch of plenty smart guys.

A detailed look at a one line calculation

The layperson is often astonished that physicists could predict the behavior of pendulums without doing a detailed calculation, but, as you probably know, our reasoning hides a lot of important, even profound, physics that took centuries to understand.

Since this is our first example, we will examine it in detail, attempting to beat it to death, so to speak. The rest of this chapter consists of a series of comments.

Occasionally but not too often,[7] a student worries unduly about all sorts of possible complications. What about air resistance? What about the stretchiness of the string, if the pendulum consists of a weight attached to a string? The list could go on.

Over time, several characters have wandered into my three previous textbooks, the most beloved appearing to be Confusio,[8] judging from comments by students and others. Worrywort here is not confused, but merely worries too much. The self-evident answer to put Worrywort at ease is that the spirit of fly by night physics is to treat the simplest possible case first. If Worrywort persists, then how about the changing mass of the pendulum as the incense burns? How about the rotation of the earth? Perhaps the best answer is that idealization is an essential part of theoretical physics.

Circular frequency over everyday frequency

Results obtained by dimensional analysis, such as (1), could easily be off by factors of 10 or more in one direction or the other. Indeed, the exact result for the pendulum is well known to be (see below) $T = 2\pi \sqrt{l/g}$.

Often, as in this case, factors of 2π and such could be fixed, or at least guessed at, by a linear combination of physics sense and experience with similar problems. In everyday usage, frequency, that is, the inverse of the period, $f = 1/T$, is defined as the number of repetitions per unit time. (For example, a hertz is 1 cycle per second.) But in physics, when we are faced with a situation periodic in time, after we set up the equation of motion, solve the differential equation, and so forth, we typically have solutions involving $\sin \omega t$, $\cos \omega t$, or for the more sophisticated, the complex exponentials $e^{\pm i\omega t}$, with the so-called circular frequency ω.

Evidently, these cyclic functions repeat themselves after a time duration $T = 2\pi/\omega$. But from experience, in the differential equation we solve, differentiation with respect to time would bring down factors of ω. So our dimensional analysis should apply, not to T, but to $\omega \sim \sqrt{g/l}$. Thus, we expect $T = 2\pi/\omega \simeq 2\pi \sqrt{l/g}$.

Take a pendulum 1 meter long. Then $T \simeq 6\sqrt{1/10}$ sec $\simeq 2$ sec. Without the 2π, we would have gotten a period of $\sim \frac{1}{3}$ sec. You do not have to be a keen experimentalist to realize that 2 sec is more reasonable.

Moral of the story: always ω, not f.

Indeed, in the rest of this book, I will drop the word "circular" and simply refer to ω as frequency. Here is an instructive exercise for the curious undergrads in my class. Glance at a textbook on an advanced subject, for example, quantum field theory. See if you can find ω or f more often.

The goal of this book is not to derive results that would agree with experimental data to n significant figures. But still, if you could obtain a better prediction, for free so to speak, by simply using a better variable, then why not?

The dull function hypothesis

As is also well known, there is actually a dimensionless parameter in the problem, namely, the initial angle θ_0 of the swinging pendulum, and so dimensional analysis, strictly speaking, only tells us that $T \simeq 2\pi f(\theta_0)\sqrt{l/g}$. The unspoken assumption is that $f(0)$ is a number of order 1.

Sometimes students ask: How do we know that $f(0)$ does not vanish? A valid question. But then you have to give a reason why $f(0)$ might vanish.

The unspoken assumption, again based on sense and experience, is that the majority of the functions appearing in physics are fairly boring functions that do not have especially wild properties. Of course, there are some exceptions.[9] But then, fairly obvious reasons usually would present themselves. Otherwise, we will end up with a major unsolved problem in physics.

In this particular example, it is fairly easy to see that the period will become independent of θ_0 in the small angle limit. Equating potential energy to kinetic energy, we have $mgl\theta_0^2 \simeq \frac{1}{2}mv^2$, thus giving $v \simeq \sqrt{gl}\theta_0 \propto \theta_0$. The pendulum swings slower and slower as $\theta_0 \to 0$, but by geometry, the distance traversed $\simeq l\theta_0$ also $\to 0$. Dividing distance by velocity, we get $T \propto \theta_0/\theta_0 = 1$, and we see explicitly that θ_0 disappears.

We will come back to the dull function hypothesis in chapter I.4.

Exact calculation, yes, but only after a back of the envelope calculation

Here is another quote:*

> Erst kommt das Denken, dann das Integral. (Roughly, "First think, then do the integral.") [Rudolf Peierls to the young Hans Bethe]

For the pendulum, the exact calculation is part of freshman physics and so is hardly hard.

A tactic that an advanced undergrad should have learned is that it is better to use energy conservation

$$E = \frac{1}{2}m(l\dot{\theta})^2 + mgl(1 - \cos\theta) \simeq \frac{1}{2}m(l\dot{\theta})^2 + \frac{1}{2}mgl\theta^2 \qquad (2)$$

rather than the equation of motion. Effectively, we have already integrated the equation of motion once to turn the second derivative in time into the square of a first derivative. This is an important theme we will come back to often, for example, in chapter VII.3.

Plugging $\theta = \theta_0 \cos\omega t$ into (2) and dividing through by $ml^2\theta_0^2$, we see that the expression on the far right, containing a $\sin^2\omega t$ and a $\cos^2\omega t$, could be constant only if $\omega^2 = g/l$, thus giving $T = 2\pi\sqrt{l/g}$.

By the way, we could also readily solve the problem for θ_0 not necessarily small; it merely involves an unappetizing and unilluminating integral. The left half of (2) allows us to solve for $\dot{\theta}$: it is proportional to the square root of some stuff plus $\cos\theta$, that is, $\dot{\theta} = (\text{stuff})\sqrt{\cos\theta - \cos\theta_0}$. (Without having to do any tedious algebra, we know the additive constant inside the square root must be $\cos\theta_0$, since $\dot{\theta}$ vanishes when θ reaches θ_0.) Thus,

$$T = \int dt = \int \frac{d\theta}{d\theta/dt} = 4(\text{stuff}) \int_0^{\theta_0} \frac{d\theta}{\sqrt{\cos\theta - \cos\theta_0}} \qquad (3)$$

Nothing much doing with this integral alluded to above, but for small θ_0, it becomes

$$\int_0^{\theta_0} \frac{d\theta}{\sqrt{\theta_0^2 - \theta^2}} = \int_0^1 \frac{\theta_0 du}{\theta_0\sqrt{1 - u^2}} \qquad (4)$$

After scaling by $\theta = u\theta_0$, we see with our very own eyes that θ_0 drops out of the problem.

Scaling is an important tactic that we will come back to again and again. One way of saying this is that we reject measuring angles following some loco Babylonian[10] idea of dividing a right angle into 90 equal parts. Instead, we measure the angle θ more sensibly in units of θ_0.

---

*We will literally apply this remark in a math interlude.

Drive your calculus teacher mad

Of course, we could also use Newton's equation of motion* $F = ma$:

$$ml\frac{d^2\theta}{dt^2} = -mg\theta \tag{5}$$

which amounts to differentiating (2) with respect to t.

Sure, you and I know how to solve differential equations like this. The fly by night physicist, however, prefers the "drive your calculus teacher mad" approach. Write $\frac{d^2\theta}{dt^2}$ out as $\frac{d}{dt}\frac{d\theta}{dt}$ and cancel the ds:

$$\frac{d}{dt}\frac{d\theta}{dt} \sim \frac{\theta}{t^2} \tag{6}$$

Note that in this respect, Leibniz[11] had better notation than Newton. Then (5) becomes, after some more canceling, $ml\frac{\theta}{t^2} \sim mg\theta$, that is, $t^2 \sim l/g$, the correct answer.

Your calculus teacher screams. But this sort of fly by night maneuver is basically correct,[12] and in essence amounts to dimensional analysis.

When we discuss the expanding universe in chapter VII.3, we will use this approach to great advantage.

The profundity behind Einstein gravity

Imagine yourself in Galileo's shoes. Observation of the pendulum led you to $T \propto \sqrt{l}$. Historically, Galileo also knew that gravity produces a universal acceleration g (recall the legend of the cannonballs dropped from the leaning tower), and he even had a reasonably good value for g, obtained by, more sensibly, rolling balls down planes inclined at various angles. But still, to construct a theory that would produce $T \propto \sqrt{l}$ would have been supremely difficult and would have to wait for the genius of Newton. Familiarity might breed contempt, but (5) contains an awful lot of conceptual profundities. We might be in a similar situation regarding some of the unsolved problems in fundamental physics, such as the cosmological constant puzzle.[13]

We now arrive at a profundity, one of the greatest in physics.

To Newton, mass measures the amount of stuff.[14] He quite naturally assumed that the mass m appearing in his law of gravity and the mass m appearing in his law of motion are one and the same.[15]

But a hair-splitting lawyer, or a habitual reader of mysteries, would surely have detected a hidden assumption here: Are the blonde[16] seen kissing the butler and the blonde caught leaving the house on the night of the murder really the same blonde? Are those two masses really the same mass?

---

*Which, curiously enough, is spoken as $F = ma$ but written as $ma = F$. Talk about my confusion when first studying physics!

Did you ever notice that the two ms appearing in (2) have different conceptual origins?

A couch potato's obligation to gravity versus his reluctance to move

To distinguish between the masses that appear in Newton's law of gravity and in Newton's law of motion, physicists called them the gravitational mass m_G and the inertial mass m_I, respectively. The former measures a couch potato's obligation to listen to gravity, the latter his reluctance to get up and move. Conceptually, they are quite distinct and could, by logic alone, not be equal.

Unlike the faculty in some other university departments, those in the physics department do not accept proofs by authority, not even a single-named giant in a likely apocryphal story about dropping stuff off a leaning tower. And thus the Hungarian Baron Loránd Eötvös, instead of doing whatever barons did in the 19th century, devoted much of his life performing ever more precise experiments establishing the equality of the gravitational mass and the inertial mass. In our days, a series of experiments, known collectively as Eötvös experiments, have established the equality of the gravitational mass and the inertial mass $m_I = m_G$ to a fantastic degree of accuracy.[17]

So, in principle, we could only say that the pendulum's period $T = 2\pi \sqrt{m_I l / m_G g}$.

It turns out that this amazing equality of the gravitational mass and the inertial mass furnishes the key[18] to Einstein's theory of gravity. See appendix Eg.

Exercises

(1) A physics major planning to become a high school physics teacher (a noble profession!) told me that in her high school, she was taught to remember the "Big 4." I asked her what those are. They turned out to be, for example, the formula for the distance Δx covered in time t by an object with constant acceleration a and initial velocity v_0: $\Delta x = v_0 t + \frac{1}{2} a t^2$. In fact, these kinematic formulas follow essentially from the definitions of various concepts, such as acceleration and dimensional analysis. Show this and argue for the factor of $\frac{1}{2}$.

(2) Look through your high school physics textbook and pick out some results that follow essentially from dimensional analysis.

(3) To ace exams on elementary physics, always keep dimensional analysis in your pocket. For instance, suppose you were confused about whether work done is force times distance traversed or force times time elapsed. What to do?

(4) When a spring with Hooke's constant k is stretched by Δl, it exerts a force $F = k\Delta l$. Find the oscillation frequency of a mass m attached to this spring.

(5) Consider the 1-dimensional motion of a particle in the potential $V(x) = g|x|^\alpha$. Let t and d denote the characteristic time and distance, respectively, for this motion. Show by dimensional analysis that

$$t \propto d^{1-\frac{\alpha}{2}}$$

Apply this result to (a) the harmonic oscillator, (b) a particle in free fall on the surface of the earth, and (c) a planet falling radially toward a star.

Notes

[1] According to a photo, most likely photoshopped, circulating on the web several years ago.

[2] For those readers unfamiliar with American culture, the name of a long defunct football team.

[3] And 1869 light years converted into feet.

[4] To be fair, Daniel Gabriel Fahrenheit (1686–1736) was actually a great physicist for his time.

[5] And any deviation from this law could be explained, namely, that one of the buses was delayed.

[6] To me this statement is so self evident that I absolutely refuse to spend 30 or so pages explaining why this must be true and orating about things like Buckingham's theorem (which, by the way, I had never heard of until I started writing this book and looking at the literature).

[7] I put in these two paragraphs at the request of a critical reader. But in fact, by the time a student reaches the course I mentioned in the preface, he or she almost invariably understands that idealization is a necessary part of physics.

[8] He first burst onto the scene in *QFT Nut*.

[9] And they sometimes open up new fields of study.

[10] Babylonians were incredibly smart. See, for example, *GNut,* p. 214.

[11] I read somewhere that in fact he "envisioned" $\frac{dy}{dx}$ as the division of one infinitesimal by another, dy by dx.

[12] The creators of LaTeX also thought so: the derivative is written as "slash frac."

[13] For example, *GNut*, chapter X.7.

[14] Newtonian physics cannot entertain the existence of massless particles.

[15] The following paragraph is taken from my book, *On Gravity*, Princeton University Press, 2018.

[16] As to what type of blonde, see the classification in the scholarly study *Blonde Like Me* by Natalia Ilyin (Simon & Schuster, 2000).

[17] In particular, an ingenious effort, led by my former colleague Eric Adelberger at the University of Washington, is fondly referred to as the Eöt-Wash experiment. See https://www.npl.washington.edu/eotwash/node/1. Nerd humor in full force here!

[18] For more details, see, for example, *GNut*. For a brief discussion, see T. P. Cheng, *Einstein's Physics*, Oxford University Press, 2013.

From Kepler's law to black holes

In anxious and uncertain times like ours, when it is difficult to find pleasure in humanity and the course of human affairs, it is particularly consoling to think of the serene greatness of a Kepler. Kepler lived in an age in which the reign of law in nature was by no means an accepted certainty. How great must his faith in a uniform law have been, to have given him the strength to devote ten years of hard and patient work to the empirical investigation of the movement of the planets and the mathematical laws of that movement, entirely on his own, supported by no one and understood by very few! [Albert Einstein, *Essays in Science*, 1934]

Newton's constant

Newton's law of universal gravity states that the force $\vec{F}$ of gravitational attraction between a mass M and a mass m is given by

$$\vec{F} = \left(\frac{GMm}{r^2}\right)\frac{\vec{r}}{r} \tag{1}$$

with G being Newton's gravitational constant and $\vec{r}$ the distance vector separating the two masses. The gravitational potential energy between two masses

$$V = -\frac{GMm}{r} \tag{2}$$

then follows.

You may have noticed that Newton's constant G never occurs by itself,* but always in combination with a mass M. For convenience, define $\kappa \equiv GM$.

---

*A slightly more sophisticated version of this statement still holds for Einstein gravity.

Using $ma = GMm/r^2$ and hence $a = GM/r^2$, we see that GM has dimension of acceleration times length squared:

$$[\kappa] = [GM] = (L/T^2)L^2 = L^3/T^2 \tag{3}$$

A quick derivation of Kepler's law: Why is Mercury called Mercury?

> It is as though I had read a divine text, written into the world itself, not with letters but rather with essential objects, saying: "Man, stretch thy reason hither, so that thou mayest comprehend these things."
> [Johannes Kepler][1]

Let $\kappa = GM$ with M the mass of the sun. To determine the period T of a planet going around the sun in a circular orbit of radius R, you would equate the gravitational force with the centrifugal force, or at a more advanced level, you would solve some differential equations[2] to determine the motion. After the dust settles, you find R as a function of T (or vice versa), but no matter what, by dimensional analysis, since $[\kappa] = L^3/T^2$, the result must have the form

$$R^3 \sim \kappa T^2 \tag{4}$$

Thus, without breaking a sweat, we obtain Kepler's law[3] stating that the orbital period of a planet scales like its distance from the sun to the $\frac{3}{2}$ power. Jupiter, about $\simeq 5.2$ times farther from the sun than the earth, orbits once every $\simeq (5.2)^{\frac{3}{2}} \simeq (140)^{\frac{1}{2}} \simeq 12$ years.

Why is the planet Mercury called Mercury? Kepler's law explains why. Being the planet closest to the sun, Mercury has the shortest orbital period. And Mercury is the Roman name for Hermes, the Greek god of speed, of travelers, and of thieves.[4] No wonder that Einstein proposed[5] the precession of Mercury, not of Jupiter, as one of the three tests of his theory of gravity.

Worrywort, whom we met in chapter I.1, might wander by and ruin a perfectly elegant derivation of Kepler's law. What about the rotation of the planet? What about the ellipticity of the orbit? Oy vey. Well, if we couldn't get away fast enough, we could always soothe him with Einstein's advice: "Physics should be made as simple as possible, but not any simpler."

You could always try to do better

The title is not a proverb, nor is it an exhortation from a self-help book. Instead, it is a statement of the possibility of finding a larger used envelope.

In deriving Kepler's law, we have essentially taken the sun to be much more massive than the orbiting planet. What if we are talking about two black holes of comparable mass orbiting each other, as was "seen" via gravity waves a

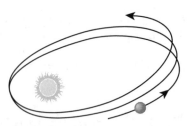

FIGURE 1. Mercury's perihelion advances (vastly exaggerated). From Zee, A. *Einstein Gravity in a Nutshell*, Princeton University Press, 2013.

while ago? We would have to modify our dimensional analysis result to $R^3 \sim GMT^2 f(m/M)$ with f an unknown function of the ratio of the two masses.

Unknown? Hah, we can apply interchange symmetry! Thus, $R^3 \sim GmT^2 f(M/m)$. Equating these two expressions, we obtain $f(x) = xf(1/x)$. This tells us that if we know f in some neighborhood around $x = a$, we immediately know f in some neighborhood around $x = 1/a$.

Don't forget that we also know that $f(0)$ is some constant, namely, the overall multiplicative constant in (4). I will let you have the fun of solving $f(x) = xf(1/x)$. See exercise (2).

This problem is treated in every textbook on classical mechanics, but here we did it sans souci. Of course, it is "merely" a matter of understanding the concept of center of mass. But notice that we could get by without ever having heard of this term.

Einstein gravity and the precession of Mercury

> When I was studying in Switzerland, I did not even know that I was a Jew. I was satisfied to know only that I was a man. [Albert Einstein, in a speech in Shanghai, 1923]

In Einstein gravity, the perihelion of Mercury does not return to the same point after one orbit, but instead moves through a tiny angle $\Delta\phi$. The orbit does not close, as it does in (idealized) Newtonian gravity (see figure 1).[6]

A precise calculation of $\Delta\phi$ may be found in standard textbooks[7] but let's use dimensional analysis to get a rough idea. We have to construct the dimensionless $\Delta\phi$ out of $\kappa = GM$, the speed of light c (since Einstein gravity is a relativistic theory), and the radius R of Mercury's orbit. With $[\kappa] = L^3/T^2$, the simplest[8] possibility is

$$\Delta\phi \sim \frac{GM}{c^2 R} \tag{5}$$

While this estimate is off by a factor of 6π, you could plug in numbers and see that this is crazy tiny, of order 10^{-8} radians per orbit.*

---

*With the 6π, this translates into the famous 43 seconds of arc per century!

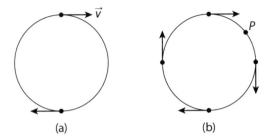

FIGURE 2. Fly by night approach to centripetal acceleration.

The reader might ask: Why didn't we also include Mercury's speed v? Well, v is not an independent variable; it could be expressed in terms of κ and R by (4).

Ever better approximation to 1

For Newton to figure out the orbits of the planets (or of the moon) and to obtain the exact version of (4), he has to know the acceleration a of an object moving around in a circle of radius r with velocity v. This many centuries later, of course any student could simply differentiate $x = r \cos \omega t$, $y = r \sin \omega t$ twice to obtain*

$$a = \frac{v^2}{r} \qquad (6)$$

But Newton has yet to invent calculus.

Try a fly by night approach. Let the object go halfway around. See figure 2(a). After time $\Delta t = (\pi r)/v$, its velocity has flipped direction, hence $\Delta v = 2v$. Thus,

$$\frac{\Delta v}{\Delta t} = \frac{2v}{\pi r/v} = \frac{2}{\pi} \frac{v^2}{r} \qquad (7)$$

Hey, $2/\pi$ is a pretty good approximation to 1.

Ambition seizes us! Let's try to do better. Now let the object go only a quarter of the way around. See figure 2(b). Now $\frac{\Delta v}{\Delta t} = \frac{\sqrt{2}v}{(\pi r/2)/v} = \frac{2\sqrt{2}}{\pi} \frac{v^2}{r}$, and $2\sqrt{2}/\pi \simeq 2.8/3.14$ is an even better approximation to 1. Note also that the acceleration $\vec{a}$ at the point P is pointing toward the center of the circle.

What you are discovering is a series of ever improving approximations to 1. You might even have, maybe, perhaps, invented calculus!

A particle moving at constant speed v around a circle has an acceleration pointing toward the circle's center. A colleague told me that people researching physics education have shown that this is one of the most difficult physics facts for the guy or gal in the street to wrap his or her head around. As a homework exercise, try to explain acceleration pointing toward the center of the circle to one of these humans.

---

*$a = r\omega^2 = r(2\pi/T)^2 = r(2\pi v/2\pi r)^2 = v^2/r$; phew, a lot of equal signs.

Free fall time for a cloud of dust

The dimension L^3/T^2 of κ, with a length cubed, almost invites us to consider a density of some kind. Consider a cloud of dust in interstellar space. The official definition of dust is that the collision between dust particles is negligible. Denote the mass density of the dust cloud by ρ with dimension M/L^3. Then $[G\rho] = [GM/L^3] = [\kappa/L^3] = (L^3/T^2)/L^3 = 1/T^2$ is an inverse time squared. The time scale for collapse τ follows immediately:

$$\tau \sim \frac{1}{\sqrt{G\rho}} \tag{8}$$

As the cloud contracts, its density increases, and so the relevant time scale becomes even shorter. A runaway process! But at some point, the collision between dust particles can no longer be neglected; the heat and pressure thus produced resist further collapse. We will come back later to this point, clearly of cosmological and astrophysical interest.

A length for every mass

That $\kappa \equiv GM$ has dimension $L^3/T^2 = (L/T)^2 L$ leads to all sorts of interesting physics besides Kepler's law. For any mass M, we can form a characteristic length $r_S \equiv 2GM/c^2$, known as the Schwarzschild radius of that mass.

What could this length be? Let us estimate the Schwarzschild radius for the most massive object in our neighborhood, namely, the sun. Even though the mass of the sun $M_\odot$ is so huge, $GM_\odot/c^2$ is still a small length, around 1.5 km, because gravity is absurdly feeble compared to the other three interactions,[9] electromagnetic, weak, and strong.

But wait, small compared to what? As a budding physicist, that phrase should always be hanging around your lips.

The only relevant length scale to compare the Schwarzschild radius of the sun with is the actual radius of the sun $R_\odot$. We find $GM_\odot/c^2 R_\odot \sim 10^{-6}$. In fact, as we will see shortly, we won't even have to look up the solar constants $M_\odot$ and $R_\odot$ to realize that this dimensionless ratio would be tiny.

Deflection of light from Newton to Einstein

> Do not Bodies act upon Light at a distance, and by their action bend its Rays? [Isaac Newton]

Many undergraduates do not realize that Newton already speculated on the bending of light, with his theory that light is made up of tiny "corpuscles" with an unknown tiny mass m (which, you realize, would cancel out in the bending angle due to the celebrated equality between inertial mass and gravitational mass). Historically, Newton's idea was swept away by the discovery that light is an electromagnetic wave.

But Einstein's theory predicted the bending of light for good. Like anything else, light is affected by the curvature of spacetime. Thus, starlight grazing the sun (so that the deflection of light would be as large as possible) would be bent, albeit still by a teeny amount.

As a naive theorist, Einstein wrote to George Hale, the director of the Mount Wilson Observatory, wanting to know "how close to the Sun fixed stars could be seen *in daylight*" (italics Einstein's). Hale explained that exploiting a solar eclipse would be more promising. Hence the famous expeditions of 1919 to Brazil[10] and Africa.

Let's write the deflection angle of light passing by the sun as*

$$\Delta\varphi = 2\eta \frac{GM_\odot}{c^2 R_\odot} \qquad (9)$$

with η a pure number. By now you understand that dimensional analysis dictates[11] this form; in Newton's, Einstein's, or your car mechanic's theory, $\Delta\varphi$ would be given by this expression regardless, unless some other relevant fundamental constants enter the theory. In Newtonian gravity, $\eta = 1$, while in Einstein gravity, $\eta = 2$. Loosely speaking, the 2 comes about because Newton curved only time, while Einstein curved space as well as time.

Einstein's luck[12]

> For the deviation of light by the sun I obtained twice the former amount. [Albert Einstein, writing to Arnold Sommerfeld, late 1915]

Of the three classic tests of Einstein gravity, it was the deflection of light[13] that made Einstein a worldwide celebrity—the general public could hardly be expected to care about the perihelion shift of Mercury. But space warp? Now that's another story!

But Einstein's road to global fame had many twists and turns, full of suspense.

First, unbeknownst to Einstein, the German physicist Johann Soldner had already obtained $\eta = 1$ decades earlier by treating light as a Newtonian "corpuscle." Second, in calculating the deflection angle, Einstein made a mistake and also obtained $\eta = 1$. Third, the expedition that set off to Crimea to verify Einstein's amazing prediction was financed by the German munitions manufacturer Krupp.

---

*Clearly, this should not be confused with (5), in which R denotes the radius of Mercury's orbit.

Fortunately for Einstein, World War I broke out, and the Russians promptly arrested the members of the expedition as spies. If you were in charge, with war clouds on the horizon, wouldn't you be suspicious of a bunch of nerdy guys paid by a foreign arms merchant wandering around with telescopes and whatnot, muttering some gibberish about the curvature of spacetime?

During this delay, Einstein discovered his error, as he mentioned in his 1915 letter to Sommerfeld. So he was spared an embarrassing disagreement between observation and his prediction.

Decades later, Nazi officials, in their antisemitic campaign, would accuse Einstein of stealing his celebrated space warp prediction from an Aryan German physicist. Little did they know that, even if that had been true, the thief would have gotten away with the wrong stuff.

A hint of black holes

Unless you are a Papuan headhunter, you would have heard of black holes by now. John Michell in 1783 and the Marquis Pierre-Simon Laplace in 1796 pointed out that even light could not escape from an object excessively massive for its size.

It's "merely" freshman physics by now. A particle of mass m at the surface of an object of mass M and radius R has a gravitational potential energy $-GMm/R$ and kinetic energy $\frac{1}{2}mv^2$. Equating these two energies gives the escape velocity $v_{\text{escape}} = \sqrt{2GM/R}$. Setting v_{escape} to c tells us that if $2GM > Rc^2$, not even light can escape, and the object is a black hole.[14] Remarkably, even though the physics behind the argument[15] is certainly not correct, this criterion, including the factor of 2, turns out to hold in Einstein's theory.

A more modern heuristic argument incorporates Einstein's $E = mc^2$. At a distance of R from an object of mass M, a small particle of mass m feels a gravitational potential energy of GMm/R. As the particle gets closer and closer to the massive object (that is, as R gets smaller and smaller), the gravitational potential energy gets larger and larger.

At what point does the particle feel that the oppression of gravity is too much to bear? Well, according to Einstein, if the particle were to be entirely converted to energy, that energy would amount to $E = mc^2$. Thus, when the gravitational potential energy gets to be comparable to this energy, the particle would not be able to stand it any more.

It is as if an oppressive boss is dumping on a cowering employee's head an amount of negative energy that exceeds the employee's entire inner reserve of energy. Then something has to give. This critical state of affairs is reached when $GMm/R \sim mc^2$. Again, m cancels out. We recover more or less the same Michell-Laplace criterion for a black hole (see figure 3):

$$GM/c^2 \gtrsim R \qquad (10)$$

One advantage of this argument is that it sows the seed for Hawking radiation, as we will see in chapter IV.3.

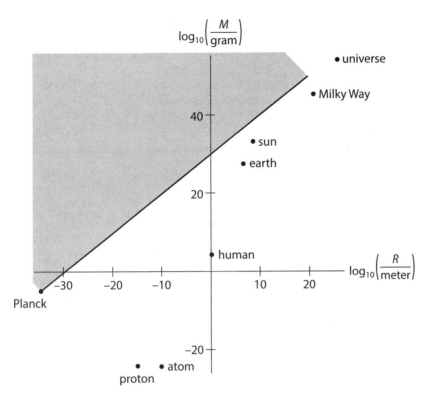

FIGURE 3. A plot of the Michell-Laplace criterion. The mass M of an object is plotted along the vertical axis and its characteristic size R along the horizontal axis. Note that this is a log-log plot, in which both mass and size are plotted in powers of 10; otherwise, it would hardly be possible to accommodate the universe and a proton in the same figure. The point labeled "Planck" stands for an object with Planck mass and Planck length, quantities to be defined in chapter IV.1. From Zee, A. *Einstein Gravity in a Nutshell*, Princeton University Press, 2013.

Indeed, the formula $\Delta\varphi \sim GM/c^2R$ for the deflection angle $\Delta\varphi$ of light around a body of mass M and radius R also points toward the Michell-Laplace criterion. If $\Delta\varphi \gtrsim 2\pi$, light is trapped.

Since light is hardly trapped around the sun, we could see, without looking up any numbers, that $GM_\odot/c^2R_\odot \ll 1$, as promised.

As Kepler said, the purpose of physics is to read the divine text.

An amusing unit

The English physicist Arthur Eddington was an ardent believer in Einstein's theory due to its aesthetic merits even before he led the solar eclipse expeditions. Afterward, he promoted Einstein by lecturing to the public and writing a popular book.[16] Eddington, as a master popularizer, expressed the result in terms that his readers in England at that time could relate to directly. He said that you were paying dearly for this newfangled electric light, but since light had been shown to weigh, you could ask how many pounds of light you were getting per pound £ of your money, and compare the cost of light with the cost of gold.

So my favorite (apparently) dimensionless unit has got to be pound per pound. I couldn't resist mentioning this historical oddity.

Exercises

(1) Continue the series to approximate 1 started in the text: $2/\pi$, $2\sqrt{2}/\pi$, ...

(2) Laboring as little as possible, obtain the exact version of $R^3 \sim GmT^2 f(M/m)$, including the factors of 2π.

(3) Show that $GM_\odot/c^2 \simeq 1.5$ km, so that the precession of Mercury is given by 1.5 km divided by the radius of Mercury's orbit, and hence is a very tiny number.

(4) Verify that r_S comes out to be about 10^{-9} (earth), 10^{-6} (sun), 10^{-4} (white dwarf), and 10^{-1} (neutron star).

Notes

¹ In the spirit of Wheeler asking freshman physics majors to pay a moment of silence to Einstein, I should ask you to do the same for Kepler. Without his devoting a lifetime to meticulous observation, physics might not have gotten off the ground when it did.

² For example, see *GNut*, pp. 28–30.

³ Historically, Kepler's law predates Newton's law of gravity.

⁴ In the United States, he delivers flowers.

⁵ Of course, this is because the effect is most pronounced for the fast moving Mercury and hence is the effect best measured by astronomers.

⁶ In the nonidealized world, perturbations from the other planets cause Mercury's orbit to precess, an effect much larger than that due to Einstein gravity. See footnote on p. 372 of *GNut*.

⁷ For example, *GNut,* p. 372.

⁸ Strictly speaking, dimensional analysis alone would allow $\Delta\phi$ to be given by some higher power of $\frac{GM}{c^2 R}$, but this begs the question of why the order G contribution to $\Delta\phi$ should happen to vanish.

⁹ See my book, *On Gravity*, Princeton University Press, 2018, p. 15.

¹⁰ I recommend the Brazilian movie "House of Sand." Once, when I taught the course on which this book is based, I mentioned that people usually associated Brazil with jungles, not a vast desert, and one student said he knew all about the desert featured in the film. Turned out that he was Brazilian.

¹¹ Endnote 8 about $\Delta\phi$ also applies here.

¹² See J. Waller, *Einstein's Luck*, cited in *GNut*, p. 370.

¹³ J. J. Thomson, the discoverer of the electron, presiding over a special meeting of the Royal Society convened to announce the result of the solar eclipse expeditions, hailed the result as the most important since Newton's work and Einstein's theory as "one of the highest achievements of human thought," which regrettably, he added, was incomprehensible.

¹⁴ Named by John Wheeler almost 200 years later.

¹⁵ This often cited Newtonian argument actually does not establish the existence of a black hole defined as an object from which nothing could escape. The escape velocity refers to the initial speed with which we attempt to fling something into outer space. In the Newtonian world, we could certainly escape from any massive planet in a rocket with a powerful enough engine.

¹⁶ Freeman Dyson told me that his father rushed out to buy this book. I am indebted to Dyson for lending me this particular copy, bearing his father's signature.

Bohr's atom and Heisenberg's uncertainty principle
opposition and compromise

Bohr atom

From Kepler's law we go over to the Bohr model[1] of the hydrogen atom, pictured as a miniature solar system with an electron going around a proton. Equating Coulomb attraction to the centrifugal force, we have

$$ma = \frac{mv^2}{r} \sim \frac{e^2}{r^2} \tag{1}$$

Multiplying by r, we find $mv^2 \sim e^2/r$. In other words, the fly by night physicist could have just as well started by saying that the kinetic energy and the potential energy are roughly equal.[2]

But no matter what, if you were Bohr, you are faced with having only one equation for two unknowns: v and r. This actually makes sense. In classical physics, you could have given the electron any velocity you wanted, and the larger the electron's velocity, the smaller its orbit. In contrast, in the unknown realm of the atom, v and r are somehow fixed.

Venturing into this terra incognita, Bohr desperately needed another equation. The clue came from dimensionally analyzing the mysterious constant $\hbar$ that Planck introduced* back in 1900. Planck's proposal (as interpreted later by Einstein) was that electromagnetic radiation consists of photons, and each photon of frequency ω carries energy of[3] $E = \hbar\omega$.

---

*We will discuss this in chapter III.5.

What is the dimension of $\hbar$? It is the dimension of energy divided by frequency:

$$[\hbar] = E/(1/T) = (ML^2/T^2)/(1/T) = ML^2/T = (ML/T)L \qquad (2)$$

Hence, equivalently, Planck's constant also has dimension of momentum times length.

But momentum times length is the dimension of angular momentum. So?

Bohr, a fly by night physicist par excellence if ever there was one, leapt off into the dark beyond and guessed that the angular momentum of the electron is given by $\hbar$:

$$mvr \sim \hbar \qquad (3)$$

This provides the second equation that was needed.

Multiplying the energy equation $mv^2 \sim e^2/r$ we had by mr^2, we obtain $\hbar^2 \sim (mvr)^2 \sim e^2 mr$ and thus

$$r \sim \frac{\hbar^2}{e^2 m} \qquad (4)$$

The energy is then given by $E \sim \frac{e^2}{r} \sim \frac{e^4 m}{\hbar^2}$.

The precise answer

$$E = -\frac{e^4 m}{2\hbar^2} \qquad (5)$$

was drummed into me as a student, and is numerically equal to 13.6 eV and known as 1 Rydberg. Note that we are sloppy about signs; E is of course negative, being a bound state energy.

Let's do a quick check to make sure that $\hbar$ is in the right places. As $\hbar \to 0$, we go over to classical physics, and $r \to 0$, $E \to -\infty$, as expected. The electron crashes into the nucleus.

Bohr moved physics forward by his wild guess

Incidentally, if we replace the $\sim$ sign by an $=$ sign in (1) and (3), we find that (4) could be written with an $=$ sign, and we obtain $E = \frac{1}{2}mv^2 - \frac{e^2}{r} = -\frac{e^4 m}{2\hbar^2}$, the precise result. This is, however, merely a happy coincidence: we now know that Bohr's inspired guess is actually wrong, wrong, wrong.

As a standard exercise in introductory courses on quantum mechanics, students routinely solve Schrödinger's equation to obtain the ground state of the hydrogen atom, showing that the wave function is actually spherically symmetric and has zero angular momentum. To set the electron's angular momentum to $\hbar$ is roundly incorrect.

But allow me to ask you: Is Bohr or the student who aces the homework problem immortalized in physics? Case closed. A wild and ultimately incorrect guess managed to move an entire area of physics forward. Face it, nobody much cares whether you and I can solve Schrödinger's equation for the hydrogen atom* almost a hundred years after it was introduced. That's essentially what I said in the preface about flying by night in total darkness.

The muttering nabobs might also complain that Bohr's fly by night approach relies on purely classical concepts, and the quantum $\hbar$ only sneaks in through (3).

The uncertainty principle

The Bohr atom dates from 1913. In 1926, Heisenberg introduced his uncertainty principle

$$\Delta p \Delta x \sim \hbar \tag{6}$$

(Note that this is consistent with the earlier remark that $\hbar$ has the dimension of angular momentum.)

The uncertainty principle could also have served, anachronistically, as the second equation Bohr needed. Write the energy equation above as $p^2/2m \sim e^2/r$. Since the electron is confined to a region of radius r, its momentum, according to the uncertainty principle, should be of order $p \sim \hbar/r$. Thus, $\hbar^2/mr^2 \sim e^2/r$, which leads to (4) and (5).

Opposition and compromise

The two approaches to the Bohr atom are arithmetically the same, but arguably, the uncertainty principle argument is physically closer to the truth of the ground state wave function being a fuzzy ball of radius r. Indeed, as just mentioned, we know that the actual ground state is spherically symmetric and has zero angular momentum. However, the theme shared by both approaches, of opposition and compromise, is common in physics. Here the potential energy $\propto -1/r$ drives $r \to 0$, while the kinetic energy $\propto +1/r^2$ wants $r \to \infty$. A compromise is reached at the Bohr radius.

Incidentally, while the Bohr atom is now standard textbook stuff, at the time there was widespread skepticism.[4] Otto Stern, famous for the Stern-Gerlach experiment, and Max von Laue took an oath that they would quit physics "if this nonsense of Bohr would turn out to be right." Pauli called this oath "Utlischwur" as a play on the traditional Swiss oath "Rutlischwur" from the rebellion against their Austrian rulers, in which figured the legend of William Tell about shooting an apple off a boy's head.

The harmonic oscillator

The uncertainty principle argument works well in many situations, in particular for the 1-dimensional harmonic oscillator. During the oscillations, kinetic

---

*Indeed, we will do this in chapter III.2.

energy is turned into potential energy and then back into kinetic energy. Setting the kinetic and potential energies roughly equal, we have $p^2/2m \sim kx^2$. Again, the smaller x is, the larger $p \sim \hbar/x$ becomes, and a compromise is reached when $p^2/2m \sim \hbar^2/mx^2 \sim kx^2$, thus giving $x^4 \sim \hbar^2/mk$ and hence a ground state energy of order

$$E \sim kx^2 \sim \hbar\sqrt{\frac{k}{m}} \sim \hbar\omega \tag{7}$$

Here we recall the frequency $\omega \sim \sqrt{\frac{k}{m}}$ of the classical harmonic oscillator.

The use of appropriate units

Of course, anybody could plug numbers into (5) and obtain the binding energy E in ergs (or BTU, British thermal units, if you prefer). But physicists in different areas of physics use different units for good reasons. The units should be appropriate for the physics at hand.

The result in (5), being proportional to m, almost shouts at us to compare E to the rest energy mc^2 of the electron, about half a million electron volts.[5] Then

$$E = -\frac{e^4 mc^2}{2\hbar^2 c^2} = -\frac{1}{2}\left(\frac{e^2}{\hbar c}\right)^2 mc^2 \equiv -\frac{1}{2}\alpha^2 mc^2 \tag{8}$$

The physics is requesting us to define the so-called fine structure constant[6]

$$\alpha = \frac{e^2}{\hbar c} \simeq \frac{1}{137} \tag{9}$$

By writing the electrostatic potential as e^2/r, we indicate that we are using Gaussian units, in which α has the stated value, as explained in appendix M.

Hence, $-E \simeq \frac{1}{2}(\frac{1}{\sqrt{2}} \times 10^{-2})^2(\frac{1}{2} \times 10^6)\, eV \simeq \frac{100}{8}\, eV \simeq 13\, eV$. Not bad!

Bohr's luck

A colleague who read this chapter harangued me about being too generous to Bohr. I was quite taken aback, saying that I did include the critical comment that the orbital angular momentum turned out not to be equal to $\hbar$ and the "Utlischwur" story. Certainly, physicists do not rank Bohr among the greats like Einstein and Schrödinger, even though his influence was enormous and the Bohr model was crucial to cracking quantum mechanics open.

But I think that bridging the mystery of the atom to the black body radiation was highly nontrivial.[7] It is also true that once $\hbar$ was brought in, much of the result followed by dimensional analysis: $[e^2] = EL$ (because of the Coulomb potential), $[\hbar] = ET$, and $[m] = M = E/(L/T)^2 = ET^2/L^2$ (because of Newton's formula for kinetic energy): To cancel out the L and T so as to obtain an expression for the binding energy, we are forced to* $me^4/\hbar^2$.

Bohr was very lucky to get the[8] $\frac{1}{2}$ in (5).

Exercises

(1) Use the uncertainty principle to find the ground state energy of a particle of mass m in an infinite square well of width a.

(2) Estimate the velocity of the electron in the hydrogen atom, and show that we do not have to worry about relativistic effects.

(3) Consider the innermost electron in an atom with atomic number Z. At what values of Z do we have to start worrying about relativistic effects?

---

*This is an example showing that sometimes it is better to use E rather than M.

Notes

[1] The Austrian A. E. Haas proposed this model 3 years before Bohr and was greeted by much ridicule. He was the first to obtain what is now called the Bohr radius. See Wikipedia https://en.wikipedia.org/wiki/Arthur_Erich_Haas.

[2] More sophisticated people say "virial theorem," which in this case states that the kinetic energy equals minus half the potential energy.

[3] We have $\hbar\omega = hf$, and hence $\hbar = h/2\pi$. Again, the 2π comes in because circular frequency is more relevant for physicists than frequency f, as explained in chapter I.1.

[4] See S. Pakvasa, "The Stern-Gerlach Experiment and the Electron Spin," physics archive.

[5] Another number worth remembering is the proton mass $m_p \simeq 1\ GeV = 10^9\ eV$, a billion electron volts. Incidentally, the large dimensionless ratio $m_p/m_e \simeq 2,000$ plays an important role in solid state physics.

[6] Eminent physicists have written about how to distinguish crackpots from normal physicists. Like many physicists, I receive my share of crackpot letters. If a letter starts out claiming to derive the integer 137, I know for sure that the theory is crackpot. First, α^{-1} is only approximately an integer. Second, in modern quantum field theory, it varies according to the energy scale. See, for example, *QFT Nut*, chapter VI.8. Long before all this was understood, Arthur Eddington, who was mentioned in chapter I.2, toward the end of his life claimed to have calculated $1/\alpha$, which at the time was measured to be 136. Later, when $1/\alpha$ turned out to be closer to 137, he tried to modify his theory. Freeman Dyson told me that, to Eddington's credit, he was always careful to say that his theory represents a wild speculation.

[7] Indeed, it never occurred to the likes of Planck and Einstein.

[8] Use equal signs. Then $mvr = \hbar$ and $F = ma$ give $e^2/r^2 = mv^2/r = \hbar^2/mr^3$, so that $r = \frac{\hbar^2}{e^2 m}$. So $E = \frac{1}{2}mv^2 - \frac{e^2}{r} = \frac{1}{2}\frac{e^2}{r} - \frac{e^2}{r} = -\frac{e^2}{2r} = -\frac{e^4 m}{2\hbar^2}$, which is the correct result.

The dull function hypothesis

Limiting behaviors: only three numbers

The fly by night physicist lives and dies by the dull function hypothesis. Most of the functions of dimensionless variables in physics are pretty blah, sort of flat, but sometimes a power law or an exponential pops up for reasons that are usually fairly self evident. Or, they are determined by symmetry, such as a spherical harmonic. Occasionally, there might be a resonance peak, approximated by a delta function, and physicists celebrate.

More often than not, if you could figure out the limiting behavior of a function, when its argument goes to zero or infinity, or some special value (such as $\frac{\pi}{2}$ for an angle), then an intelligent interpolation might work. We look at an elementary example in the next section. Another example of this will be given in chapter VIII.1 on water waves.

Early in my studies, an eminent professor told me that in experimental physics, there were an infinite number of numbers, but in fundamental physics, only three numbers: 0, 1, and ∞. What he meant was of course that after you had employed the appropriate units so the number of interest is dimensionless[1] and had done your dimensional analysis, the result could only come out to be either (a) much less than you expected, (b) more or less what you expected, or (c) much larger than you expected. You understand that, in fact, there are only two numbers, 0 and 1, since ∞ is inversely related to 0.

Thus, in physically interesting limits, a function would either go to a constant or to 0 (or equivalently, ∞). The default guess is a constant. If you guess 0 or ∞, then you have to provide an explanation.

You are of course expected to adapt these general considerations to the specific situation at hand.

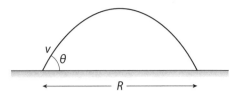

FIGURE 1. Range R of a projectile as a function of the angle θ.

Artillery range

Here is a freshman physics level problem. A cannonball is launched with speed v at an angle θ. What is its range R? See figures 1 and 2.

While the problem is elementary, it has kept countless physicists in the favor of kings and generals throughout human history. Many eminent physicists have served as artillery officers.[2]

Since $[v] = L/T$ and $[g] = L/T^2$ (once again, the mass of the cannonball cancels out), it follows almost instantly that

$$R = \frac{v^2}{g} f(\theta) \tag{1}$$

with some unknown function $f(\theta)$.

Well, if you fire horizontally, the cannonball would just plop down on the ground. And if you are so idiotic to fire straight up, watch out! You are eliminated from the gene pool. Thus, $f(0)$ and $f(\pi/2)$ vanish. A simple guess would be that $f(\theta)$ equals the product of a function that vanishes at $\theta = 0$ and another that vanishes at $\theta = \pi/2$, and if you have any physics sense, you would reject something like $\theta(\frac{\pi}{2} - \theta)$. Trigonometric functions enter when you decompose a vector into its vertical and horizontal components. Hence you guess[3] $\sin\theta\cos\theta$, or up to some overall factor of 2, $\sin 2\theta$.

Know thy function at both ends

In numerous situations, we know an otherwise unknown function at both ends of its range. We will encounter an example in the quantum wave function for the Coulomb problem in chapter III.2. The wave function vanishes exponentially as some radial variable r goes to infinity, and it vanishes linearly as $r \to 0$. Can you guess the wave function without solving the Schrödinger equation? Give it a try.

Worrywort speaks up, as we all knew that he would. "Knowing a function at two ends does not determine the function uniquely," he says. Yes yes, we know that. So, let's go on.

FIGURE 2. The author fires a Civil War cannon, twisting his body to the left to yank the leather firing cord as instructed by his crew. The camera man is barely visible in the lower right together with the winter hat of the director of the TV film.

Capturing the Ceryneian hind: the third labor of Hercules[4]

Hercules has to[5] capture the Ceryneian hind, sacred to the goddess Artemis and the speediest of all deer.[6] The hind plans to come down from Mount Artemisus (treated as a point here right on the bank of the river; see figure 3), swim (with speed v_w) across the Ladon River (with width w) to the point P, and then run along the bank (with speed v_l) to shelter in the forest of Arcadia. Where should P be for the hind to get to shelter ASAP? Assume that $\gamma \equiv v_l/v_w > 1$, of course. Neglect the flow speed of the river. (Note also that the problem as formulated here has nothing to do with Hercules as such, in case you are wondering.)

One reason I use this problem is because you could easily fly by day to solve this problem, using Pythagoras's theorem and differentiating to find the minimal time.[7] We will also invoke an instructive but elementary point about dimensional analysis.

So, fly by night!

Let x denote the distance between P and C, the point directly across the river from Mount Artemisus. We are to determine x in terms of v_l, v_w, w, and the distance to the forest of Arcadia.

At first sight, it would appear that dimensional analysis will not be able to help, since two distances are available for x to depend on. But a moment's

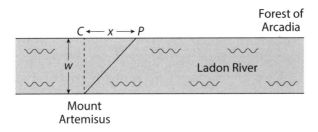

FIGURE 3. The Ceryneian hind swims across the Ladon River to the point P and then runs along the bank to the forest of Arcadia.

reflection indicates that the distance to the forest of Arcadia is irrelevant. One way to see this is to consider moving the forest 10 kilometers, say, to the right in figure 3. That just adds some extra time it takes the hind to get to the forest, but does not affect the optimal location of P.

Another way is to let the hind run backward from P back to C, and then from C to the forest. The time consumed going from P back to C should count as a negative contribution to the hind's time budget. The time it takes the hind to get from C to the forest is fixed and does not depend on x at all. The distance to the forest is irrelevant.

The lesson is that we should not blindly count the number of dimensional variables in the problem. Thus, dimensional analysis implies that $x = f(\gamma)w$. In other words, the width w of the river sets the scale for x.

As I said, it is easy[8] enough for us to fly by day and determine $f(\gamma)$, but in the spirit of this chapter, let us guess and argue instead.

As $\gamma = v_l/v_w$ goes to infinity, that is, if the hind could run much faster than she could swim, then x should go to zero. The simplest guess would be $f(\gamma \to \infty) \sim 1/\gamma$.

As $\gamma \to 1$, the hind could swim as fast as she could run. To her, the river might as well be paved over.[9] The hind could just proceed in a straight line to the forest, and so x equals the undefined distance to the faraway forest, effectively infinitely far away compared to w. We expect $f(\gamma)$ to blow up as $\gamma \to 1$.

Indeed, $\gamma = 1$ imposes a limit on this problem. If the hind could swim faster than she could run on land, the problem no longer makes sense. In elementary physics (and mathematics), the simplest way for a calculation to go haywire is for the result to turn imaginary. Some kind of square root, perhaps. So, a "reasonable" guess, taking into account all the preceding considerations, might be that $f(\gamma) \sim 1/\sqrt{\gamma^2 - 1}$.

No mathematical rigor here, to say the very least. Indeed, the fly by night guy here may have forgotten in which direction the math department is located.

Try your hand at interpolating

Here is a more difficult challenge for you. Guess a function with the properties $g(x) \to x$ as $x \to 0$ and $g(x) \to e^x$ as $x \to \infty$.

Invariably, some undergraduate would tell me that an infinite number of functions behave like this. Well, I understand that. But in the spirit of fly by night physics, what I would like to see is the simplest function with these properties. No, I do not want to discuss what "simple" means; let's just say a function you could write down with the least number of arithmetical symbols.

The "correct" guess, which will be revealed in chapter III.5, is of ginormous importance in the history of physics. Indeed, it ushered in a veritable revolution. Undergraduates typically get the impression that physics progresses logically, and every result is derived step by step. Surely, dear reader, you know that this is often far from the truth. A truly profound advance in physics involves a leap of faith, a wild but educated guess.[10]

The pendulum function revisited

We have already analyzed, in chapter I.1, the pendulum function, defined by $f(\theta_0) \equiv \int_0^{\theta_0} \frac{d\theta}{\sqrt{\cos\theta - \cos\theta_0}}$ in the limit $\theta_0 \to 0$, whereby it goes to a constant. What about a generic value of θ_0, such as $47°$? From everyday experience, we know that nothing much happens. Let's make sure that the integral is not misbehaving. Near the upper limit, $\cos\theta \simeq \cos\theta_0 - (\theta - \theta_0)\sin\theta_0$, and we have a soft square root singularity $\sim \int^{\theta_0} \frac{d\theta}{\sqrt{\theta_0 - \theta}}$, and the integral is perfectly fine, and boring.

The dull function hypothesis, combined with good sense and a flying guess, will get you far in physics.

Exercises

(1) Verify the guess $f(\theta) \propto \sin 2\theta$, and compute the overall numerical factor.

(2) Guess a function such that $f(x) \to 1/x$ as $x \to 0$ and $f(x) \to e^{-x}$ as $x \to \infty$.

Notes

[1] Instead of something with dimension of, say, erg fathoms per hour.

[2] For example, Erwin Schrödinger, and also Rudolf Thun, one of my fellow physics students in college. I understand that Laplace was able to keep his head during the French Revolution partly because of his service to the French artillery corps. D. I. Duveen and R. Hahn, "Laplace's Succession to Bézout's Post of Examinateur des Elèves de l'Artillerie," *Isis* 48, no. 4 (Dec. 1957): 416–427, https://doi.org/10.1086/348608.

[3] A physicist friend of mine told me that in artillery training, when he tried to explain sine and cosine to the sergeant, the man bellowed, "Shove your sinus and conus up you know where!" So much for my friend's erudition. (You could deduce that this anecdote did not occur in the US Army.)

[4] Adapted from M. Huber, *Mythematics: Solving the Twelve Labors of Hercules*, Princeton University Press, 2009, p. 22.

[5] Why? We don't really care to know, but if you must, the ten labors constitute his punishment for throwing his children and his brother's children into the fire.

[6] See https://en.wikipedia.org/wiki/Ceryneian_Hind.

[7] Indeed, the Greek story reminds me of a similar, but more modern story, involving a real hero of 20th century physics. See *GNut*, p. 3.

[8] Indeed, $\frac{d}{dx}\left(v_w^{-1}\sqrt{w^2+x^2} - v_l^{-1}x\right) = 0$ gives $f(\gamma) = 1/\sqrt{\gamma^2-1}$.

[9] As is often the case in modern cities.

[10] A tip of the hat to the young Bohr of chapter I.3.

Match wits with Einstein
over diffusion and dissipation

The reproachful question: why don't you get the Nobel Prize? can no
longer be posed to me. (I would reply each time: Because I am not the
one who awards the prize.) [A. Einstein, in a letter[1] to S. Arrhenius on
learning that he had been awarded the prize]

Diffusion

Put a drop of ink into a glass of water. The ink spreads, and with the passage of
time, the density of ink $n(\vec{x}, t)$ as a function of position $\vec{x}$ and time t will settle
down to a constant value through diffusion.[2]

Diffusion is described by a current $\vec{J}(\vec{x}, t)$ that flows from a region of high
density to neighboring regions of low density. Phenomenologically, the current
is given by

$$\vec{J} = -D\vec{\nabla}n \tag{1}$$

The diffusion constant $D > 0$ is a phenomenological parameter that depends on
various properties of the fluid and so on. To see that the sign in (1) is correct,
we need only check the case in which n depends on the x coordinate only.
Then $J_x = -D\frac{dn}{dx}$, and indeed, for n decreasing with x, the current flows in the
direction of increasing x.

The description of diffusion in (1) evidently applies quite generally. Instead
of two fluids (ink and water), it could apply to a single fluid or gas whose
density varies throughout space.

Friction and dissipation

Imagine applying a force $\vec{F}$ on a particle in a fluid. (The particle may be a molecule of the fluid.) Due to the surrounding fluid, the particle is not going to accelerate as it would in empty space. Rather, the force will impart a velocity[3] given by

$$\mu\vec{v} = \vec{F} \tag{2}$$

with μ a coefficient of friction[4] characteristic of the fluid and of the particle.

Three approaches to the Einstein relation

Pause to think. Ready to match wits with Einstein? Do you feel that there might be a relation between the two pieces of physics described above, diffusion and friction?

Well, microscopically, they both involve molecular collision. Let us imagine how μ and D would be affected by our cranking collision up or down.

Fewer molecular collisions would imply less friction and hence a smaller μ.

However, without collisions with the surrounding particles, the diffusing particle would sail right through, corresponding effectively to an infinitely large diffusion constant D. Thus, a smaller μ should correspond to a larger D.

Also, without thermal agitation, the diffusing particle might simply sit there. Diffusion is due to the jostling of the crowd and so should cease at zero temperature. Conversely, we expect D to increase as temperature goes up.

Yes, there is a relation between D and μ, known as the Einstein relation.[5]

Once you think that a relation might exist, there are three approaches I can think of that you could follow:

(1) The macho serious scholar approach, in which you would write down some fancy equations (who knows what, perhaps Fokker-Planck or something) describing molecular collisions et cetera.
(2) The fly by night guess supplemented by dimensional analysis.
(3) Cleverly setting up a situation in which the two effects under study fight each other. Recall opposition and compromise!

Actually, to tell you the truth, I like all three approaches, but (1) does not belong in this book.

A fly by night guess

So first, a fly by night guess. The heuristic discussion above indicates that with increasing collisions, D decreases while μ increases, suggesting that[6] $D \propto 1/\mu$. Since thermal agitation favors D, let's guess $D \sim T/\mu$.

Next, use dimensional analysis. From (1), since current is given by the number of particles going through a unit area per unit time, and density is the number of particles in a unit volume, we have $[J] = 1/(L^2 T)$ and $[n] = 1/L^3$. Therefore, $[D] = LL^3/(L^2 T) = L^2/T$.

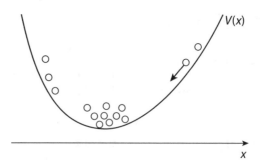

FIGURE 1. Einstein relation between diffusion D and friction μ.

Since according to (2), the friction coefficient μ has dimension of force divided by velocity, we have* $[\mu]=[F]/[v]=[F]T/L=[FL](T/L^2)$. Thus, μ has dimension equal to that of D^{-1} times something with the dimension of force times a distance. But force times a distance is an energy, and thus $[\mu]=ET/L^2$. That energy could only be the ambient temperature† T. Our guess $D \sim T/\mu$ is dimensionally correct.

Thermal equilibrium

Our next approach is to set up a situation involving both diffusion and friction. So put the fluid or gas in an external potential $V(\vec{x})$. The molecules are then pushed by the force $\vec{F}=-\vec{\nabla}V$, leading to a current $\vec{J}=n\vec{v}=-n\vec{\nabla}V/\mu$. In the last step, we used (2). (Again, to keep track of the sign, it suffices to go to one dimension and a region in which $\frac{dV}{dx}>0$, so that $J<0$.)

On the other hand, according to Boltzmann, in thermal equilibrium, particles tend to congregate in regions of low potential energy, with the density of particles given by $n \propto e^{-V/T}$. Aha! This is how temperature gets into the problem. We have $\vec{\nabla}n \propto -(\vec{\nabla}V/T)e^{-V/T}$ and thus

$$\vec{\nabla}n \sim -(\vec{\nabla}V/T)n \tag{3}$$

Note that we did not need to know the proportionality factor to obtain this relation. Most students in my undergraduate class know that $\frac{df}{dx}/f=\frac{d\log f}{dx}$ is known as the logarithmic derivative of the function f and that the logarithm turns the proportionality factor into an additive piece of junk to be annihilated by differentiation. Evidently, $\vec{\nabla}n/n=\vec{\nabla}\log n$ is a logarithmic gradient. We will come back to this "highly forgiving" operation when we discuss gases in chapter V.1.

The diffusion current is then $\vec{J}=-D\vec{\nabla}n \sim +D(\vec{\nabla}V/T)n$. (For $\frac{dV}{dx}>0$, we have $\frac{dn}{dx}>0$ and hence the sign.) In equilibrium, the two currents must fight

---

*This exemplifies my earlier remark that in dimensional analysis, sometimes it is better to use something other than M, L, T. Here we use F, L, T and E, L, T.

†Again, if you are confused by the two uses of T, go back to square one.

each other to a standstill.[7] One current moves to the left, the other to the right, and hence, equating the magnitude of the two currents, we obtain the Einstein relation (see figure 1)

$$D \sim T/\mu \tag{4}$$

Notice that n and $\vec{\nabla}V$ both cancel out. The relation confirms that diffusion ceases at zero temperature.

Incidentally, I regard this relation* as one of Einstein's prettiest contributions to physics.

Random walk

When we combine the diffusion equation (1) with the equation of continuity[8]

$$\frac{\partial n}{\partial t} + \vec{\nabla} \cdot \vec{J} = 0 \tag{5}$$

which states that particles are conserved (that is, the change in density with time is determined by the net influx of particles), we obtain

$$\frac{\partial n}{\partial t} = D\nabla^2 n \tag{6}$$

This equation, which indicates clearly that D has dimension of L^2/T, is also called the diffusion equation.

From dimensional analysis, how far a diffusing particle has gotten to from its starting point in time t is given by[†]

$$r^2 \sim Dt \tag{7}$$

This square root law $r \propto \sqrt{t}$ could be readily derived invoking the well-known random walk or drunkard's walk.[9] Consider a drunkard taking a step of length l, and after every step, he moves off with equal probability in every direction. Denote the displacement of the ith step by (x_i, y_i). Then after a large number N of steps, the average (indicated by the symbol $[\cdot]$) of the distance squared that the drunkard has gotten to from the starting point equals to, according to Pythagoras's theorem,

$$r^2 \equiv \left[\left(\sum_{i=1}^{N} x_i \right)^2 + \left(\sum_{i=1}^{N} y_i \right)^2 \right]$$
$$\simeq \sum_{i=1}^{N} x_i^2 + \sum_{i=1}^{N} y_i^2$$

---

*It ultimately culminated in what is known to the cognoscenti as the fluctuation-dissipation theorem.

[†]Just solve (6) by the "drive your calculus teacher mad" method.

$$= \sum_{i=1}^{N}(x_i^2 + y_i^2) = Nl^2 \tag{8}$$

The key observation here is that the $\simeq$ sign holds, since the cross terms in the sum $\sum\sum_{i\neq j} x_i x_j$ (and similarly $\sum\sum_{i\neq j} y_i y_j$) average approximately to zero, as each term can be either positive or negative. Thus, indeed, $r \propto \sqrt{N}$ follows the square root of N rule. The diffusion constant D is determined by the number of collisions per unit time and the mean free path between collisions.

Notice that, interestingly, the diffusion equation (6) has the same form, up to a factor of $(-i)$ and Planck's constant, as the Schrödinger equation for a free particle: $i\hbar\frac{\partial\psi}{\partial t} = -\frac{\hbar^2}{2m}\nabla^2\psi$.

No k

You might have also noticed that I do not see the need for Boltzmann's constant k in physics. Do you? More on this in chapter IV.1.

Exercise

(1) In a typical star like the sun, the mean free path of a photon, in between scattering off an electron, is about 1 mm. (This could be estimated from the Thomson cross section discussed in chapter II.3. Actually, other interactions may also be important.) Estimate how long it will take a photon to get out of the sun.

Notes

[1] Z. Rosenkranz, *The Travel Diaries of Albert Einstein*, Princeton University Press, 2018, p. 257.

[2] Examples abound in daily life. If you are not in the habit of dropping ink into water, you could always pour milk into coffee.

[3] We could say, somewhat tongue in cheek, that Aristotelian dynamics, rather than Newtonian dynamics, hold here.

[4] For what it's worth, the inverse of μ is known as the mobility.

[5] The Matthews principle is at work here: The same result was also derived by W. Sutherland and by M. Smolukowski, independently.

[6] A fly by day reader asserts that I can only say that D goes like some inverse power of μ. Well, yes, of course. This endnote could be inserted throughout this book.

[7] In chapter I.3, we spoke about opposition and compromise.

[8] See appendix ENS.

[9] I first learned about this by reading G. Gamow's popular physics book, *One Two Three ... Infinity*, Bantam Books, 1971.

Energy released in the first atomic bomb test

State secrets from a picture magazine

The atomic bomb was first tested in 1945 in New Mexico. I know of two stories regarding the energy released.

At the moment of detonation, Fermi let drop a piece of paper. As it fluttered to the ground, pushed by the blast, Fermi was able to estimate the energy released.

The second story, which I will tell here, involves a popular American picture magazine publishing, a few years later, a series of photos showing the expanding fireball[1] from the blast. The actual time after detonation was conveniently stamped in one corner of the photos (see figure 1).

When the photos were published, the energy released in the explosion was still classified information. I read that the British government had asked for this information but was refused. Yet the British physicist Geoffrey Ingram Taylor was able to deduce the energy released, to the consternation of both intelligence establishments. Once again, physicists rule!

Can you obtain state secrets by looking at a few photos? Try it before reading on.

Assume a spherical fireball

Assume that the fireball is spherical, that is, the effect of the ground is negligible. The energy is essentially liberated instantaneously from a point, because the size of the bomb is tiny compared to the size of the fireball. Let us find R as a function of t. Denote by ρ_0 the density of quiet air, that is, the air just outside the fireball (see figure 2).

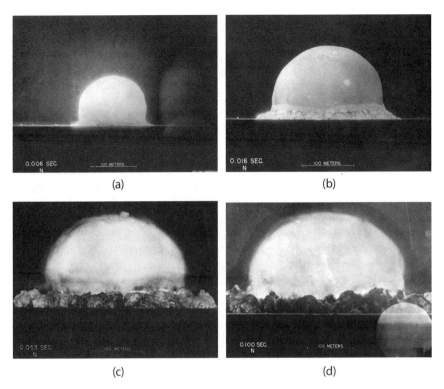

FIGURE 1. Expanding fireball from an atomic bomb. Courtesy Los Alamos National Laboratory, via the Federation of American Scientists web site.

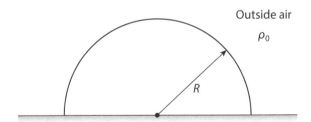

FIGURE 2. The fireball from an atomic explosion pushing the ambient air outward.

Note that $[E] = ML^2/T^2$, and $[\rho_0] = M/L^3$. We want a relation between R and t, and so we need to get rid of M. Thus, note that $[E/\rho_0] = L^5/T^2$. We conclude that

$$R \sim \left(\frac{Et^2}{\rho_0}\right)^{\frac{1}{5}} \tag{1}$$

A quick check: R grows with time, and the larger E is, the larger R will be at a given time. The larger ρ_0 is, the harder the fireball would have to push, and hence a smaller R. All sensible.

Estimating the bomb yield

Numerically, make a log-log plot of the growth of the fireball with time: $\log R \simeq \frac{2}{5} \log t + \frac{1}{5}(\log E - \log \rho_0)$. The slope of $\sim \frac{2}{5}$ allowed Taylor to check that the analysis is more or less correct. The energy released pops out almost instantly.

Alternatively, rewrite (1) as

$$E \sim \frac{\rho_0 R^5}{t^2} \tag{2}$$

The density of air ρ_0 is about 1.2 kg/m³. From the first photo, we see that at $t = 0.006$ sec, $R \sim 90$ m, and so $E \sim 2 \times 10^{14}$ joules. Looking up the conversion 1 kiloton TNT $\simeq 4 \times 10^{12}$ joules, we get a bomb yield of about 50 kilotons TNT. Note that since E depends on R to the 5th power, the result depends sensitively on our estimate of the radius.* To obtain a better estimate, we should thus use all four photos instead of just one, and go through the log-log analysis outlined in the preceding paragraph.

The pressure P exerted by the fireball

We could use the same approach to determine other quantities, such as the pressure P exerted by the fireball. I will let you show (in exercise 1) that

$$P \sim \left(\frac{E^2 \rho_0^3}{t^6}\right)^{\frac{1}{5}} \tag{3}$$

Pressure decreases with time and increases with the energy released and the density of air the fireball has to push against, which makes sense.

Suppose you are standing at a distance R from the detonation. By time t, the blast reaches you. What is the pressure exerted on you? From (1), we have $t^2 \sim \rho_0 R^5 / E$. Plugging that into (3), we obtain $P \sim E/R^3$. The pressure is just the energy density contained in the fireball if the energy released were distributed more or less uniformly in a sphere of radius R.

Incidentally, in elementary physics, pressure is introduced as a force per unit area, but it is often useful, as we will see in this book, to think of it as an energy per unit volume: $[P] = [F/A] = (ML/T^2)/L^2 = M(L/T)^2/L^3 = [E/V]$, using a more or less self-evident notation. In words, energy is force times length.

Were you surprised that ρ_0 drops out of the result $P \sim E/R^3$? Once we reached the correct dimension for P, there is, so to speak, no more room for ρ_0 to come in.

With an energy release of about $E \sim 2 \times 10^{14}$ joules, the pressure 10 km away from the bomb blast is about $E/R^3 \sim 200$ newton/m².

---

*For instance, if we had taken $R \sim 80$ m, we would have gotten a yield of about 25 kilotons, since $(9/8)^5 \simeq 2$.

Lessons to fly by night with

The story of Taylor thumbing his nose at the military censors teaches fly by night physicists useful lessons.

Upon first hearing "atomic bomb blast," we might think that we need to master some nasty bit of nuclear physics. But in fact, the job of nuclear physics is to produce the energy E, and once E is produced, its job is done. The problem at hand only requires that E is produced almost instantaneously in a region much, much smaller than R.

Next, we might think that all kinds of complicated fluid dynamics, shock waves and the like, would be involved. That may be so, but ultimately, the blast just generates a ball of hot gas whose purpose in life is to push the ambient air outward.

Exercises

(1) Find the pressure P exerted by the fireball.

(2) The following interesting problem is discussed in the book[2] by P. Krapivsky, S. Redner, and E. Ben-Naim. Consider a classical gas consisting of hard spheres (called particles) all sitting at rest (that is, the temperature of the gas equals zero) and filling all of space. What happens if a particle suddenly starts moving at high speed inside this gas? It collides with particles in the gas, which in turn collide with other particles, and so on. This collision cascade may be thought of as an explosion. Numerical simulation (see p. 76 of the book just cited) showed that eventually the developing cascade is roughly spherical. Estimate by dimensional analysis the radius R as a function of time t and the energy E injected by the high speed particle. Assume that R depends only weakly on the radius a.

By the way, if the gas is initially confined to half of space, say, $x \geq 0$, and a particle is shot into the gas at high speed along the x direction, some particles are ultimately ejected into the $x < 0$ region. Details of this backsplatter (shown on the cover of the book cited) are not completely understood.

Notes

[1] A colleague expert in explosions cautioned that the commonly used term "fireball" may be misleading: There is no fire inside the ball! The term "spherical blast wave" would be more accurate.

[2] P. Krapivsky, S. Redner, and E. Ben-Naim, *A Kinetic View of Statistical Physics*, Cambridge University Press, 2010.

Interlude

Math medley 1

I introduce this math medley partly to interrupt the physics narrative, but, more importantly, partly to show that the tools we have been developing are often applicable to "purely mathematical" problems physicists might encounter. In my course, students liked these brief digressions. Here we will enjoy the combined power of cyclic symmetry, dimensional analysis, taking limits, and so forth.

Calculating without calculating

Given a triangle with sides of length a, b, and c. What is its area A?

The brute force approach would be to pick the side c and call it the base. From the base, erect a perpendicular line to the opposite vertex and call this line the altitude (just draw a figure as you read along; it is elementary high school stuff). The length of the altitude, namely the height h of the triangle, is determined by applying Pythagoras's theorem twice, once to each of the two right triangles the altitude divides the original triangle into. After some tedious algebra, we obtain the altitude $h = h(a, b, c)$ as a function of a, b, and c. Then $A = \frac{1}{2}ch$. The resulting formula is expected to be highly asymmetrical and ugly, since we have privileged c, ruining the inherent symmetry between a, b, and c.

Instead, let's argue as follows. If $a + b = c$, the triangle collapses and $A = 0$. Thus A should be proportional to $(a + b - c)$, and hence by the principle of democracy, proportional to the product $Q \equiv (a + b - c)(b + c - a)(c + a - b)$. But this has dimension of length cubed. By dimensional analysis alone, you might argue that A must be given by $Q^{\frac{2}{3}}$.

But this is surely wrong. Your math sense, the analog of physical intuition, has kicked in. No way a cube root could emerge from the brute force method outlined above! Pythagoras's theorem involves only squares and square roots. An interesting case of imagining the result of a calculation without calculating!

Thus, by dimensional analysis, the sought-for formula should be for A^2, not for A, and involves multiplying Q by a linear function of the three given lengths, which by symmetry must be $(a+b+c)$. The area squared A^2 should be proportional to $\Pi \equiv (a+b-c)(b+c-a)(c+a-b)(a+b+c)$.

The proportionality constant could now be determined by considering the extreme (isosceles) case of $a=b\to\infty$ with c fixed. The area should go as $A\to\frac{1}{2}ca$, so that $A^2\to\frac{1}{4}c^2a^2$, but by inspection, $\Pi\to(2a)cc(2a)=4c^2a^2$.

Hence we obtain the pleasingly symmetric result

$$A^2 = \frac{1}{16}(a+b-c)(b+c-a)(c+a-b)(a+b+c) \tag{1}$$

known as Heron's formula[1] since antiquity.

It is fun to check (1). For example, take the 45° right triangle with $a=b=1$, $c=\sqrt{2}$. Then $A^2=\frac{1}{16}(2-\sqrt{2})\sqrt{2}\sqrt{2}(2+\sqrt{2})=\frac{1}{16}(4-2)2=\frac{1}{4}$. Hey, math works.

By the way, when I said that, by dimensional analysis, the sought-for formula should be for A^2, you could have objected and said, "Why couldn't the formula be for A and involve dividing Q by $(a+b+c)$?" But then it won't pass the $a=b\to\infty$ test. You could think of other nitpicking objections, but they are all easily answered.

That A is given as a square root of the expression in (1) provides another check. It is clever enough to "tell" us that when $c>a+b$, for instance, the triangle does not exist.

Incidentally, the quantity $Q/(a+b+c)$ you proposed in your objection has dimension L^2 and thus defines the square of a length. That length in fact has a nice geometrical meaning, yours to find out by doing exercise (2).

This strikes me as a wonderful example of the wrong answer of a problem that may actually be the right answer for another problem, when the answer is based on general considerations such as symmetry and dimensional analysis, rather than on some despicable approximation.

Yet another proof of Pythagoras's theorem

The world is not lacking in proofs of Pythagoras's theorem. (One proof was even given by an American president.[2]) Applying (1) to a right triangle, we obtain yet another proof. I let you have the fun of working it out.

Keeping things as symmetric as possible

Throughout physics and mathematics, it is generally advantageous to keep calculations as symmetric as possible, as strikingly illustrated by the elegant derivation of Heron's formula given here. Interchange, or more generally,

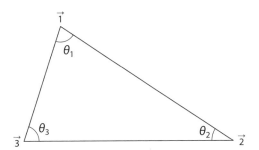

FIGURE 1. An arbitrary triangle.

permutation symmetry, is an important theme in theoretical physics (we saw a particularly simple example in chapter I.2). Here we give another example.

Consider an arbitrary triangle whose vertices we locate with three 2-dimensional vectors $\vec{1}, \vec{2}$, and $\vec{3}$, referred to some origin O. See figure 1. Denote the length of the side joining vertices 1 and 2 by $(12) \equiv \sqrt{(\vec{1} - \vec{2}) \cdot (\vec{1} - \vec{2})}$ and similarly for (23) and (31). The sine of the angle θ_1 at vertex 1 is then, recalling the definition of the vector cross product (note that it is a rotational scalar here, since we are in two dimensions), given by

$$\sin \theta_1 = \frac{(\vec{1} - \vec{2}) \times (\vec{1} - \vec{3})}{(12)(31)} = \frac{(\vec{2} \times \vec{3}) + (\vec{3} \times \vec{1}) + (\vec{1} \times \vec{2})}{(12)(31)} \tag{2}$$

Check that, almost by definition, geometrical quantities such as $\sin \theta_1$ remain unchanged as we move O around, by letting $\vec{1} \to \vec{1} + \vec{x}$, $\vec{2} \to \vec{2} + \vec{x}$, and $\vec{3} \to \vec{3} + \vec{x}$ with $\vec{x}$ an arbitrary vector.

We notice wistfully that while the numerator of this expression is cyclically invariant under $1 \to 2$, $2 \to 3$, $3 \to 1$, the denominator is not. Sigh! But we could readily make it cyclically invariant: divide it by the length (23). In other words,

$$\frac{\sin \theta_1}{(23)} = \frac{(\vec{2} \times \vec{3}) + (\vec{3} \times \vec{1}) + (\vec{1} \times \vec{2})}{(12)(23)(31)} \tag{3}$$

is cyclically invariant.

It follows immediately that

$$\frac{\sin \theta_1}{(23)} = \frac{\sin \theta_2}{(31)} = \frac{\sin \theta_3}{(12)} \tag{4}$$

Surely you recognize this as the law of sines[3] taught to many unsuspecting school children.[4]

Theorems about three lines intersecting at the same point

As a kid, I was amazed by all those theorems in geometry about three lines intersecting at a single point. You know the ones[5] I mean. Let us derive one of them here.

Keep the notation we just used. Consider the vertex $\vec{3}$. The midpoint of the side opposite this vertex is located by $\frac{1}{2}(\vec{1}+\vec{2})$. The line joining $\vec{3}$ to this midpoint is thus described by

$$\vec{3} + 2\lambda_{12}\left(\frac{1}{2}(\vec{1}+\vec{2}) - \vec{3}\right) = \lambda_{12}(\vec{1}+\vec{2}) + (1 - 2\lambda_{12})\vec{3} \tag{5}$$

with a parameter λ_{12} running from $-\infty$ to $+\infty$.

By cyclic symmetry, the other two lines are described by $\lambda_{23}(\vec{2}+\vec{3}) + (1 - 2\lambda_{23})\vec{1}$ with λ_{23} another parameter, and so on.

By inspection and by symmetry, we see that the lines intersect when $\lambda_{12} = \lambda_{23} = \lambda_{31} = \frac{1}{3}$. The simultaneous intersection of the three median lines of a triangle occurs at* $\frac{1}{3}(\vec{1}+\vec{2}+\vec{3})$. My amazement turned into a deep respect for the power of symmetry.

Exercises

(1) Use Heron's formula (1) to prove Pythagoras's theorem. (By its very nature, this is a fly by day exercise.)

(2) Show that the radius of the circle (known as the incircle) inscribed inside a triangle with sides a, b, c is given by

$$R = \sqrt{\frac{(a+b-c)(b+c-a)(c+a-b)}{4(a+b+c)}}$$

(3) The radius of a circle circumscribing a triangle (that is, passing through all 3 vertices) with sides a, b, c is given by

$$R = \frac{abc}{4A}$$

with A the area. Argue by invoking cyclic symmetry, dimensional analysis, and so forth.

(4) Show that the common quantity in (4) is equal to $1/2R$ with R the radius of the circle circumscribing the triangle.

---

*This expression also tells how far this intersection point is from the three vertices.

Notes

[1] Named after Hero of Alexandria. This formula can assume many forms. For example, denoting by $s = \frac{1}{2}(a+b+c)$ half of the perimeter of the triangle, then $A^2 = s(s-a)(s-b)(s-c)$. According to Wikipedia, this formula was discovered independently by the Chinese and published in *Shushu Jiuzhang* ("Mathematical Treatise in Nine Sections").

[2] James Garfield, assassinated at age 50 after $6\frac{1}{2}$ months in the White House.

[3] According to Wikipedia, this law was discovered in the 10th century by Abu-Mahmud Khojandi, Abu al-Wafa' Buzjani, Nasir al-Din al-Tusi, or Abu Nasr Mansur.

[4] Another easy way is to calculate the area of a triangle three ways, using each of the three sides in turn as the base. For example, using (23) as the base, we have $A = \frac{1}{2}(12)(23)\sin\theta_2$.

[5] For an elegant proof of the theorem stating that the three altitudes of a triangle intersect at a single point, see the pseudo-reference cited in endnote 8 on p. 202 of *Group Nut*.

Part II

Telecommunication is possible

Electromagnetism

strange dimensions

The dimension of electromagnetism

I have to assume that you are familiar with electromagnetism. In appendix M, I offer you a quick review of Maxwell's equations. But for us fly by night physicists, the details do not matter too much; here we care only about the broad outline. Indeed, in this chapter, we are mostly concerned with the dimensions of the relevant quantities and symmetries.

Since the Coulomb potential e^2/r is an energy, we have

$$[e^2] = EL = ML^3/T^2 \tag{1}$$

Hence

$$[e] = E^{\frac{1}{2}}L^{\frac{1}{2}} \quad \text{or} \quad [e] = M^{\frac{1}{2}}L^{\frac{3}{2}}/T \tag{2}$$

It follows that the electric field $\vec{E}$ has dimension

$$[E] = [e/r^2] = (M/L)^{\frac{1}{2}}/T \tag{3}$$

In the convention used in this book (see below), we insist that $\vec{E}$ and $\vec{B}$ have the same dimension $[E] = [B]$, so that the Lorentz force acting on a particle with charge e moving with velocity $\vec{v}$ is given by

$$\vec{F} = e(\vec{E} + \frac{\vec{v}}{c} \times \vec{B}) \tag{4}$$

A quick check: $[eE] = ML/T^2$ has dimension of force.

Peculiar dimensions

Dyson has remarked that since electric charge e and field E have peculiar fractional dimensions, they are not "physical." He asked rhetorically, how can an experimentalist measure the square root of a centimeter? True, they can't. But I think that this fractional dimension is simply because e never occurs by itself but is always effectively in the combination e^2 (as implicit in quantities such as eE), just as Newton's gravitational constant G always occurs multiplied by a mass, as was pointed out in chapter I.2. Still, it would be good for students to realize that the electric field can be measured only by the effect it has on charged particles.

Indeed, the dimension $[e^2] = ML^3/T^2$ should remind you of something. Yes, you encountered the combination L^3/T^2 as the dimension of $\kappa \equiv GM$ back in chapter I.2. This all makes sense, since according to Newton's and Coulomb's laws, GM_1M_2 and e^2 have the same dimension.

Electromagnetic duality

When you studied electromagnetism, you probably had a vague sense of some kind of interchange symmetry between the electric and magnetic fields. Indeed, you could check that in empty space, the set of Maxwell's equations is invariant under the operation $\vec{E} \to \vec{B}$ and $\vec{B} \to -\vec{E}$. For instance, the equation $\vec{\nabla} \times \vec{E} + \frac{1}{c}\frac{\partial \vec{B}}{\partial t} = 0$ is transformed into $\vec{\nabla} \times \vec{B} - \frac{1}{c}\frac{\partial \vec{E}}{\partial t} = 0$, and vice versa. This mysterious symmetry, not fully understood to this day, is known as electromagnetic duality.[1] The caveat "in empty space" is necessary because, while the electric charge is commonplace, the corresponding magnetic charge (namely, the fabled magnetic monopole theorized by Dirac) has not yet been experimentally discovered.

Energy density and energy flow

In a course on electromagnetism, you have surely derived formulas for the energy density ε contained in an electromagnetic field and the energy flux $\vec{S}$ carried by an electromagnetic wave.

The fly by night physicist could easily write down these formulas. By rotational invariance, the scalar ε can depend only on the magnitude but not on the direction of $\vec{E}$. Thus, it should be expressed in terms of $\vec{E}^2$. By electromagnetic duality, ε should not discriminate between $\vec{E}$ and $\vec{B}$. Furthermore, from (3), we see that $\vec{E}^2$ and $\vec{B}^2$ have dimension $[E^2] = [B^2] = M/(LT^2) = (ML^2/T^2)/L^3$, namely, the dimension of an energy density. From these considerations, the energy density of an electromagnetic field could only be

$$\varepsilon \sim (\vec{E}^2 + \vec{B}^2) \tag{5}$$

You might wonder about the scalar $\vec{E} \cdot \vec{B}$. Since it also has the dimension of an energy density, you might naively think that it could be added[2] to (5) with some coefficient. One reason that this possibility is ruled out is that $\vec{E} \cdot \vec{B}$ flips sign under duality. (When we get to chapter VI.1, we will see that this is also ruled out by space reflection and time reversal.)

The energy flux carried by an electromagnetic wave, namely, the energy going through a unit area per unit time, is described by a vector $\vec{S}$, known as the Poynting vector.[3] By the powerful combination of rotational symmetry and dimensional analysis, we can write down almost immediately[4]

$$\vec{S} \sim c\vec{E} \times \vec{B} \tag{6}$$

We have $[S] = (L/T)[E]^2 = (L/T)(M/L)/T^2 = M/T^3 = (ML^2/T^2)/L^2T$, which is indeed energy per unit area per unit time. Note that it is invariant under duality.

As I explained in the preface, the time to worry about the 2s and πs is after you have obtained (5) and (6), not before. Indeed, it is not all that difficult to show[5] that the coefficient is $\frac{1}{8\pi}$ in (5) and $\frac{1}{4\pi}$ in (6).

A word, and only a word, about units in electromagnetism

As you know, arguments between physicists about the proper choice of units[6] in electromagnetism often risk descending into physical violence, but to each his or her own choice.* De gustibus non disputandam.[7] Pace.

As the informed reader can see, I tend to switch back and forth between Gaussian and Heaviside-Lorentz (also known as rationalized) systems of units,[8] whichever is more convenient for my purpose at hand.

Guessing the electromagnetic Lagrangian

For later use, I need to tell you about the electromagnetic Lagrangian. This is not needed here in part II, so readers totally unfamiliar with Lagrangians could skip this section. In fact, I will need the Lagrangian only when we get to the emission of gravitational waves by orbiting black holes in chapter VII.4, and even then, not in an essential way. Lagrangians will be crucial in part IX, however.

One step at a time. First, the Lagrangian of free Newtonian point particle is just its kinetic energy $L = \frac{1}{2}m\dot{q}^2$, which evidently has dimension of energy. This is also clear if we recall that the Lagrangian L and the Hamiltonian H are

---

*Standard textbooks, such as Jackson, Garg, and Zangwill, all offer conversion tables. Very helpfully, Garg also writes all important equations and formulas in two different systems of units. For further remarks about units, see appendix M.

related by a Legendre transformation.[9] Next, the energy, or the Hamiltonian, of a point particle in a potential equals (with $p = m\dot{q}$) $H = \frac{p^2}{2m} + V(q) = \frac{1}{2}m\dot{q}^2 + V(q)$, namely, the sum of the kinetic energy and potential energy. In contrast, the Lagrangian $L = \frac{1}{2}m\dot{q}^2 - V(q)$ equals the difference of the kinetic energy and potential energy. The cognoscenti would know that I am glossing over some technicalities here.[10]

Electromagnetism is the first field theory in physics, thanks to Faraday. In contrast to a particle, the electromagnetic field exists at each and every point of space. In other words, the electric and magnetic fields $\vec{E}$ and $\vec{B}$ are functions of $\vec{x}$ and t. The electromagnetic Lagrangian L is given by an integral over 3-dimensional space. Hence, for a field theory, it is more sensible and convenient to speak of a Lagrangian density $\mathcal{L}$, defined self-evidently by $L = \int d^3x \, \mathcal{L}$.

Perhaps the reader already knows the Lagrangian density of electromagnetism. But we could also easily guess. Again, since a Legendre transformation relates the Lagrangian L and the Hamiltonian H, the Lagrangian density $\mathcal{L}$ has the same dimension as energy density $\varepsilon \sim (\vec{E}^2 + \vec{B}^2)$. By rotational symmetry, parity, and time reversal, $\mathcal{L}$ can only be a linear combination of $\vec{E}^2$ and $\vec{B}^2$. But the sum of the two is already taken, so it can only be the difference:

$$\mathcal{L} \sim (\vec{E}^2 - \vec{B}^2) \tag{7}$$

As I said, we don't need this stuff until much later.

To be honest, perhaps like you, I also thought the Lagrangian was some useless formal repackaging when I first encountered it as an undergraduate.[11] Indeed, decades later, whenever I mentioned the Lagrangian, a Nobel certified condensed matter theorist would pretend that he had never heard of it. He would ask rhetorically, "What do I need the Lagrangian for? Give me the Hamiltonian and I will try to find you its eigenvalues and eigenfunctions." Bad attitude.

Exercise

(1) Now that you know the coefficient in (6) to be $1/8\pi$, deduce the coefficient in (5). Hint: Differentiate ε with respect to time and use Maxwell's equations.

Notes

[1] For a glimpse of how vast this subject has become, see J. M. Figueroa-O'Farrill, "Electromagnetic Duality for Children," https://www.maths.ed.ac.uk/~jmf/Teaching/Lectures/EDC.pdf.

[2] We certainly cannot simply have $\varepsilon \sim \vec{E} \cdot \vec{B}$, since ε is manifestly positive while $\vec{E} \cdot \vec{B}$ could take on either sign.

[3] When as an undergraduate I learned about this vector pointing in the direction of the electromagnetic wave, I thought that J. H. Poynting with his family name was surely born to discover this vector. Now I am puzzled that Maxwell didn't know about it; he probably did. Poynting's other great contribution was to propose that human activity would increase global temperature through the greenhouse effect.

[4] Compare and contrast with the fly by day derivation. See, for example, Jackson pp. 189–190.

[5] A typical student might square the Coulomb field around a point charge, integrate, and promptly crash into the notorious infinite self-energy $\sim \int d^3x (e/r^2)^2 = \infty$. Since this issue of the self-energy could not be resolved without quantum electrodynamics, the correct procedure to follow here is to ignore it and consider a large number of charges sitting around. Then the electrostatic energy equals $E = \frac{1}{2} \sum_i \sum_{j \neq i} q_i q_j / |\vec{x}_i - \vec{x}_j| = \frac{1}{2} \sum_i q_i \phi(\vec{x}_i)$, where $\phi(\vec{x}_i) = \sum_{j \neq i} q_j / |\vec{x}_i - \vec{x}_j|$ is the electrostatic potential at the charge i due to all the other charges. The factor of $\frac{1}{2}$ is needed because the double sum counts both the energy due to 7 acting on 9 (know the children's joke about 7 and 9?) and due to 9 acting on 7, while the restriction $j \neq i$ is to ignore self-energy. Go to the continuum limit, and relate the electrostatic potential ϕ to the charge density ρ via $\nabla^2 \phi = -4\pi \rho$; thus $E = \frac{1}{2} \int d^3x \rho(\vec{x}) \phi(\vec{x}) = -\frac{1}{8\pi} \int d^3x \phi(\vec{x}) \nabla^2 \phi(\vec{x}) = \frac{1}{8\pi} \int d^3x (\vec{\nabla}\phi(\vec{x}))^2 = \frac{1}{8\pi} \int d^3x \vec{E}^2$. QED. By the way, the $\frac{1}{4\pi}$ for the Poynting factor, rather than $\frac{1}{8\pi}$, is because the identity used in appendix M involving two cross products contains two terms.

[6] Not to mention units such as statcoulomb, tesla, and volts actually used by experimental, applied, and other real physicists. I studied from Jackson when I was a student; I cannot stand ε_0, μ_0 all over the place. One of my colleagues calls the SI units, much preferred by experimentalists and engineers, a "curse on physics." A brief history of electromagnetic units is given by Garg, p. 15, explaining the roots of all this confusion.

[7] Or in Yiddish, if you prefer: So sue me! Or in Italian as spoken in New Jersey: Would you like a knuckle sandwich?

[8] See appendix M for more.

[9] That is, $H(p,q) = p\dot{q} - L(\dot{q},q)$, with the momentum defined by $p = \frac{\delta L}{\delta \dot{q}}$.

[10] We will be content to let the fly by day people rant and rave about symplectic structures.

[11] A typical stumbling confusion of students on first exposure is thinking that q corresponds to the x in $\vec{E}(x)$. See the table on p. 19 of QFT Nut.

The emission of electromagnetic waves ▬

An apparent paradox

Next time you run into a physics student, or even better, a physics professor, ask him or her this question.

The electric field around a charge falls off rapidly, like $1/r^2$, with r the distance from the charge. OK, then how is electromagnetic radiation possible? How can an electromagnetic wave propagate over long distances, even granted that some amplification is usually needed after detection?

Many professors of theoretical physics, I believe, would be unable to answer this off the top of their heads. (Notice that I exclude experimentalists; they deal with electromagnetism all the time.) Somehow, shaking the charge makes all the difference in the world. Do you know the resolution?

Let us try to quantify this apparent paradox a bit more. Using "merely" rotational symmetry and dimensional analysis, we deduced in chapter II.1 that the energy per unit area per unit time transmitted by an electromagnetic wave is given by $\sim c\vec{E} \times \vec{B}$. We do not know how B falls off around a radiating charge, but at least we have no reason to think that it would fall off less rapidly than E (so that far from the charge, the magnetic field would be stronger than the electric field). So suppose the B also falls off like $1/r^2$. In fact, this is implied by electromagnetic duality. Consider a large sphere of radius R enclosing the radiating charge. Then the energy per unit time going through the sphere would be $\sim 4\pi R^2 (1/R^2)^2 \sim 1/R^2$, which vanishes rapidly with increasing R.

Clearly, this line of reasoning is wrong, and we will soon see why.

Maxwell's equations

Before I resolve this apparent paradox, let me remind you that (in a certain gauge) Maxwell's equations lead to the wave equation (for those who need a quick review, see appendix M):

$$\left(\nabla^2 - \frac{1}{c^2}\frac{\partial^2}{\partial t^2}\right) A_\mu(t,\vec{x}) = -\frac{1}{c}J_\mu(t,\vec{x}) \tag{1}$$

The electrostatic potential and the vector potential are packaged into $A_\mu = (A_0, A_i) = (\varphi, \vec{A})$; charge and current into $J_\mu = (J_0, J_i) = (c\rho, \vec{J})$.

The solution of (1) is given in appendix Gr:

$$A_\mu(t,\vec{x}) = \frac{1}{c}\int d^3x' \int dt' \frac{\delta(t - t' - \frac{1}{c}|\vec{x} - \vec{x}'|)}{|\vec{x} - \vec{x}'|} J_\mu(t',\vec{x}') \tag{2}$$

This may look formidable, but it really isn't, as noted in appendix Gr. Also, for the problem at hand, it simplifies enormously.

With no loss of generality, we can assume that the source has been oscillating steadily for a long time at a single frequency ω (since we can always superpose solutions). Writing $J_\mu(t',\vec{x}') = e^{-i\omega t'}J_\mu(\vec{x}')$, we integrate over t' instantly:

$$\int dt' \delta\left(t - t' - \frac{1}{c}|\vec{x} - \vec{x}'|\right) J_\mu(t',\vec{x}') = e^{-i\omega t}e^{i\frac{\omega}{c}|\vec{x}-\vec{x}'|}J_\mu(\vec{x}') \tag{3}$$

The delta function merely instructs us to set t' equal to $t - \frac{1}{c}|\vec{x} - \vec{x}'|$.

The electromagnetic potential far away from the source

The factor $e^{-i\omega t}$ in (3) tells us that the detected electromagnetic wave also oscillates at frequency ω. Indeed, we do not need math to tell us this intuitive result: after a period $T = 2\pi/\omega$, the source has returned to the same point, so the world should not have changed.[1]

So write $A_\mu(t,\vec{x}) = e^{-i\omega t}A_\mu(\vec{x})$, and define $k \equiv \omega/c$. Then

$$A_\mu(\vec{x}) = \frac{1}{c}\int d^3x' \frac{e^{ik|\vec{x}-\vec{x}'|}}{|\vec{x} - \vec{x}'|} J_\mu(\vec{x}') \tag{4}$$

To address the apparent paradox we started this chapter with, we take the observer to be far away, at a distance much larger than the size a of the radiating system and the wavelength λ of the wave emitted. In other words, $|\vec{x}| = r \gg a, \lambda$. Then $|\vec{x} - \vec{x}'| \simeq r(1 + O(a/r))$, and so (4) simplifies drastically to

$$A_\mu(\vec{x}) \sim \frac{e^{ikr}}{r}\frac{1}{c}\int d^3x' \, J_\mu(\vec{x}') + \cdots \tag{5}$$

The paradox is on the verge of being resolved. Do you see how?

Paradox resolved

The crucial point is the appearance of the factor e^{ikr} in (5).

As you know, in electromagnetism, we have to deal with gauge issues. It is best to focus on gauge invariant quantities, namely, the magnetic field $\vec{B} = \vec{\nabla} \times \vec{A}$ and the electric field $\vec{E}$. Once we know $\vec{B}$, the electric field $\vec{E}$ is determined by one of Maxwell's equations in empty space (far from the radiating system): $\frac{1}{c}\frac{\partial \vec{E}}{\partial t} = \vec{\nabla} \times \vec{B}$.

To obtain the magnetic field $\vec{B} = \vec{\nabla} \times \vec{A}$, all we have to do is evaluate $\vec{\nabla}\,(e^{ikr}/r)$ in (5). (The integral manifestly does not depend on $\vec{x}$.)

When $\vec{\nabla}$ hits $1/r$, it produces the $1/r^2$ dependence characteristic of static electromagnetic fields. In contrast, when $\vec{\nabla}$ hits e^{ikr}, it merely produces a factor of ik. By assumption, $k \sim 1/\lambda \ggg 1/r$. We are much farther away from the source than one wavelength. One of the two terms in $\vec{\nabla}\,(e^{ikr}/r)$ wins, the other loses. The fields $\vec{B}$ and $\vec{E}$ fall off like $1/kr$, much more slowly than $1/r^2$.

So, ta da, we have solved the mystery of why an electromagnetic wave can propagate, while the Coulomb field falls off rapidly like $1/r^2$.

The all-crucial factor e^{ikr} oscillating in space makes our civilization,* such as it is, possible!

With the benefit of hindsight, you could argue that we do not even need all this "math." We are talking, after all, about a wave, duh. Hardly surprising that a factor e^{ikr} must appear. However, many physicists would not have been satisfied with this explanation.

Anupum Garg, in his textbook on electromagnetism, asserted forcefully from the start that the electromagnetic field is as real as a rhino,[2] presumably in response to a student's question.

Not only is the electromagnetic field real, it could leave home and go off on its own, living independently from the charges and currents that generated it.

Let's summarize. After learning that the electric field produced by a charge diminishes like $1/r^2$, we might pessimistically conclude that telecommunication would be impossible. By dimensional analysis, our only hope would be to trade one of the two factors of r for another length, but there is not a length in sight near a charge lazily sitting still. We need to create a length, and it turns out that Nature obliges. Shaking a charge with frequency ω produces a wave with wavelength $\lambda = 2\pi c/\omega$. Instead of $1/r^2$, we have $1/\lambda r \sim \omega/cr$, quite a gain! Crucially, we live in a universe with a finite c.

Were this an electromagnetism textbook ... but it is not

If this were an electromagnetism textbook,[3] we could now go on for tens of pages calculating various aspects of electromagnetic radiation. But it is not. Still, let's push (5) a tiny bit forward.

---

*Not only that, it also allows us to find out about the inconceivably vast universe out yonder.

For convenience, define

$$i\omega\vec{p} \equiv \int d^3x' \; \vec{J}(\vec{x}'),$$
(6)

a vector characteristic of the radiating system. Note that dimensionally, the current is something something per unit time, and so $[\vec{J}] \propto 1/T$. Taking out the factor $i\omega$ renders $\vec{p}$ dimensionally independent of time and easier to think about physically as a dipole moment, as we will see shortly. It also makes later formulas cleaner.

Now differentiate $\vec{A}(\vec{x}) \sim \frac{e^{ikr}}{cr}(i\omega\vec{p})$ in (5) to obtain the magnetic field $\vec{B}$. Recall, from appendix Del, the useful formula* $\partial_i r = x^i/r = \hat{r}^i$, with $\hat{r}$ the unit vector in the radial direction. Define the wave vector $\vec{k} \equiv k\hat{r}$.

Thus, at large r, the leading term for the magnetic field is

$$\vec{B} \sim \omega(\vec{k} \times \vec{p})/cr$$
(7)

You couldn't have it any simpler!

To recognize what $\vec{p}$ represents, we use current conservation,[4] $\partial_0 J_0 = \partial_i J_i$, to eliminate the current J_i in favor of the charge distribution J_0. Since $\partial_j x^i$ vanishes unless $j = i$ (in which case it is equal to 1), we can write

$$i\omega p_i = \int d^3x \; J_j(\vec{x})\partial_j x^i$$
$$= -\int d^3x \; (\partial_j J_j(\vec{x}))x^i = -\int d^3x \; (\partial_0 J_0(\vec{x}))x^i$$
$$= i\omega \int d^3x J_0(\vec{x})x^i$$
(8)

Hence the vector $\vec{p} = \int d^3x \; \rho(\vec{x}) \; \vec{x}$ is just the electric dipole moment of the radiating system.

We have derived the famous theorem that an electric monopole cannot radiate; an electric dipole is needed. Intuitively, this result follows immediately from the fact that electric charge (namely, the monopole) is conserved and hence constant, while time variation is required for radiation. (We will see later that for gravitational radiation, a quadrupole moment is needed.)

According to (7), the magnetic field is perpendicular to the direction of propagation $\vec{k}$ (and to the dipole moment $\vec{p}$). As remarked earlier, the electric field can be determined in terms of the magnetic field using one of Maxwell's equations, $\frac{1}{c}\frac{\partial \vec{E}}{\partial t} + \vec{\nabla} \times \vec{B} = 0$, which in this context says $\omega\vec{E} = -c\vec{k} \times \vec{B}$. Thus, $\vec{E}$ is perpendicular to both $\vec{k}$ and $\vec{B}$ and equal to $\vec{B}$ in magnitude.

To summarize, for the total power transmitted through a large sphere of radius r centered on the radiating system to be constant, $4\pi r^2 S \sim$ constant (with $\vec{S} = c\vec{E} \times \vec{B}$ being the Poynting vector derived in chapter II.2), we need $S \sim 1/r^2$, $E \sim 1/r$, and $B \sim 1/r$. From energy conservation (and from democracy between

---

*Notation: $\partial_i = \frac{\partial}{\partial x^i}$, $\partial_0 = \frac{\partial}{\partial x^0} = \frac{\partial}{\partial t}$.

$\vec{E}$ and $\vec{B}$), we see that in a propagating electromagnetic wave, the fields must fall off like $1/r$. Indeed, (7) tells us that this is the case.

Physics works.

The fly by night physicist admires the sunny blue sky

When a molecule or an atom is shaken by an electromagnetic wave, it radiates. The incoming electromagnetic wave is effectively scattered. From (7), $B \sim \omega k p / c r = k^2 p / c r$, and thus $S \sim cEB = cB^2 \sim ck^4 p^2 / r^2$. The total power radiated equals

$$P \sim ck^4 p^2 \sim \frac{p^2}{c^3} \omega^4 \tag{9}$$

A quick dimensional check: $[p^2] = [e^2 d^2] = [e^2]L^2 = (EL)L^2 = EL^3$, and so $[P] = EL^3 / ((L/T)^3 T^4) = E/T$. Check.

The dependence of the power radiated on the 4th power[5] of frequency exhibited in (9) leads us to one of the famous celebrated triumphs of late 19th century physics. Humans must have wondered since time immemorial why the sky is blue.[6] Lord Rayleigh[7] gave the answer: in the visible spectrum, which the human eye has evolved to tune to, blue scatters a lot more than red.

A corollary is that sunsets look red.

Newton's second derivative versus Aristotle's first derivative

This section lays the groundwork for Einstein gravity. I want to tell you that the dynamical variables in electromagnetism are not the electromagnetic fields $\vec{E}$ and $\vec{B}$, but the potentials A_μ, whose spacetime derivatives determine the electromagnetic fields.

When you first encounter the electrostatic potential and the vector potential, they seem to be invented solely for mathematical convenience. This first impression is incorrect. At a deeper level, quantum electrodynamics is formulated in terms of A_μ, not $\vec{E}$ and $\vec{B}$. In quantum physics, the A_μ are more real[8] than $\vec{E}$ and $\vec{B}$. For our purposes here, you don't need to know all this, but I would like to ask you a question.

Physics began with Newton's $ma = F$, which involves a second derivative, namely, the acceleration $a = \frac{d^2 q}{dt^2}$ (with q the position of the particle), instead of Aristotle's $mv = F$, which involves a first derivative, namely, the velocity $v = \frac{dq}{dt}$. Here comes my question. When you moved on from classical mechanics to electromagnetism, did it strike you as odd that Maxwell's equations involve the first (partial) derivatives $\frac{\partial}{\partial t}$ and $\vec{\nabla}$?

It should, but I can't imagine many of us would ask this question when we were struggling to learn electromagnetism. As it turns out, the underlying dynamics of electromagnetism also involves second derivatives, $\frac{1}{c^2}\frac{\partial^2}{\partial t^2}$ and ∇^2, as in (1). In the 19th century, the A_μ seemed like mathematical luxuries, but with the advent of the quantum, they became necessities of life.[9]

Exercise

(1) Why is the sky not violet?

Notes

[1] The detected frequency could be some integer times ω, but this is possible only in a nonlinear theory.

[2] To which I responded that quantum fields are as real as quantum rhinos.

[3] For example, Jackson, Garg, and Zangwill.

[4] For those unfamiliar with Einstein's repeated index summation convention, $\partial_i J_i = \sum_i \partial_i J_i = \vec{\nabla} \cdot \vec{J}$.

[5] For those who know some quantum field theory, a one-line derivation of this fact is given on p. 457 of *QFT Nut*.

[6] The problem challenged numerous intellects, from Da Vinci and Newton to Tyndall and Lorenz. See Pedro Lilienfeld, *Optics & Photonics News*, 2004, p. 32. L. V. Lorenz also obtained the same ω^4 as Lord Rayleigh but was ignored because he published only in his native Danish. He was also responsible for the Lorenz gauge; see appendix M.

[7] Since I often berate students for not knowing more physics history, which to me only adds to the enjoyment of physics, I told my class at this point that one question on the final exam will be "What is Lord Rayleigh's given name?" Incidentally, I was surprised to read that, contrary to what I had assumed ever since I started studying physics, he was not made a lord because of his achievements in physics.

[8] I am referring to the Aharonov-Bohm effect here.

[9] See *QFTNut*, 2nd edition, p. 474.

Electromagnetic radiation from moving point charges and Compton scattering

Radiation from an accelerating charge

In chapter II.2, we determined that the electromagnetic wave from a source nicely oscillates within bounds at a single frequency ω. But in particle physics and astrophysics, we often have to deal with the electromagnetic radiation from a point charge not simply oscillating and playing nice, but screaming by like a bat out of hell and moving out to spatial infinity.

Let the observer sit at $\vec{x}$ and observe at time t. The charge e particle is located at $\vec{r}(t')$ at time t'. As t' varies from $-\infty$ to $+\infty$, the particle traces out a trajectory in space described by $\vec{r}(t')$. Call the distance between the observer at $\vec{x}$ and the particle $R(t') \equiv |\vec{x} - \vec{r}(t')|$. See figure 1. For convenience, refer to the electromagnetic radiation observed at $\vec{x}$ at time t as the signal.

Finite speed of light

We are interested in the position of the particle not at any old time t', but at the specific time t' when the signal observed at $\vec{x}$ at time t was emitted. Call that time t_e. Due to the finite speed of light c, it takes the electromagnetic wave time R/c to cover the distance R. Thus, t_e is the solution of the equation

$$t - t_e = \frac{1}{c}|\vec{x} - \vec{r}(t_e)| \equiv \frac{R(t_e)}{c} \qquad (1)$$

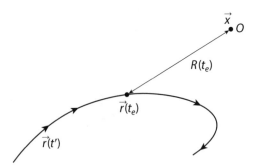

FIGURE 1. The charged particle's trajectory is indicated by $\vec{r}(t')$ as t' varies. The observer at $\vec{x}$ observes a signal at time t. The signal was emitted at time t_e.

Somehow, some undergrads expect a simple formula for $R(t_e)$, but there cannot possibly be an explicit formula, since you are free to specify any trajectory $\vec{r}(t')$ you like. But once you specify the trajectory, we can solve (1) for t_e and then evaluate $R(t_e) \equiv |\vec{x} - \vec{r}(t_e)|$ to tell you what $R(t_e)$ is. Note that t_e is a complicated function of t, $\vec{x}$, and the trajectory $\vec{r}(t')$.

The electromagnetic field observed at $\vec{x}$ and t

Now that we have gotten that straight, let us determine the observed electromagnetic field. But in some sense, we already have solved that problem. Go back to the central result (which allegedly contains all of electromagnetism) stated in appendix Gr and in chapter II.2:

$$A_\mu(t, \vec{x}) = \frac{1}{c} \int d^3x' \int dt' \frac{\delta(t - t' - \frac{1}{c}|\vec{x} - \vec{x}'|)}{|\vec{x} - \vec{x}'|} J_\mu(t', \vec{x}') \tag{2}$$

Just plug in the given J_μ, integrate to obtain A_μ, and then differentiate to obtain the electromagnetic field $\vec{E}$ and $\vec{B}$. Can't get more algorithmic than that!

Let us focus on the $\mu = 0$ component, with the potential $\varphi \equiv A_0$ and charge density $\rho \equiv J_0$. We will then come back for $\vec{A}$.

For our problem, $\rho = e\delta^3(\vec{x}' - \vec{r}(t'))$. So, plug this into (2) to obtain φ. The integral over[1] $\vec{x}'$ can be done instantly:

$$\int d^3x' \delta\left(t - t' - \frac{1}{c}|\vec{x} - \vec{x}'|\right) e\delta^{(3)}(\vec{x}' - \vec{r}(t'))$$

$$= e\delta\left(t - t' - \frac{1}{c}|\vec{x} - \vec{r}(t')|\right) = e\delta\left(t - t' - \frac{R(t')}{c}\right) \tag{3}$$

We have merely set $\vec{x}' = \vec{r}(t')$, as mandated by the 3-dimensional delta function.

From my experience, occasionally a student is inexplicably befuddled by these delta function gymnastics, but it really is trivial. The delta function in (3) just says that t' is what you think it should be. Thus, we obtain a fairly simple result:

$$\varphi(t, \vec{x}) \sim e \int dt' \, \frac{1}{R(t')} \, \delta\left(t - t' - \frac{R(t')}{c}\right) \sim e \int \frac{1}{R} \, \delta \qquad (4)$$

It couldn't be any simpler, as Einstein would say

Indeed, it couldn't be any simpler, as Einstein would say. The result is just Coulomb's law $\varphi \sim e/R$, except that R is not the distance from us to where the particle is now, but the distance from us to where the particle was when it emitted the electromagnetic wave at t' that is just getting to us now. Read this sentence again if it is not clear to you. Makes total sense, yes?

The delta function in the integral over t' simply mandates that we are to solve (1) to determine t_e and then evaluate $R(t_e)$.

Now that we have obtained φ and understood what it means, we can practically write down the vector potential $\vec{A}$. Just imagine plugging into the $\mu = i$ component of (2), and remember that the current $\vec{J}$ contains an extra factor of $\vec{v}$ when compared to the charge density ρ. We simply insert this factor into (4) to obtain

$$\vec{A}(t, \vec{x}) \sim e \int dt' \, \frac{\vec{v}(t')}{cR(t')} \, \delta\left(t - t' - \frac{R(t')}{c}\right) \sim \frac{e}{c} \int \frac{\vec{v}}{R} \, \delta \qquad (5)$$

Again, advice to undergraduates as in the preface. Occasionally, some undergrads, used to neat explicit formulas in introductory courses, expect "the" solution to the problem being discussed, something they could highlight. But a neat formula is just not possible: you have to specify the trajectory $\vec{r}(t')$ first.

Just imagine differentiating instead of actually doing it

At this point, you merely have to differentiate away to obtain the electric and magnetic fields, for example, $\vec{E} = -\vec{\nabla}\varphi - \frac{1}{c}\frac{\partial \vec{A}}{\partial t}$. A quick dimension check of (4) and (5): $[\varphi] = [A] = [e]/L$, which implies $[eE] = [e^2]/L^2$, indeed a force according to Coulomb's law. But things, while conceptually straightforward, could get hairy in a hurry, as you can see by looking at a standard textbook[2] on electromagnetism.

Now, like a zen swordsman, the fly by night physicist merely imagines differentiating rather than breaking a sweat actually differentiating. To obtain the electric field, you hit

$$\varphi \sim e \int \frac{1}{R} \, \delta \qquad (6)$$

and

$$\vec{A} \sim \frac{e}{c} \int \frac{\vec{v}}{R} \delta \qquad (7)$$

with a ∂, which generically could be a $\frac{\partial}{\partial t}$ or a $\frac{\partial}{\partial x}$. Note that if you hit the delta function, you can always integrate by parts.

So we merely need to hit $1/R$ and $\vec{v}$ in turn.

Pick the winner

Now the crucial observation is that if you hit the $1/R$, you get $1/R^2$, and you are dead in the water. As explained in chapter II.2, this is a real loser at large distances. Thus, to get a propagating electromagnetic wave, we have to hit $\vec{v}$. As we can see as clear as day from (6) and (7), only $\vec{A}$ offers us the opportunity to do that.

Physically, this makes a lot of sense, since we expect that it is acceleration that causes a charged particle to radiate. Readers conversant with relativity would know immediately that a uniformly moving charged particle cannot radiate, since we can always transform to a frame in which the particle is at rest doing nothing.

Hitting $\vec{v}(t')$ produces the needed acceleration: $\vec{a}(t') \equiv \frac{d\vec{v}(t')}{dt'}$.

Indeed, even more simply, we could argue by dimensional analysis. To win, you have to get $1/R$. The goal is to avoid the loser $1/R^2$, and instead end up with a $1/R$. Since the winner and the loser must have the same dimension, we need to multiply $1/R$ by an inverse length that is not $1/R$. The only available inverse length that makes physical sense is the acceleration divided by c^2: dimensionally, $[a] = L/T^2$ and so $[a/c^2] = (L/T^2)/(L/T)^2 = 1/L$.

The fly by night physicist argues that the electric field, by comparison with the electrostatic field e/R^2, must be

$$E \sim \frac{e}{R}\left(\frac{1}{c^2}\frac{dv}{dt}\right) \simeq \frac{ea}{c^2 R} \qquad (8)$$

The magnetic field follows from one of Maxwell's equations, $\vec{B} = c\vec{k} \times \vec{E}/\omega$. As in chapter II.2, with $E \sim 1/R$ and $B \sim 1/R$, $E \times B \sim 1/R^2$, and the power radiated through a sphere of radius R is $\sim 4\pi R^2/R^2 \sim 1/R^0$, as desired.

Instead of the slogan "just do it," the fly by night physicist just imagines doing it.

What do we miss with our fly by night approach?

Differentiating more carefully, we would encounter all sorts of relativistic factors.[3] Conceptually, these are of course more important than the various factors of 2 and π we dropped, and we will come back to some of them shortly and in chapter II.4.

Larmor formula for the power radiated by an accelerating charge

But now we can put things together to obtain the Larmor formula for the power radiated by an accelerating charge in the nonrelativistic regime. I need not belabor how energy radiated away per unit time is of basic importance to many areas of physics (and science and engineering).

So, plug $E \sim B \sim \frac{ea}{c^2 R}$ into the Poynting vector $S \sim c(\vec{E} \times \vec{B})$. Then $S \sim \frac{e^2 a^2}{c^4 R^2}$, which when multiplied by the area $4\pi R^2$ of the sphere of radius R gives power radiated as

$$P \simeq \frac{e^2}{c^3} a^2 \tag{9}$$

a remarkably simple formula. I emphasize again that this is a nonrelativistic result.

Since P is by definition positive, but e can take on either sign, we are hardly surprised that e^2 appears. Similarly, rotational invariance requires[4] $\vec{a}^2$. Thus, the only feature in (9) that merits a comment is the c^3.

Every time an odd power appears, it seems a bit, well, odd. So let's check it by dimensional analysis. From chapter II.1, $[e^2] = EL$, and thus $[\frac{e^2}{c^3} a^2] = EL(L/T^2)^2(T/L)^3 = E/T$, which is indeed energy per unit time.

A curious fact. Since power P is proportional to energy and hence $[P]$ contains M, we might have thought that the mass of the radiating particle might come in. But mass does not appear in Maxwell's equations. The puzzle is resolved by recalling that an M is hidden in e^2!

A couple of useful points about dimensional analysis

Of course, this little check also implies that we could have obtained the Larmor formula by dimensional analysis if we could convince ourselves (fairly easily, I think) that, for example, the mass of the charged particle does not come in. Indeed, I think that the Larmor formula offers a couple of valuable lessons about dimensional analysis.

We would like to obtain the power P with $[P] = (ML^2/T^2)/T = ML^2/T^3$. In some discussions of dimensional analysis I have seen, the answer is first expressed in arbitrary powers of the relevant quantities, in the form $e^\alpha a^\beta c^\gamma \cdots$ with α, β, γ, et cetera to be determined.

But I would prefer to inject some physics from the start. We know that the electric field is proportional to e, and so $P \propto e^2$. As already mentioned, we also know that acceleration $\vec{a}$ is a vector and by rotational invariance, it must appear as an even power, most likely 2, unless we can think of a reason otherwise. Another way to see this is the following. Recall that $\vec{E}$ and $\vec{B}$ are obtained by acting with ∂ once on A_μ, and so they can only be linear in a. So $P \propto e^2 a^2$, with

$[e^2a^2] = (ML^3/T^2)(L^2/T^4) = ML^5/T^6$. That we can get it to be ML^2/T^3, the dimension of P, by dividing by c^3 offers a consistency check.

Indeed, suppose we mistakenly thought that a uniformly moving charge could radiate. Then we only have its velocity v, not a, to construct P. With $[e^2] = ML^3/T^2$ and $[v] = [c] = L/T$, that would be impossible. There will always be one more power of L than of T. So, dimensional analysis could arguably tell us a bit about relativity.

You might ask: Why didn't we include v in the Larmor formula? That's because we restricted ourselves to the nonrelativistic regime. Yes, we should include v and write $P \simeq \frac{e^2}{c^3} a^2 f(v/c)$. Evidently, dimensional analysis cannot fix the form of the unknown function[5] f, which could well (and indeed does) contain factors like $\sqrt{1 - \frac{v^2}{c^2}}$. What we did do is to appeal to the dull function hypothesis and assume that $f(0) \sim 1$: A particle sitting at rest could radiate if suddenly accelerated.

For future use, it is more convenient to write the acceleration as $\vec{a} = \frac{d\vec{v}}{dt} = \frac{1}{m}\frac{d\vec{p}}{dt}$, so that

$$P \simeq \frac{e^2}{m^2 c^3} \left(\frac{d\vec{p}}{dt}\right)^2 \qquad (10)$$

Larmor from the dipole and vice versa

Now the fly by night physicist wants to soar. Let's skip the entire discussion starting from (2) and see if we could obtain the Larmor formula (9) by shimmying from the dipole radiation formula $P \sim ck^4p^2$ in chapter II.2.

But you object. The dipole radiation formula was for a dipole obediently oscillating in a confined space at a given frequency ω. In contrast, the Larmor formula is for an accelerating point particle tracing out a possibly unbounded trajectory.

Well, maybe, but let's forge ahead before we get too worried. Picture the dipole as an electron oscillating over a characteristic distance d around a fixed positive neutralizing charge, so that $p \sim ed$. With the position of the electron given by $x \simeq d \sin \omega t$, the acceleration $\ddot{x}$ equals $a = \ddot{x} \simeq \omega^2 d \sin \omega t$ and so, on average, $p \sim ed \sim ea/\omega^2$. Plugging this into the dipole radiation formula derived in chapter II.2, we obtain

$$P \sim ck^4p^2 \sim ck^4e^2a^2/\omega^4 \sim e^2a^2/c^3 \qquad (11)$$

precisely the Larmor formula!

And of course we could go in the other direction and derive the dipole radiation formula from the Larmor formula.

Incidentally, we see the origin of the odd power of c: The Poynting vector has an explicit power of c.

Causality: discrepancy between inside and outside observers

That a charged particle radiates only when accelerated[6] suggests a more physical understanding. A charge just sitting there emanates electric field lines radially outward, as everybody knows. For a uniformly moving charge, we simply Lorentz transform the electric field, which leads to a magnetic field as well. The field lines are smooth and nice, as portrayed these days even in textbooks for nonphysics majors.[7] In contrast, if the particle is subject to a sudden jerk (I picture a charge involved in a car crash), we would expect the shock to be transmitted to the field lines, in a manner vaguely reminiscent of shock waves[8] in fluid dynamics. We are familiar with the wake "radiated" by a speed boat and the sound wave "radiated" by a passing jet. Speaking very loosely, we expect an electromagnetic wave to be "radiated" also.

Ed Purcell worked out an elegantly pictorial and quantitative determination of electromagnetic radiation[9] based on this notion. I said pictorial, so it is crucial to examine figure 2 in detail. The key point is that the field lines must reconcile the expectation of the "outside observer" with that of the "inside observer," as will be introduced presently.

Consider a nonrelativistic particle with charge e moving with velocity v significantly less than c (say, $v \sim c/10$ to be definite). At time $t = 0$, when the particle is at $x = 0$, it is brought to rest in a short time interval τ (car crash) with uniform deceleration $a = v/\tau$. Elementary physics tells us that the particle stops at $x_s = \frac{1}{2} v \tau$.

The situation at time $t = T \gg \tau$ is depicted in figure 2a. The figure shows the arcs of two circles, one centered at $x = 0$ with radius $R = cT$, and the other centered at $x_s = \frac{1}{2} v \tau$ with a slightly smaller radius $c(T - \tau) \lesssim cT$. The two radii ($\sim cT$) are much larger than the separation between the two arcs ($\sim c\tau$). This separation is in turn much larger than x_s. So the figure is hardly to scale.

Consider an observer far away, outside the arc centered at $x = 0$. In his innocence, he does not know that the particle has stopped. By causality, that information, traveling at the speed of light, could not have reached him yet. Instead, he thinks that the particle continues to move with velocity v and is now at $x_e = vT \gg x_s$, emitting the radiation he now sees. (The subscript e stands for "emit," and s stands for "stop.") He expects to see that a field line makes an angle θ with the x-axis and intersects the larger circle at the point denoted by B, as shown. See figure 2b.

Now consider an observer inside the smaller arc centered at x_s, aware that the particle has stopped. Focus on an electric field line extending from x_s at an angle θ from the x-axis, as shown. This field line intersects the smaller arc at the point denoted by A.

By continuity, the field line has to quickly jog over from A to B. Since $T \gg \tau$, the geometry of the situation indicates that the field going from A to B has to be enhanced. The angular component of the electric field E_θ is much larger than the radial component of the electric field E_r. Referring to figure 2c, we

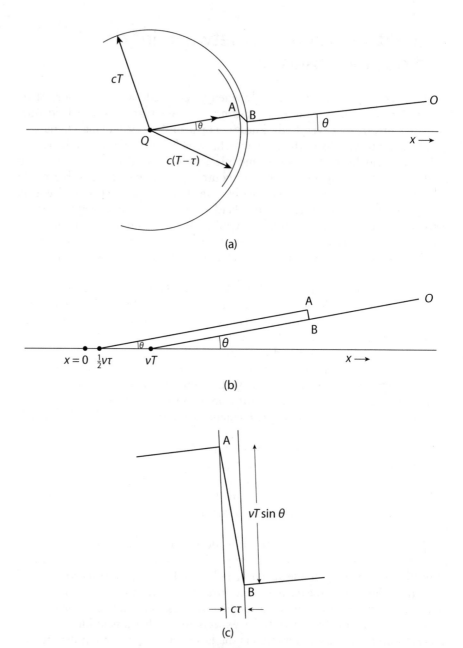

FIGURE 2. Purcell's argument for electromagnetic radiation. The figure is wildly not to scale. (a) A charge (denoted by Q in the figure) is brought to rest abruptly. The figure depicts what an observer at O sees. Details are given in the text. (b) The location of the charge when it comes to rest, and the alleged location of the charge when the observer observes it. Details are given in the text. (c) The electric field line has to make a quick jog from A to B.

see that the enhancement factor is determined geometrically by

$$\frac{E_\theta}{E_r} = \frac{vT\sin\theta}{c\tau} = \left(\frac{v}{c}\right)\left(\frac{T}{\tau}\right)\sin\theta \tag{12}$$

For $T/\tau \gg c/v$, this enhancement factor could be huge. Indeed, since $T = R/c$, we have the enhancement factor $\frac{vT}{c\tau} = \frac{vR}{c^2\tau} = \frac{a}{c^2}R \propto R$, proportional to the large distance to the observer.

At large distances, the feeble electrostatic field $E_r = e/R^2$ is enhanced by a factor $\propto R$ to

$$E_\theta \sim \left(\frac{a}{c^2}R\sin\theta\right)E_r \sim \left(\frac{a}{c^2}\right)\left(\frac{e}{R}\right)\sin\theta \propto \frac{1}{R} \tag{13}$$

thus enabling telecommunication and other wonders of our age!

To obtain the energy radiated, we have to square the electric field $E^2 = E_\theta^2 + E_r^2 \simeq E_\theta^2$, average $\sin^2\theta$ over the spherical shell, integrate, et cetera, all kinds of fun stuff you could do in the privacy of your own home. We obtain precisely the Larmor result for the power radiated: $P = 2e^2a^2/3c^3$. (You could maybe even see the $\frac{2}{3}$ from averaging $\sin^2\theta$. Hint: two transverse directions in 3-dimensional space.)

I find this physical derivation extremely appealing, particularly when backed up by the more mathematical derivation involving (7). This argument also shows that in an electromagnetic wave, the electric field $\vec{E}$ is perpendicular to the direction of propagation.

Once again, the finite speed of light is crucial.

Scattering cross section: Thomson

I assume that you are familiar with the concept of a scattering cross section, a concept already present in classical physics, which is what we have been doing thus far in this chapter. It measures the effective area a target particle presents to a scattering particle. We will be discussing the scattering of an electromagnetic wave off of an electron, evidently a phenomenon of basic importance in the development of modern physics. So, instead of a scattering particle, we have an incident wave.

The classical picture consists of an electromagnetic wave shaking an electron, thus causing it to radiate (see figure 3). The scattering cross section σ is defined to be equal to the energy radiated by the electron per unit time divided by the energy carried by the incident electromagnetic wave per unit time per unit area, which is manifestly an area: $\sigma = $ (energy radiated per unit time)/(incident energy per unit area per unit time).

For the numerator, plug in the Larmor formula: $P \simeq \frac{e^2a^2}{c^3}$. Newton told us how to calculate the acceleration a on the very first day we started studying physics: $a = F/m = eE/m$, with E the electric field of the incident wave. Thus, the electron radiates away $P \simeq \frac{e^2}{m^2c^3}E^2$.

For the denominator, plug in the power per unit area of the incident wave. Since this $\sim c(\vec{E} \times \vec{B}) \sim cE^2$, we see that E^2 cancels out in the cross section σ,

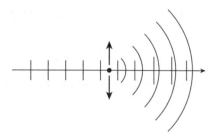

FIGURE 3. A plane wave shaking an electron, causing it to radiate.

as it should. Dividing, we obtain what is known as the Thomson cross section

$$\sigma_{\text{Thomson}} \sim \left(\frac{e^2}{mc^2}\right)^2 \tag{14}$$

Interesting how the mass comes in! Incidentally, the exact result equals $8\pi/3$ times this result. You can almost see how this factor could come in. The Thomson cross section for the electron comes out to be $\sim 6.6 \times 10^{-25}$ cm^2, that is, about[10] 0.66 barn.

It is worth remarking that the electromagnetic scattering cross section for a charged particle as derived here starts out of order ω^0, namely, that it is frequency independent at low frequency, in sharp contrast to the electromagnetic scattering cross section of a neutral object (such as an atom or a molecule), as discussed in chapter II.2. Rayleigh scattering starts as $\propto \omega^4$.

I need hardly remark that we could also obtain the Thomson cross section by dimensional analysis. Since $[e^2]$ has the dimension of an energy times a length, we note that $\frac{e^2}{mc^2}$ is a length, known as the classical radius of the electron. Note the classical radius of the electron is the length scale r over which the electron's Coulomb energy e^2/r is on the order of its rest energy mc^2. So σ_{Thomson} is literally the classical cross sectional area of an electron.

The photon discovers that the electron is a wave

When does this classical picture break down?

Well, when the electromagnetic wave listens to Einstein and reveals its quantum character, that it is actually made up of a stampeding herd of photons, each with energy $\hbar\omega$.

Interestingly, the frequency ω of the wave does not even come into the classical discussion above. The energy $\hbar\omega$, being proportional to $\hbar$, evidently measures how quantum the wave is. What should we compare it to? The electron's rest energy is the only candidate around. So, we expect the classical result (14) to break down when $\hbar\omega$ exceeds mc^2, or equivalently, when the wavelength of the electromagnetic wave $\lambda \lesssim \hbar/mc$.

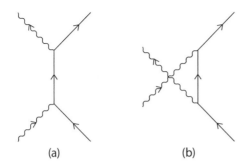

FIGURE 4. Feynman diagrams for Compton scattering: (a) A photon (wavy line) is absorbed by an electron (straight line), which subsequently emits a photon. (b) The outgoing photon observed by the experimentalist was actually emitted by the incoming electron, which subsequently absorbs the incoming photon. According to quantum field theory, both possibilities must be included.

The length $\hbar/mc$ is known as the Compton wavelength[11] of the electron. Recall that de Broglie taught us that the wavelength of a material particle is $\simeq \hbar/p$, an insight that broke the quantum puzzle open. Interestingly, here mc plays the role of the target electron's momentum p.

Quantum mechanics started when Planck proposed that an electromagnetic field actually consists of discrete particles called photons. Here the classical Thomson scattering breaks down when the photon discovers that, hey, the electron is a wave also. What irony!

Note the hierarchy of lengths.[12] The Bohr radius is 137 times larger than the Compton wavelength of the electron: $(\hbar^2/me^2)/(\hbar/mc) = \hbar c/e^2 = 1/\alpha$, and the Compton wavelength of the electron is in turn 137 times larger than the classical radius of the electron: $(\hbar/mc)/(e^2/mc^2) = \hbar c/e^2 = 1/\alpha$. Here is a mnemonic that may or may not be useful to you. With a tiny electron zinging around the proton, the Bohr atom is big. Growing some fuzz, the quantum electron in turn dwarfs the teeny classical thingy it emerges from.

Compton scattering

The experimentalist Arthur Compton won the Nobel Prize in 1927 for the eponymous scattering of a photon γ off an electron e^-: namely, $\gamma + e^- \to \gamma + e^-$, which established definitely the particle nature of light.

The calculation of the cross section for Compton scattering is most easily done using Feynman diagrams (see figure 4).[13] We will discuss what these little pictures mean in part IX.

By the way, I am very impressed by the fact that Oskar Klein and Yoshio Nishina,[14] using a rudimentary formulation of relativistic quantum mechanics, were able to calculate the scattering cross section in 1928 (when Richard Feynman was 10 years old).

At frequencies $\hbar\omega \gg 2mc^2$, the highly energetic photon is capable of producing pairs of electrons and positrons in processes such as $\gamma + e^- \rightarrow \gamma + e^- + e^+ + e^-$.

Let's write the Klein-Nishina cross section for the process $\gamma + e^- \rightarrow \gamma + e^-$ as $\sigma_{KN} = \sigma_{Thomson} f(\frac{mc^2}{\hbar\omega})$ with some function f,[15] as dictated by dimensional analysis. By definition, at low frequencies, $f(x) \rightarrow 1$ as $x \rightarrow \infty$. I will let you guess the high frequency behavior of $f(x)$, as $x \rightarrow 0$.

Intellectual continuity

One fascinating aspect of theoretical physics is that, while the Klein-Nishina calculation (I have in mind the modern version using Feynman diagrams) is based on concepts (for example, photon, Dirac equation, fermion propagator, and electron photon coupling, as explained in any textbook on quantum field theory) totally alien to the calculation of Thomson scattering given here, based entirely on classical concepts (electromagnetic wave, Newton's force law, Larmor radiation, Poynting vector, and the like), $\sigma_{KN}(\omega)$ must approach $\sigma_{Thomson}$ as $\omega \rightarrow 0$. Such is the intellectual continuity of physics that other fields can only envy.

Who is afraid of electromagnetic radiation?

I summarize what we discussed in this chapter as follows. Essentially with only dimensional analysis, perhaps backed up by a rudimentary understanding of relativity, an astute student could write down Larmor's formula, and from that jump over to the dipole radiation formula. Thomson scattering follows readily. Not a few physics students stumble over electromagnetic radiation, but as you can see here, the essential physics can be understood without much effort.

The one restriction we had to impose is that the radiating particle has to be nonrelativistic. I will show in the following chapter how to lift this restriction with almost no work.

Exercises

(1) Find the numerical value of the Thomson cross section.

(2) Guess the behavior of $f(x)$ in the Klein-Nishina formula as $x \rightarrow 0$.

Notes

[1] Recall that in chapter II.2, we integrated over t' first.

[2] For example, Jackson.

[3] These factors can be viewed in their full glory in, for example, Jackson, pp. 466–467.

[4] I have already said in earlier chapters that the fly by night spirit is founded on our faith in simplicity. So, please, no more asking why not use $(\vec{a}^2)^6$.

[5] You might ask why f can't depend on a. We can exclude that by dimensional analysis. Also, $\vec{E}$ and $\vec{B}$ are obtained from A_μ by differentiating once, and so P can depend on a only quadratically.

[6] For the resolution of the apparent contradiction with Einstein's equivalence principle, see *GNut*. As an undergraduate, I was deeply troubled by this for quite a long time.

[7] See R. Freedman, T. Ruskell, P. R. Kesten, and D. L. Tauck, *College Physics*, W. H. Freeman, 2017.

[8] Of course, the physics is entirely different; the electromagnetic wave is still perfectly sinusoidal.

[9] This result is explained in a number of sources; here I follow the textbook by E. Purcell, *Electricity and Magnetism*, Cambridge University Press, 2011, p. 459.

[10] Invented during World War II by midwestern physicists at Purdue University for the sake of secrecy. The barn wins my award for coolest unit in physics. See *Physical Review Letters* 3 (1959), p. 161.

[11] Many authors define the Compton wavelength using h instead of $\hbar$ as is done here; they refer to $\hbar/mc$ as the "reduced Compton wavelength."

[12] We are using Gaussian units here.

[13] See, for example, *QFT Nut*, pp. 152–155.

[14] His students include Hideki Yukawa, Sin-Itiro Tomonaga, and Shoichi Sakata.

[15] Given in the form of an integral in, for example, Bjorken and Drell, *Relativistic Quantum Mechanics*, McGraw-Hill College, 1965, vol. 1, p. 132.

Relativistic effects by promotion and completion

Leaping from the nonrelativistic to the relativistic domain

Undergraduates are well versed in exploiting rotational symmetry. Having worked out a physical process in the plane, they can readily extend it to 3-dimensional space. They recognize, for example, that they should promote the 2-dimensional momentum vector (p_x, p_y) to the 3-dimensional vector $\vec{p} = (p_x, p_y, p_z)$, and complete an expression like $p_x^2 + p_y^2$ to $p_x^2 + p_y^2 + p_z^2 = \vec{p} \cdot \vec{p} = \vec{p}^2$.

Lorentz symmetry is not significantly more difficult. To render a result Lorentz invariant, simply extend the result to 4-dimensional spacetime. Promote the momentum 3-vector (with a slight change of notation) $\vec{p} = (p^1, p^2, p^3)$ to the momentum 4-vector $p = (p^0, p^1, p^2, p^3)$, with p^0 being the energy.[1] Replace the scalar product $\vec{p}^2$ by the Lorentz scalar product $p^2 = \vec{p}^2 - (p^0)^2$: It is like going from 2- to 3-dimensional space, but going from 3-dimensional space to $(3+1) = 4$-dimensional spacetime requires putting in a minus sign to distinguish space from time. Promotion and completion[2] is the name of the game.

Recall from chapter II.3 that the Larmor formula for the power radiated* by an accelerating, but slowly moving, charge is

$$P = \frac{2e^2}{3m^2 c^3} \left(\frac{d\vec{p}}{dt} \right)^2 \tag{1}$$

---

*We did not obtain the factor $\frac{2}{3}$ there but might as well write it here.

To work out the corresponding expression for a relativistic particle would require considerably more work, but the fly by night physicist could get away with almost no heavy lifting.[3]

As promised, I will now show you that we can extend the Larmor formula that we obtained for a nonrelativistic particle to the relativistic domain. A self-evident remark: I have to assume that you are conversant with special relativity.

Larmor's power

The first step is to recognize that the power P is the amount of energy dE radiated in the infinitesimal time interval dt: $dE = Pdt$. Next, note that dE and dt each represent the time component of a Lorentz 4-vector, respectively, $dp^\mu = (dE, d\vec{p})$ and $dx^\mu = (dt, d\vec{x})$. Thus, the power P transforms like a scalar under a Lorentz transformation. In other words, P does not transform at all.

The fly by night physicist's task is to generalize (1) in such a way that P goes from a rotational scalar to a Lorentz scalar.

We promote the 3-vector $\frac{d\vec{p}}{dt}$ to the spatial components of the 4-vector $\frac{dp^\mu}{d\tau}$, where[4]

$$d\tau \equiv +\sqrt{dt^2 - d\vec{x}^2} = dt\sqrt{1 - v^2} \tag{2}$$

defines the Lorentz invariant proper time of the particle.[5] Of course, at the risk of repeating myself, I have to assume that you know special relativity and have a nodding acquaintance with this formalism. If so, then you can write down the generalization of (1) instantaneously:

$$P = \frac{2e^2}{3m^2c^3} \left(\frac{dp}{d\tau}\right)^2 \tag{3}$$

Here the Lorentz scalar product of the 4-vector $\frac{dp^\mu}{d\tau}$ with itself is implied: $(\frac{dp}{d\tau})^2 = (\frac{d\vec{p}}{d\tau})^2 - (\frac{dE}{d\tau})^2$. Remarkably, this result was obtained by the French physicist Liénard in 1898, years before special relativity!

Natives in the relativistic world

We are less interested in the relativistic Larmor formula and its many important applications[6] than in the method, namely, promotion and completion that you can apply to other situations.

The key is to use concepts native to the world you want to move to, rather than those native to the world you are leaving behind.

In the situation at hand, know that momentum $\vec{p} = m\vec{v}$ is more natural than velocity $\vec{v}$, and the relativistic 4-momentum p^μ is more natural than $\vec{p}$. Similarly, proper time τ is more natural than t. An important first step, already taken in chapter II.3, was replacing $\vec{a}$ by $\frac{1}{m}\frac{d\vec{p}}{dt}$. Here we promote $d\vec{p}$ to dp^μ and dt to $d\tau$. With practice, you could jump from (1) to (3) almost instantly without much thought.

Not surprisingly, formulas and results look more at ease in their natural habitat. That is one "secret" usually not taught to undergraduates (at least not enough, in my opinion). A higher symmetry normally renders an expression simpler. For example, if you insist on expressing (3) in terms of $\vec{v}$, you would do best to first define (after putting c back) $\vec{\beta} \equiv \vec{v}/c$ and $\gamma = 1/\sqrt{1 - \vec{\beta}^2}$, as is done in textbooks on special relativity. After some straightforward but tedious algebra, you could verify that (3) becomes

$$P = \frac{2e^2}{3c}\gamma^6 \left(\dot{\vec{\beta}}^2 - (\vec{\beta} \times \dot{\vec{\beta}})^2 \right) \tag{4}$$

Talk about whether (3) or (4) looks more at ease! Simply at the practical level, is it easier for you to remember (3) or (4)?

Indeed, for you to appreciate the fly by night approach better, I strongly urge you to derive (4) step by step from the general result $A_\mu(t, \vec{x}) = \int d^3x'$ $\int dt' \delta(t - t' - \frac{1}{c}|\vec{x} - \vec{x}'|) J_\mu(t', \vec{x}') / |\vec{x} - \vec{x}'|$ given in appendix Gr. Okay, flying by night, I cannot get the $\frac{2}{3}$ in (3), but if I really need to,[7] I could do it by working out the simplest possible case of a particle accelerating slowly from rest.

A summary of electromagnetic radiation

Let us summarize what we have covered in the last three chapters.

In electrostatics, the electric field around a charge is given by $\frac{e}{R^2}$, which diminishes rapidly for large R.

For a monochromatic source oscillating with frequency $\omega = ck$, one of the Rs in the inverse square law $1/R^2$ is replaced by the wavelength $\lambda \sim 1/k$. Thus, the electric field in the outgoing wave is given by $\frac{e}{\lambda R}$, which decreases like $1/R$ instead of $1/R^2$.

For an accelerating nonrelativistic charged particle with acceleration a, one of the Rs in the inverse square law $1/R^2$ is replaced by c^2/a. Thus, the electric field in the outgoing wave decreases like $1/R$ instead of $1/R^2$. What does the length c^2/a mean? Write it as $(c/a)c$. Then c/a is the time it takes the accelerating particle to (formally) attain the speed of light. Thus, the length c^2/a is the distance light would have traveled in that time. We win only if $a \gg c^2/R$, that is, for R large.

While the derivation of the two results, dipole radiation and Larmor radiation, look quite different, the fly by night physicist is able to move from one to the other with ease.

The restriction, in the fly by night derivation, of the accelerating charged particle to the nonrelativistic regime is easily lifted through promotion and completion.

Notes

[1] Lest a student in the back row squeal, let me say that I am talking about a method, not a formula for you to compute with, so I am dispensing with the cs here.

[2] See *GNut*, p. 218.

[3] I follow Jackson, p. 470.

[4] Again, I set $c = 1$ unless the factor is explicitly needed.

[5] Indeed, the velocity $\frac{d\vec{x}}{dt}$ is an extremely awkward concept in special relativity. See *GNut*, p. 218.

[6] Such as radiation loss in circular versus linear accelerators.

[7] When has that ever happened?

Part III

Quantum physics: tunneling in stars, scaling,
atoms, and black holes

From drawing the Schrödinger wave function to tunneling in stars ▬▬▬

Undergraduates in quantum mechanics courses spend too much time solving the Schrödinger equation exactly; certainly I did. In contrast, the ability to produce a quick sketch of the wave function is not emphasized enough.

I expect the reader to be familiar with at least some of what I say in this chapter.

Piecewise constant potentials

The 1-dimensional Schrödinger equation

$$-\frac{\hbar^2}{2m}\frac{d^2\psi}{dx^2} + V(x)\psi = E\psi \tag{1}$$

cannot be solved in general.

But for $V(x)$ equal to a constant, it is easily solved.

For $E > V$, write

$$\frac{d^2\psi}{dx^2} = -\frac{2m}{\hbar^2}(E - V)\psi = -k^2\psi \tag{2}$$

with oscillatory solutions ψ variously equal to $\cos kx$; $\sin kx$; or equivalently, $e^{\pm ikx}$. For $\psi > 0$, $\frac{d^2\psi}{dx^2} < 0$, and so ψ is convex upward (like $-x^2$). For $\psi < 0$, $\frac{d^2\psi}{dx^2} > 0$, and so ψ is convex downward (like $+x^2$).

For $E < V$, write, in contrast to (2),

$$\frac{d^2\psi}{dx^2} = \frac{2m}{\hbar^2}(V - E)\psi = \kappa^2\psi \tag{3}$$

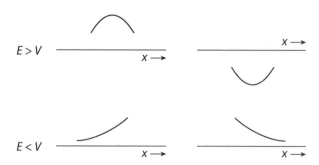

FIGURE 1. Segments of wave function; for $E > V$, the wave function can be positive or negative, but for $E < V$, the overall coefficients of the increasing and decreasing exponentials are taken to be positive, merely to minimize the figure.

with exponentially growing or decaying solutions ψ equal to $e^{\kappa x}$ or $e^{-\kappa x}$. See figure 1.

For a $V(x)$ that is piecewise constant, in each region we can use the solution to either (2) or (3) and then match ψ and $\frac{d\psi}{dx}$ across the regions. The reader should be familiar with all this. As a quick review, I go through the celebrated finite square well in appendix Fsw.

Solving the Schrödinger equation by drawing the wave function

If $V(x)$ varies slowly as a function of x, with regions in which it is more or less constant, we can obtain a back of the envelope sketch of the wave function $\psi(x)$. See figure 2.

Various theorems help a great deal.

If the potential is even ($V(x) = V(-x)$), then the wave function must be[1] either even ($\psi(x) = +\psi(-x)$) or odd ($\psi(x) = -\psi(-x)$).

We will restrict ourselves to even potentials.[2]

Zeros of the wave function

Another helpful theorem states that, for a generic attractive potential, the ground state wave function does not have a zero.* The proof provides a beautiful example of pictorial back of the envelope thinking. See figure 3.

---

*For a free particle, or a particle in a repulsive potential, the wave function goes like $\cos kx$, $\sin kx$ with energy $\propto k^2$, and strictly speaking, the ground state is not defined. As $k \to 0$, the infinite number of zeros become spaced farther and farther apart.

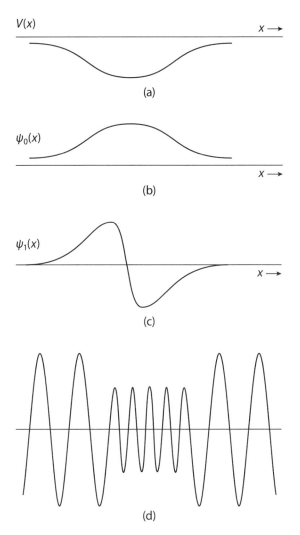

FIGURE 2. Sketching the wave function: (a) the potential, (b) the ground state wave function, (c) the first excited state, and (d) a typical scattering state.

Suppose somebody tries to sell us a ground state wave function, $\psi(x)$, with a zero as shown in figure 3(a). (Again I do the simple 1-dimensional case with an even potential.) She insists that we aren't going to find another wave function with a lower price, oops, energy. Do we buy it?

No, we first reflect the portion of ψ that is negative by letting $\psi \to -\psi$, while keeping the portion of ψ that is positive untouched. We now have the wave function shown in figure 3(b). The energy, given by

$$E = \int dx \left(\frac{\hbar^2}{2m} \left(\frac{d\psi}{dx} \right)^2 + V(x)\psi^2 \right) \bigg/ \int dx\psi^2 \qquad (4)$$

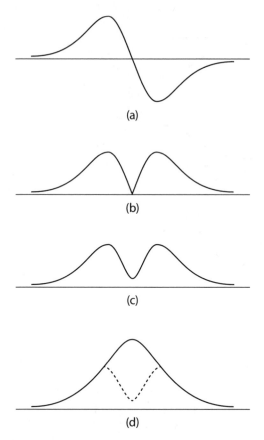

FIGURE 3. (a) An alleged ground state wave function; (b) the alleged wave function "folded over;" (c) the wave function with cusp smoothed out; (d) a wave function with energy lower than that of the alleged wave function.

is unchanged. (Here we have written E without assuming that the wave function ψ is normalized.)

We could remove the cusp in ψ by rounding it off a little, as shown in figure 3(c). Furthermore, if the potential is negative in the region of the cusp, we could "flip" the wave function to the form shown in figure 3(d) and lower the energy E, since we now have more ψ^2 in the region where $V(x) < 0$.

We have found a wave function with energy lower than that of the wave function the shady gal was trying to sell us.

This proves that the ground state wave function cannot have any zeros.

In three dimensions, we apply this sort of reasoning to the radial wave function $u(r) \equiv r\psi(r)$, as explained in appendix L.

That the ground state wave function minimizes[3] the energy E forms the basis of the variational principle. The wave function "tends to go to" where the potential is the lowest, while trying to keep $(\frac{d\psi}{dx})^2$ in (4) as small as possible.

Orthogonality and the first excited state

A third theorem is that two wave functions ψ_a and ψ_b, corresponding to different energy eigenvalues, must be orthogonal to each other: $\int dx\, \psi_a^* \psi_b = 0$, or simply $\int dx\, \psi_a \psi_b = 0$ if we are talking about bound states. Since the ground state wave function has no zero, then the wave function for the first excited state must have one zero,[4] the wave function for the second excited state must have two zeros, and so on.

Again, I expect the reader to have seen all these arguments in an exposure to elementary quantum mechanics.

Once we know the ground state wave function, orthogonality more or less fixes what the excited states look like. We give an example below for the harmonic oscillator.

Sketching the wave function

Equipped with these observations, we can readily sketch what the wave function ψ should look like, as already done in figure 2. In short, if a given potential resembles a square well, then its spectrum should not be too far off from that of a square well of similar width and depth.

Slowly varying potentials are best treated by the standard WKB approximation, as the reader probably knows. While the underlying idea is simple, the actual implementation often requires some work. In particular, a rather tedious (for my taste)[5] discussion of matching conditions is involved. Since this is treated in almost all textbooks, I will stay away from it. Instead, I will discuss a more fly by night approach, crudely replacing "reasonably" slowly varying potentials by piecewise constant potentials. I illustrate this with a "real life" problem encountered in stellar burning.

Stellar burning

For stellar burning, we have to go from 1-dimensional to 3-dimensional space. Let us define $u = r\psi$, with $r > 0$ the radial variable in spherical coordinates. As you learned in school, and as I remind you in appendix L, u satisfies a 1-dimensional Schrödinger equation.

In nuclear processes, the electric repulsion between two nuclei works to prevent them from coming close enough together to react via the strong interaction (or the weak interaction). The potential $V(r)$ looks something like what is sketched in figure 4 and consists of a Coulomb potential $\propto 1/r$ plus a short range potential approximated by a deep finite square well.

This potential $V(r)$ is important in the theory of stellar structure, as the process $p + p \to d + e^+ + \nu_e$ (turning two protons into a deuteron, a positron, and an electron neutrino) controls the first step in energy production in a typical star like the sun. The emission of a positron is mandated by charge conservation,

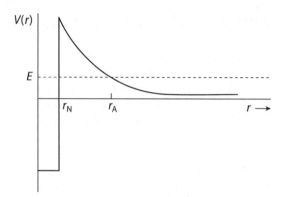

FIGURE 4. Sketch of the potential between two nuclei involved in stellar burning; vastly not to scale.

while the emission of a neutrino indicates that the weak interaction has to be involved. (See chapter IX.2.) This remarkable fact plays an important role[6] in stellar burning, and, of particular importance to us, solar burning.

Note the typical energy scale of nuclear reactions (about a few MeV) and the range (about a fermi $= 10^{-13}$ cm $= 1$ femtometer and denoted by r_N). Thus, the figure is wildly not to scale.

In contrast, the typical temperature in stellar interiors is about 1 keV. An approaching proton with energy $E \sim 1$ keV is turned around at a radius of closest approach* $r_A \gg r_N$.

For the following, it is convenient to write the Coulomb potential as $V_C = E r_A/r$, since by definition the potential is equal to E at $r = r_A$. Since the distance of closest approach r_A is determined by equating the Coulomb potential $\propto 1/r$ to E, we note and should keep in mind that $r_A \propto 1/E$. The higher your energy, the closer you get.

So, you see the problem if you were a nucleon inside a star: to ignite a reaction you need to get close, but a powerful electric repulsion keeps pushing you away. What to do?

Gamow tunneling in nuclear physics

You tunnel! Gamow invented tunneling to solve this problem (and originally that of the radioactive emission of α particles, explaining how the α particles manage to get out[7] of the parent nucleus).

Look at the potential sketched in figure 4. Let's simply replace it by a piecewise constant. Indeed, a commonly used[8] approximation is to replace the Coulomb potential by a constant potential $= W$ for $r_N < r < r_A$ and $= 0$ for $r_A < r$. Now we have an easily solved piecewise constant potential. See figure 5.

"But wait, what is W?" the attentive reader asks.

---

*A for "approach."

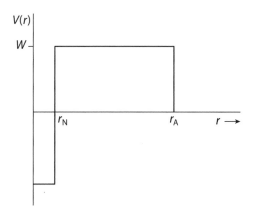

FIGURE 5. Replacing the true potential between two nuclei involved in stellar burning by a piecewise constant potential.

A plausible fly by night guess would be to equate W to the volume average of the Coulomb potential $V_C = E r_A / r$ between r_N and r_A (so let the rigorous types weep):

$$W = \int_{r_N}^{r_A} \left(\frac{E r_A}{r} \right) r^2 \, dr \Big/ \int_{r_N}^{r_A} r^2 \, dr \simeq E r_A \left(\frac{1}{2} r_A^2 \right) \Big/ \left(\frac{1}{3} r_A^3 \right) = \frac{3}{2} E \quad (5)$$

We used $r_A \gg r_N$.

One amusing feature of this problem is that the height of the effective potential is proportional to E. The higher the energy of the approaching nucleus, the higher and narrower (since $r_A \propto 1/E$) the repulsive square well!

Not quite your standard textbook Schrödinger equation. Instead, the (radial s-wave) Schrödinger equation (namely, the 3-dimensional version of (3)) in the relevant region $r_A > r > r_N$ becomes (with m some reduced mass of the two nuclei)

$$\frac{d^2 (r\psi)}{dr^2} = \frac{2m}{\hbar^2} (V - E) r\psi = \frac{2m}{\hbar^2} \left(\frac{1}{2} E \right) r\psi = \kappa^2 r\psi \quad (6)$$

The solution is $r\psi \propto e^{\sqrt{\frac{mE}{\hbar^2}} r}$.

We are interested in the ratio $\mathcal{R}$ of the probability $P(r_N)$ of finding the nucleus at r_N, where we could get down to business, to the probability $P(r_A)$ of finding the nucleus at r_A, the place of closest approach. Thus,

$$\mathcal{R} = \frac{P(r_N)}{P(r_A)} = \frac{r_N^2 |\psi(r_N)|^2}{r_A^2 |\psi(r_A)|^2}$$

$$= e^{2\sqrt{\frac{mE}{\hbar^2}}(r_N - r_A)} \simeq e^{-2\sqrt{\frac{mE}{\hbar^2}} r_A}$$

$$= e^{-\sqrt{\frac{E_G}{E}}} \quad (7)$$

In the second step, remember the geometrical factor $\propto r^2$! In the third step, we plugged in the solution $r\psi \propto e^{\sqrt{\frac{mE}{\hbar^2}} r}$ and in the fourth step, we again used

$r_A \gg r_N$. In the crucial final step, recall what I asked you to keep in mind: that the distance of closest approach r_A is proportional to $1/E$.

One nice fly by night feature is that we didn't have to bother keeping track of the mass of the nucleon, the electrostatic coupling e, Planck's constant, and so forth; all these gather themselves together to form E_G, by dimensional analysis a characteristic energy (G for Gamow).

The interesting result is the unusual dependence on E, with $1/\sqrt{E}$ in an exponential, not something you see every day. The relative probability $\mathcal{R}$ vanishes as $E \to 0$, and it approaches 1 as $E \gg E_G$, as we expect.

Physics makes sense. (I love this kind of physics!) Unhappily, this kind of approximation is becoming a lost art: people either try to solve the relevant potential exactly or, as is more likely these days, simply fire up their computers. In either approach, it might be hard to fully grasp what is going on.

Since this is not a text on stellar nucleosynthesis, I will merely sketch the rest and encourage the reader to soldier on. First, multiply the relative tunneling probability $\mathcal{R} \simeq e^{-\sqrt{\frac{E_G}{E}}}$, which increases with increasing energy E, by the Boltzmann factor $e^{-\frac{E}{T}}$ governing the distribution of the energy E and which decreases with increasing E. We see that the rate of nuclear reaction peaks at a characteristic energy E_*, known as the Gamow peak, with a spread in energy Γ. Next, integrate over E using the steepest descent approximation, which involves figuring out the minimum $\sqrt{\frac{E_G}{E}} + \frac{E}{T}$, and so on and so forth.[9] The technical details are left for our fly by day friends, ha!

The harmonic potential as piecewise constant?

Even potentials that are manifestly far from being piecewise constant can often be profitably treated as piecewise constant in the fly by night approach.

Suppose we don't know how to solve the harmonic oscillator problem. In chapter III.2, we will discuss how to clean up the Schrödinger equation by absorbing various constants. Anticipating a tiny bit, I write the clean version here:

$$\frac{d^2\psi}{dy^2} = (y^2 - \varepsilon)\psi \tag{8}$$

Here ε is the energy measured in units of $\frac{1}{2}\hbar\omega$, with ω the classical frequency of the oscillator. For a region of y around Y, where Y is fixed and large compared to $\sqrt{\varepsilon}$, we can approximate the potential by the constant Y^2. In other words, replace the potential y^2 by a piecewise constant potential, sort of like those terraced rice fields on steep hillsides seen in some photos of China. See figures 6 and 7.

This is admittedly a pretty wild approximation, but the fly by night physicist presses on. She solves the trivial equation $\frac{d^2\psi}{dy^2} = Y^2\psi$ around $y \sim Y$, and writes $\psi \sim e^{-\sqrt{Y^2}y}$, which is sort of maybe like $\sim e^{-y^2}$, at least among friends.

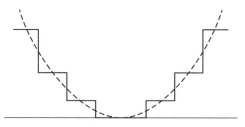

FIGURE 6. Replacing a curved potential by a piecewise constant potential.

FIGURE 7. Terraced rice fields. Photograph by Anna Frodesiak.

In words, if the potential goes to a constant at V as $y \to \infty$, then we expect $\psi \sim e^{-\sqrt{V}y}$ to decrease exponentially, but here the potential is growing fast as $y \to \infty$ and so we guess that ψ should vanish much faster than an exponential.

To allow ourselves some wriggle room, let us try $\psi = e^{-ay^2}$ with a a fudge factor, which Schrödinger will determine for us.

Thus differentiating twice, we obtain $\frac{d^2\psi}{dy^2} = (4a^2y^2 - 2a)\psi$, which equals $(y^2 - \varepsilon)\psi$ if $a = \frac{1}{2}$, thus implying that the energy $\varepsilon = 1$.

You could now carry on, applying the theorems mentioned earlier. The first excited wave function has a zero, is odd, and hence must have the form $\psi = ye^{-\frac{1}{2}y^2}$. Note that the exponential is the same as in the ground state wave function, because in this problem, the potential overwhelms the energy as

$y \to \infty$. Differentiating twice, we obtain $\frac{d^2 \psi}{dy^2} = (y^2 - 3)\psi$, and thus find easily that $\varepsilon = 3$. Orthogonality provides a check on the arithmetic.

The second excited wave function, since it is even, must equal $\psi = (y^2 - c)e^{-\frac{1}{2}y^2}$, with the constant c determined either by the Schrödinger equation or by orthogonality to the ground state wave function. Indeed, this is how you can determine the Hermite polynomials iteratively.

Exercises

(1) Draw the ground state wave function for the double well potential sketched in figure 8, with a the width of each well and $2L$ the separation between them. Assume that $V(x) = V(-x)$. Take $V(x)$ to have the form $W(x + L) + W(x - L)$, with $W(x)$ an attractive potential of width a. Do this for $L/a \simeq 10$, 1, 1/10, and 0. The energetic reader could check his or her sketches by solving the Schrödinger equation for $W(x) = -W\theta(x)\theta(a - x)$, an attractive square well.

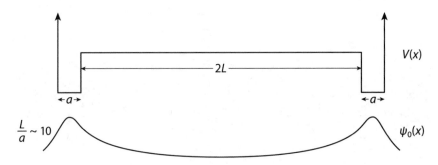

FIGURE 8. Double well potential for exercise (1).

(2) Show that the Gamow peak is located at $E_* \propto T^{\frac{2}{3}}$ and has width $\Gamma \propto T^{\frac{5}{6}}$.

(3) Determine the energy of the second excited state of the harmonic oscillator.

(4) For the hydrogen atom, write $u = r\psi$ as in the text, clean up, and obtain the binding energy up to an overall numerical factor by dimensional analysis.

Notes

[1] The proof is simple. If the wave function $\phi(x)$ is a solution of Schrödinger's equation, then $\psi(x) = \phi(x) \pm \phi(-x)$ is also a solution. The sophisticated would say that this is because the group Z_2 has only two irreducible representations. See *Group Nut*.

[2] You are cordially invited to extend the results here to more general potentials.

[3] You can prove this by inserting $\psi + \delta\psi$ into (4), expanding E to first order in $\delta\psi$, and setting its coefficient to 0.

[4] Clearly, if neither $\psi_0(x)$ nor $\psi_1(x)$ has a zero, then the sign of $\psi_0(x)\psi_1(x)$ stays the same, so that $\int dx \psi_0(x)\psi_1(x)$ cannot vanish. This proof only shows that the wave function for the first excited state has at least one zero, but it is not difficult to extend this proof.

[5] The following story, told by Feynman about a discussion he had with Fermi after the war, relates to how I feel about the WKB approximation:

> (Fermi) came to Caltech.... Somewhere along the line he said, "Let's see, what is the criterion for the WKB approximation?" That's a technical point. I said, "You should know." He said, "But I don't remember," and he hands me the chalk—"Professor, so let's see—." And I couldn't get it straightened out, so I said, "The chalk, Professor." He took it back, "I think I know now," and he starts to explain, and he couldn't get it straightened out and he gives me back the chalk. And this went on for fifteen minutes. Now, that's a very elementary proposition that we always expect all the students to know right away, and we were both confused. Finally we caught on, and then of course it was obvious and we both felt very silly. (https://www.aip.org/history-programs/niels-bohr-library/oral-histories/5020-4).

[6] The weak interaction is so named because it proceeds much more slowly than the strong and electromagnetic interactions. In more massive stars later in the universe, helium fusion into carbon takes over. A detailed and correct discussion of stellar burning is far beyond the scope of what I intend to cover here.

[7] I recommend George Gamow's autobiography, *My Worldline*, in which he describes his escape from the Soviet Union and his discovery of quantum tunneling. Perhaps the two are related subconsciously? While *My Worldline* wins my prize for best title of a physics autobiography or biography, *Tunneling* would have been a close runner-up.

[8] Maoz, *Astrophysics in a Nutshell*, p. 41.

[9] See D. Clayton, *Principles of Stellar Evolution*, p. 301ff.

Scaling and the importance of being clean

Clean up first

When confronting a physics equation to be solved, it is a good idea to clean up first. The Schrödinger equation for the simple harmonic oscillator with spring constant k offers a handy illustration:

$$-\frac{\hbar^2}{2m}\frac{d^2\psi}{dx^2} + \frac{1}{2}kx^2\psi = E\psi \tag{1}$$

Change variable: $x = \lambda y$. Then $-\frac{\hbar^2}{2m\lambda^2}\frac{d^2\psi}{dy^2} + \frac{1}{2}k\lambda^2 y^2\psi = E\psi$.

We are free to choose λ to make life as easy for ourselves as possible. A good choice is to set the coefficients of the two terms on the left hand side equal: $\frac{\hbar^2}{m\lambda^2} = k\lambda^2$. So $\lambda^2 = (\frac{\hbar^2}{mk})^{\frac{1}{2}}$, and hence the common coefficient is given by

$$\frac{1}{2}k\left(\frac{\hbar^2}{mk}\right)^{\frac{1}{2}} = \frac{1}{2}\hbar\sqrt{\frac{k}{m}} = \frac{1}{2}\hbar\omega \tag{2}$$

Here ω denotes the classical frequency of the oscillator. (See chapter I.1.)

Dividing by $\frac{1}{2}\hbar\omega$, we obtain the cleaned up version of (1),

$$-\frac{d^2\psi}{dy^2} + y^2\psi = \varepsilon\psi \tag{3}$$

where ε gives the energy measured in units of $\frac{1}{2}\hbar\omega$. Note also that (3) determines the length λ by $\frac{1}{2}k\lambda^2 = \frac{1}{2}\hbar\omega$, not surprisingly. Your mother always told you to clean up. Good advice in physics, too: compare (3) with (1).

Since (3) does not contain any dimensional parameters, we deduce that the spectrum of the harmonic oscillator is given by a series of pure numbers times $\frac{1}{2}\hbar\omega$.

Different fields of physics use units most convenient for that particular field. Cleaning up informs us that for the simple harmonic oscillator, the right units to use are $\frac{1}{2}\hbar\omega$ for energy and $(\frac{\hbar^2}{mk})^{\frac{1}{4}}$ for length. The spring stretched by this length has potential energy equal to $\frac{1}{2}\hbar\omega$, as you can see from (2).

Scaling in this context corresponds to a fancier version of dimensional analysis. Of course, dimensional analysis would yield the same result. Since we want to construct an energy, it is slightly more efficient to give dimensions not in terms of M, T, and L, but rather in terms of (the generic) E, T, and L. Thus, $[k] = E/L^2$, and $[m] = E/(L/T)^2 = ET^2/L^2$. To eliminate L, we have to form the combination k/m with $[k/m] = 1/T^2$, and thus the energy spectrum is given by dimensionless numbers times $\hbar\sqrt{k/m}$, since $[\hbar] = ET$.

Even better, we should have noticed that in simple quantum mechanics problems, m and $\hbar$ always appear in the combination[1] $\hbar^2/m$. From (1), we see that $[\frac{\hbar^2}{m}] = EL^2$ and $[k] = E/L^2$. Multiplying, we deduce that dimensional analysis implies that $E^2 \sim \hbar^2 k/m$, in agreement with the above, of course.

Anharmonic terms and interpolation

As long as the potential involves a simple power (for example, $V(x) = gx^4$), you can do the same sort of clean up.

Evidently, we do not obtain a result as simple as (3) when the potential has mixed powers, for example, $V(x) = \frac{1}{2}kx^2 + gx^4$. But we can, and should, still clean up. We conclude that $E = \frac{1}{2}\hbar\omega f(\gamma)$, where $\gamma \equiv (g/k^2)\hbar\omega$ is a dimensionless measure of the anharmonicity. A simple dimensional check: $[g] = E/L^4$, $[k] = E/L^2$, and so $[g/k^2] = 1/E$ is an inverse energy.

In physics, when we can't do anything better, interpolation often works, as discussed in chapter I.4 on the dull function hypothesis. Suppose a problem is readily solved in two limits. Then interpolate! In this example, you could deduce something about the unknown function $f(\gamma)$ by taking the limits $g \to 0$ and $k \to 0$, corresponding to $\gamma \to 0$ and $\gamma \to \infty$.

Scaling the hydrogen atom

Let us apply scaling to the zero angular momentum states of the hydrogen atom. Recall that if we set $\psi(r) = u(r)/r$ (see appendix L if you need to be reminded), then $u(r)$ satisfies the 1-dimensional Schrödinger equation with $r > 0$:

$$-\frac{\hbar^2}{2m}\frac{d^2u}{dr^2} - \frac{e^2}{r}u = Eu \tag{4}$$

Set $r = \lambda\rho$ in (4). Determine λ by requiring that the coefficients of the two terms on the left hand side be equal: $\frac{\hbar^2}{2m}\frac{1}{\lambda^2} = \frac{e^2}{\lambda} \implies \frac{1}{\lambda} = \frac{2me^2}{\hbar^2}$. Thus, the equal coefficient comes out to be $\frac{e^2}{\lambda} = \frac{2me^4}{\hbar^2}$.

Dividing by this quantity, we obtain the cleaned up version of the Schrödinger equation:

$$\frac{d^2u}{d\rho^2} + \frac{1}{\rho}u = \varepsilon u \tag{5}$$

with $E = -2\varepsilon(\frac{me^4}{\hbar^2})$. We reach the well-known result that the zero angular momentum states of the hydrogen atom have energies given by negative pure numbers times the Bohr energy $E_B \equiv \frac{me^4}{\hbar^2}$, as was last seen in chapter I.3.

Just as for the harmonic oscillator, this result for the hydrogen atom also follows from dimensional analysis, since we have three dimensional parameters, e, m, $\hbar$, to form into an energy. Indeed, $[\hbar] = ET$, $[e^2] = EL$, $[m] = M = E/(L/T)^2 = ET^2/L^2$, hence $[m/\hbar^2] = 1/EL^2$ to get rid of T, and so the only combination with dimension of energy is $me^4/\hbar^2$, as expected. This is an example of the advantage of using E instead of M.

Note that in contrast to the harmonic oscillator, E_B blows up as $\hbar \to 0$. Without any kinetic energy, the harmonic oscillator simply sits still at the origin, and so $\frac{1}{2}\hbar\omega \to 0$ as $\hbar \to 0$. Here, the wave function plunges into the origin $r = 0$. This was of course the catastrophe that the Bohr atom was invented to avoid.

We see that the uncertainty principle argument used in chapter I.3 to solve the hydrogen atom is just a version of the scaling argument used here. The scaling equality $\frac{\hbar^2}{2m}\frac{1}{\lambda^2} = \frac{e^2}{\lambda}$ used, with $\lambda \to r$, can be interpreted as kinetic and potential energies being of the same order: $\frac{p^2}{2m} \sim (\frac{\hbar}{r})^2\frac{1}{2m} \sim \frac{e^2}{r}$.

Spacing for highly excited states

The energy spectrum of the infinite square well

$$E = \frac{\hbar^2}{2m}\left(\frac{n\pi}{a}\right)^2 \tag{6}$$

consists of energy levels that are increasingly farther spaced apart. Although you should be able to derive this result, let me quickly do it here to illustrate a fairly important point. Normally, we want to choose coordinates so that things are as symmetric as possible, but for the infinite square well, the choice $V(x) = V(-x)$ is actually not so good, since the wave functions then alternate between cosines and sines. Better to choose $V(x)$ to be infinite for $x < 0$ and $x > a$. Then $\psi = \sin n\pi x/a$, and the energy spectrum follows.

In contrast, one much appreciated feature of the simple harmonic oscillator is that its energy spectrum

$$E = \left(n + \frac{1}{2}\right)\hbar\omega \tag{7}$$

consists of equally spaced energy levels. Indeed, this "beautiful" fact is exploited mercilessly in the founding of quantum field theory: the quantum field is "nothing much more" than an uncountably infinite number of coupled simple harmonic oscillators, one at each point in space.[2] Equal spacing in energy is key.

There are many ways to understand the equal spacing; that is why it is called harmonic! In the next two sections, I give not one, but two fly by night approaches to obtain the n dependence in (7), for n largish. (We already know how to nail down the spectrum for $n = 0$, 1 as shown in chapter III.1.)

First fly by night approach to the harmonic spectrum

The first is by brute force, like driving a truck at night. We have already written down, in chapter III.1, the wave function $\psi = e^{-\frac{1}{2}y^2}$ for the ground state, and $\psi = y e^{-\frac{1}{2}y^2}$ and $\psi = (y^2 - c)e^{-\frac{1}{2}y^2}$ (with some constant c we did not bother to determine) for the first two excited states. Subsequent wave functions are completely determined by orthogonality to all preceding wave functions (and by the attendant theorem on the number of zeros). Thus, the wave function for the nth excited state has the form $\psi = (y^n + a_2 y^{n-2} + a_4 y^{n-4} + \cdots)e^{-\frac{1}{2}y^2}$.

Plug this wave function straight into $\frac{d^2\psi}{dy^2} = (y^2 - \varepsilon)\psi$, the cleaned up Schrödinger equation (3). Since we just want to understand the behavior $\varepsilon \propto n$ for largish n, we are not going to worry about various numerical coefficients, factors of 2, and other such stuff.

Differentiate and use the product rule $(fg)' = f'g + fg'$ to obtain $(fg)'' = f''g + 2f'g' + fg''$. Now evaluate $\frac{d^2\psi}{dy^2}$, with ψ a product of a polynomial and an exponential. Every time we hit the polynomial $(y^n + \cdots)$, the power of y goes down, but a power of n pops up. In contrast, every time we hit $e^{-\frac{1}{2}y^2}$, a $(-y)$ descends from the exponent to raise the power of y by one. So, for largish n, $\frac{d^2\psi}{dy^2} \sim (y^{n+2} + (-n + a_2)y^n + (n^2 + \cdots)y^{n-2} + \cdots)e^{\cdots}$.

Schrödinger told us to match this to $(y^2 - \varepsilon)\psi$. The potential term goes like $y^2\psi \sim y^{n+2} + a_2 y^n + \cdots)e^{\cdots}$, while the energy term looks like $(-\varepsilon\psi) \sim -\varepsilon(y^n + \cdots)e^{\cdots}$. The $y^{n+2}e^{\cdots}$ match "automatically" (that is, by design). The $y^n e^{\cdots}$ match gives us, by eyeball, $\varepsilon \propto n$, the desired result. By the way, if I were showing this on the blackboard to my class, it could go faster with less verbiage.

Second fly by night approach to the harmonic spectrum

The second fly by night approach (and the one I like) is to replace the harmonic potential by an infinite square well. See figure 1.

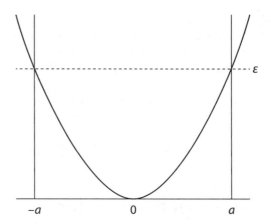

FIGURE 1. Replacing the harmonic potential by an infinite square well.

What!? you say. The two spectra are totally different, as we just saw. What about the curvature of the harmonic potential, now totally lost?

Well, for n largish, the wave function oscillates like crazy for y such that $(\varepsilon - y^2) > 0$, but then dies faster than exponentially for y such that $(\varepsilon - y^2) < 0$. So, as far as the wave function is concerned, it can't get far outside $y \sim \sqrt{\varepsilon}$, and it doesn't spend much time near $y \sim 0$ and so doesn't much care about the curvature there.

A fly by night physicist thinks that an infinite square well with width $a \sim \sqrt{\varepsilon}$ would be just fine. Therefore (weep, you rigorous guys in the jungle patrol!),

$$\varepsilon \sim \left(\frac{n\pi}{a}\right)^2 \sim \left(\frac{n}{\sqrt{\varepsilon}}\right)^2 = \frac{n^2}{\varepsilon} \implies \varepsilon \sim n \tag{8}$$

Put another way, the simple harmonic oscillator is like an infinite square well that swells with energy like $\sqrt{\varepsilon}$. The fly by night physicist likes to make statements that, strictly speaking, make no sense.

Replacing the Coulomb potential by a square well

We now have a fly by night understanding of the energy levels going like n^2 for the infinite square well, and going like n for the simple harmonic oscillator. The energy spectrum of the hydrogen atom displays yet another behavior: the energy levels, going like $1/n^2$ for large n, get squeezed closer and closer together. This is of course due to the infinitely long tail of the Coulomb potential. Let us give a fly by night derivation of this well known fact (Balmer's guess).

Set $E = -B$ in the Schrödinger equation (4) with B positive and approaching 0. Then the electron gets out to a radial distance a given roughly by $B \sim e^2/a$. So replace the Coulomb potential by an attractive square well of width $a \sim e^2/B$. See figure 2.

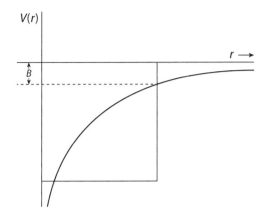

FIGURE 2. Replacing the Coulomb potential by a square well.

An immediate objection is how deep a well are we going to use: The Coulomb potential dives all the way down to $-\infty$. Turns out that the depth of the well does not matter much, just assign to it some large finite value, nominally infinite (whatever that means). We are explicitly interested in the large n states, in which the electron wanders around far from the proton. So, invoking the known spectrum of the infinite square well, we write

$$B = \frac{\hbar^2}{2m}(n\pi)^2\frac{1}{a^2} \sim \frac{\hbar^2}{2m}(n\pi)^2\frac{B^2}{e^4} \tag{9}$$

Solving, we obtain $B \sim \frac{me^4}{\hbar^2}\frac{1}{n^2}$, the right answer. Weep some more, rigorous types!

Effectively, the Coulomb potential is like a well whose width grows like the $1/B$ as the binding energy $B \to 0$.

Sommerfeld's fine structure constant and interpolation in space

It is instructive to write the Bohr energy in terms of Sommerfeld's fine structure constant[3,4]

$$\alpha \equiv \frac{e^2}{\hbar c} \simeq \frac{1}{137} \tag{10}$$

(We are using Gaussian units here; see appendix M.) The Bohr energy $me^4/\hbar^2 = m(\hbar c\alpha)^2/\hbar^2 = \alpha^2 mc^2$ is about $\alpha^2 \simeq (10^{-2}/1.4)^2 \simeq \frac{1}{2} \times 10^{-4}$ times the rest energy of the electron $\sim \frac{1}{2}$ MeV. We obtain ~ 25 eV, off by a factor of 2, which I already alluded to in chapter I.3.

Even flying by night, we can easily track down this factor of 2, using the cleaned Schrödinger equation (5). We know the boundary condition: $u(\rho) \to 0$ as $\rho \to 0$. (See appendix L.) On the other hand, as $\rho \to \infty$, we have $\frac{d^2u}{d\rho^2} \simeq \varepsilon u$, and so $u(\rho)$ decays exponentially $\sim e^{-\sqrt{\varepsilon}\rho}$. Knowing u at both ends and knowing that u has no zero (other than $\rho = 0$), the fly by night physicist immediately guesses that $u(\rho) = \rho e^{-\sqrt{\varepsilon}\rho}$. Plugging this into (5), we find that Schrödinger equation is solved if $\varepsilon = \frac{1}{4}$. Thus, $E = \frac{1}{2}\alpha^2 mc^2$. Factor of 2 caught!

Earlier, we spoke about interpolation as a coupling constant in the potential varies from 0 to ∞; here we are interpolating in space, between what we know at $\rho = 0$ and at $\rho = \infty$.

Incidentally, here we could also relate to Dyson's remark I mentioned in chapter II.1 about the bizarre dimension appearing in electromagnetism. We see that e has the dimension $[\sqrt{\hbar c}] = (M(L^2/T)L/T)^{\frac{1}{2}} = M^{\frac{1}{2}}L^{\frac{3}{2}}/T$.

Einstein misled

That the combination $\frac{e^2}{\hbar c}$ of fundamental constants is dimensionless suggested to Einstein and others a possible theory of the relativistic electron that would somehow produce Planck's constant: perhaps $\hbar \sim e^2/c$ would pop out of such a theory. Einstein had high hopes that quantum physics could emerge from classical physics.

You must admit that this possibility would have been extremely tempting at the time. In the atomic realm, the only known interaction was electromagnetism. Furthermore, Einstein himself had introduced the concept of the photon, whose energy is given by $\hbar\omega$. After the successes of the Schrödinger equation, Einstein naturally tried to introduce $\hbar$ into the Maxwell equations. Ironically, we now know that in quantum electrodynamics, the Maxwell equations stay unchanged and without an $\hbar$ in sight. What changed was the promotion of the electromagnetic field to a quantum operator capable of creating and annihilating photons.

To me, this offers a prime example of how dimensional analysis can lead some of the greatest minds in physics astray. I wonder if some of the suggestively intriguing relations now considered in fundamental physics might not turn out to be similar red herrings.

Exercises

(1) Repeat what is done in the text for the potential $V(x) = gx^4$. How do the energy eigenvalues depend on g?

(2) How much more can you say about the energy eigenvalues for the potential $V(x) = \frac{1}{2}kx^2 + gx^4$? In particular, you could do first order perturbation theory in g. How does the correction to the energy grow with n? Give a fly by night argument for the behavior.

(3) Invoking orthogonality, guess that the first excited state in the hydrogen atom has the form $u = (\rho - b\rho^2)e^{-\sqrt{\varepsilon}\rho}$. Sketch this wave function and the ground state wave function. Plug this into Schrödinger's equation to determine the energy of this state. Verify orthogonality.

(4) Obtain the large n behavior of the energy spectrum of the hydrogen atom $\varepsilon \sim 1/n^2$ (Balmer's guess), using the first fly by night approach mentioned in the text.

(5) The Thomas-Fermi model of atoms with a large number Z of electrons leads to the messy looking equation

$$\frac{1}{r}\frac{d^2(r\phi)}{dr^2} = \frac{4\pi e(2me)^{\frac{3}{2}}}{3\pi^2\hbar^3}\phi^{\frac{3}{2}} \tag{11}$$

with the boundary conditions $\lim_{r\to 0} r\phi = Ze$ and $\lim_{r\to\infty} r\phi = 0$. We will sketch how this is derived in chapter V.4; for now, this serves as a good cleaning exercise. Show that this equation can be cleaned up to have the form[5] $x^{\frac{1}{2}}\frac{d^2\zeta}{dx^2} = \zeta^{\frac{3}{2}}$.

(6) Given the equation $x^{\frac{1}{2}}\frac{d^2\zeta}{dx^2} = \zeta^{\frac{3}{2}}$ with the the boundary conditions $\zeta \to 1$ as $x \to 0$ and $\zeta \to 0$ as $x \to \infty$, analyze it and obtain the behavior of ζ in the two limits.

Notes

[1] Since writing this chapter, I have learned that some people call $\frac{\hbar^2}{2m}$ Schrödinger's constant.

[2] This is why the electromagnetic field can emit or absorb a photon at any point in space, as we will see in part IX. And note that Yang-Mills theory, namely non-abelian gauge theory (which underlies the strong, electromagnetic, and weak interactions), in some sense involves an infinite number of coupled spinning tops.

[3] I once stayed at a physics institute in Munich, where Sommerfeld held court and where a commemorative metal plaque inscribed with his formula $\alpha \equiv \frac{e^2}{\hbar c}$ was set in the lobby. Some friends who were not physicists came to visit, and I asked them what the funny symbol $\hbar$ meant. The craftsman carved the plaque in such a way that the bar in "h bar" was a short horizontal line, which crossed the long vertical line that forms the spine of the letter h. A German woman, evidently an antinuclear and peace activist, immediately responded that physicists were contrite about inventing nuclear reactors: the "rounded arch" in the h represented a nuclear reactor, right next to which was erected a Christian cross memorializing all the people physicists had killed indirectly. Very creative deconstruction of $\hbar$!

[4] When Pauli was dying of pancreatic cancer in 1958, his last assistant, Charles Enz, visited him in the hospital. Pauli asked him: "Did you see the room number?" It was number 137. Throughout his life, Pauli had been preoccupied with the question of why the fine structure constant, a dimensionless fundamental constant, has a value nearly equal to 1/137. Pauli died in that room on December 15, 1958. Thanks to the notion of running coupling constants in quantum field theory, we now understand that there is no special significance to the integer 137. My generation was amazed by this obsession with numerology.

[5] For various attempts to solve this equation, see https://en.wikipedia.org/wiki/Thomas.

The Landau problem in quantum mechanics

When I was learning quantum mechanics, I dutifully mastered the methods leading to exact solutions of the Schrödinger equation, such as the use of Hermite or Laguerre polynomials. These methods are, without question, beautiful in their own ways, but—how should I say it—they lack soul, or at least feel. Later, I came across an an interesting exercise involving the 1-dimensional Schrödinger equation in the quantum mechanics textbook by Landau and Lifshitz. It made a deep impression on me. I will describe the problem and the solution from my own perspective this many years later. (Since I don't know the history of the problem, I have always called it the "Landau problem" in my mind.)

A potential with bumps all over

Consider a particle in an attractive potential $gV(x)$ that vanishes rapidly as $x \to \pm\infty$. The function $V(x)$ is "bumpy" and rather irregular, as sketched in figure 1. Think of g as setting the strength of the potential.

Can you determine the bound state energy spectrum exactly?

At first sight, your reaction would be "of course not"! And you would be right. For a specific $V(x)$, you would just have to turn on the good old computer and grind away numerically.

You could also draw the wave function $\psi(x)$, as I exhorted you to do in chapter III.1. It also would be bumpy and irregular, slavishly following $V(x)$, being larger where the potential is deep and smaller where the potential is shallow. But, as we will see presently, in the limit $g \to 0$, we can in fact say something exact about the bound state spectrum. Perhaps you can work it out before reading on?

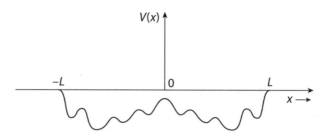

FIGURE 1. A very bumpy potential.

Setting up the problem

So start with Schrödinger's equation:

$$\left(-\frac{\hbar^2}{2m}\frac{d^2}{dx^2}+gV(x)\right)\psi = E\psi \tag{1}$$

For simplicity, we will take a symmetric potential $V(x) = V(-x)$ that is everywhere negative in the range $L > x > -L$, and is zero outside of it. You could generalize this as an exercise.

When I first saw this problem, I was impressed by the elegance of the solution offered by Landau and Lifshitz. The problem was also a refreshing change from the standard problems solved exactly in any generic run-of-the-mill textbook.[1] The solution calls on our "feel" for the Schrödinger equation.

More importantly, it is not what we feel, but what the wave function feels. For general g, the wave function ψ inside the range $L > x > -L$ feels the bumps in $V(x)$ and it "wants to go" where $V(x)$ is deepest. Outside $L > x > -L$, we have $\psi \simeq e^{\mp\kappa x}$, with κ related directly to the binding energy.

But as $g \to 0$, the bound states evaporate one by one and become unbound. Let us focus on the lowest bound state: its wave function ψ becomes flatter and flatter, so that it is less and less sensitive to the bumps in $V(x)$. See figure 2. This notion gives us hope that something can be said exactly about the ground state energy as $g \to 0$.

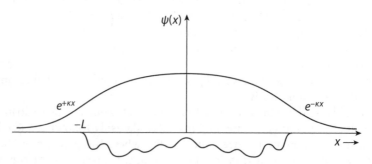

FIGURE 2. As the potential gets shallower, the wave function gets flatter inside the potential.

Clean up and intuit

First, clean up by multiplying (1) by $\frac{2m}{\hbar^2}$ (there ought to be a special symbol, say,[2] ξ, in quantum mechanics for $\frac{2m}{\hbar^2}$) to write

$$\frac{d^2\psi}{dx^2} = (\varepsilon + \lambda V)\psi \tag{2}$$

with $\lambda = \xi g$ and $\varepsilon = -\xi E > 0$ for a bound state.

Before we calculate, let us ask what our physical intuition might tell us about how $\varepsilon \to 0$ as $\lambda \to 0$. A classical ball, in the presence of some slight friction, will roll down to some local or possibly the global minimum. Regardless, its energy would be of order $\varepsilon \sim O(\lambda)$. But with quantum fluctuations, we might expect the particle to be more loosely bound, that is, to have a binding energy that vanishes faster than λ, perhaps like $O(\lambda^2)$.

Pedagogically, perhaps it would be best for me to ask you to solve for ε, assuming that it is $O(\lambda)$. You will soon give up in frustration. At the end of this chapter, I will reveal to you an easy way to see why the naive guess $\varepsilon \sim O(\lambda)$ does not work.

So, suppose[3] that $\varepsilon \sim O(\lambda^2)$. Fly by night, and drop ε in (2) as small compared to the potential λV. At first sight, this seems like a fairly dangerous mutilation of an eigenvalue equation, getting rid of the eigenvalue!

Solving the problem

Nevertheless, press ahead and integrate (2) from $-L$ to $+L$ to obtain

$$\frac{d\psi}{dx}\Big|_{-L}^{L} = -2\kappa e^{-\kappa L}$$
$$\simeq \int dx\, \lambda V \psi \simeq \lambda\left(\int dx\, V\right) e^{-\kappa L} \tag{3}$$

The first equality follows from the exponential decay of ψ outside the range $L > x > -L$. The first approximate equality comes from integrating (2) with ε dropped. The second approximate equality comes from the insight that ψ is essentially constant inside the range $L > x > -L$. We thus obtain $\kappa = -\frac{1}{2}\lambda \int dx\, V$. Note that V is negative, and so $\kappa > 0$. Since V is assumed to vanish outside the range $L > x > -L$, the integration range could be extended to $\pm\infty$.

When we dropped ε from Schrödinger's equation, you might have been alarmed and worried about how we were ever going to get ε back. Well, do you hear it just now sneaking back through the exponentially decaying ψ far away?

The scaled binding energy is then given by $\varepsilon = \kappa^2 = \frac{\lambda^2}{4}(\int_{-\infty}^{\infty} dx\, V(x))^2$. Indeed, ε is $O(\lambda^2)$. The binding energy equals

$$E = -\frac{g^2}{4}\left(\frac{2m}{\hbar^2}\right)\left(\int_{-\infty}^{\infty} dx\, V(x)\right)^2 \tag{4}$$

The moral of the story and a possible strategy

I find this result[4] remarkable: The binding energy is determined by the "area" of the potential $V(x)$ independently of its bumpy details. Each to his or her own taste, but I like this kind of simple and elegant result. It is not to say that the result[5] (4) is of much importance; it is not.

A Feynman biographer[6] wrote that "Feynman seemed to possess a frightening ease with the substance behind the equations, like Einstein at the same age, like the Soviet physicist Lev Landau—but few others." To me, the memorable phrase "the substance behind the equations" is exemplified by (4). Almost anybody can solve a second order differential equation, but few can feel what a wave function feels, trying to get low and going flat as its energy flags.

Imagine being confronted by this problem. Instead of playing dead, one possible strategy would be to solve the finite square well and observe how the bound state energies vanish as the well depth goes to zero. You might notice the wave function becoming flatter and flatter as the potential becomes shallower and shallower. Keep in mind that an analogous strategy might serve you well in some other context.

The power of dimensional analysis

Actually, the elegant result in (4) can also be obtained by dimensional analysis. Imagine showing (2), $\frac{d^2\psi}{dx^2} = (\varepsilon + v)\psi$, with the specified boundary condition, to a mathematician. (To save writing, I defined $v(x) \equiv \lambda V(x)$. By now it should be clear that λ and V do not occur separately.)

Some undergraduates think that dimensional analysis is not relevant to a pure exercise in solving differential equations, which the present problem is. But I trust that after math medley 1, you no longer think so.

Suppose the mathematician does not even know what M, L, and T are. Still, he or she could assign the dimension $[x] = K$ to the variable x, without having to specify what K is. Then $[v] = K^{-2}$. The mathematician's task is to determine ε, with dimension K^{-2}, The key is that ε is not a function of x, and hence conjectures like $\varepsilon \sim v(x)$, $\varepsilon \sim xv(x)$, or $\varepsilon \sim (xv(x))^2$ (that some students might actually write down) make no sense (the second one is not even dimensionally correct). For the first two erroneous choices, ε is indeed of $O(\lambda)$, our first naive guess mentioned earlier in this chapter, but at what value of x should $v(x)$ be evaluated? By playing around a bit, you will soon see that the only possibility is $\varepsilon \sim (\int dx v(x))^2$, with dx providing the needed dimension. We get the Landau result up to an overall numerical factor.

Exercise

(1) Repeat the derivation for $V(x)$ not necessarily symmetric.

Notes

[1] Over the decades, I have never seen this exercise in any other textbook besides Landau and Lifshitz.

[2] Or lowercased upsilon υ, as a student in one of my classes suggested, but it looks too much like script v. Later, I read somewhere that the reciprocal of this is known as the Schrödinger constant, but the term is certainly not in general use.

[3] This is best justified a posteriori.

[4] A challenge for the more mathematically inclined reader: What if $V(x)$ does not fall off fast enough, so that the integral $\int_{-\infty}^{\infty} dx\, V(x)$ does not exist? This has been worked out rigorously by mathematical physicists paid to worry about such things.

[5] This reminds me of Feynman's quip about physics and getting results.

[6] J. Gleick, *Genius: The Life and Science of Richard Feynman*, Pantheon Books, 1992.

Atomic physics

Quick, which atom is the smallest? Think, and then write down your answer, please.

In school we learn about exact solution of the hydrogen atom, but once you get past the hydrogen atom, Laguerre polynomials are not going to help you much. Approximation methods, notably the Hartree-Fock approach and the Thomas-Fermi model, were developed and are treated in detail in standard textbooks. Instead, we will explore a particular fly by night approach to atomic physics* a bit in this chapter. We will start by recovering the Balmer series for the hydrogen atom, move on to the helium atom, and finally treat low Z atoms. I will offer some critical comments at the end.

Undergraduates are veterans at taking exams, and the way the question is posed, they know that it is not going to be the hydrogen atom. If you answered hydrogen, you are way off. The hydrogen atom is not even the second smallest. The helium atom is the smallest, and the neon atom is the second smallest.

Ground states of the hydrogen atom

In treating the hydrogen atom back in chapter I.3, we spoke of the opposition between the kinetic energy and potential energy and the compromise eventually reached. To recap, the total energy is given by

$$E = \frac{p^2}{2m} - \frac{e^2}{r} \simeq \frac{\hbar^2}{2mr^2} - \frac{e^2}{r} \equiv \frac{A}{2r^2} - \frac{B}{r} \tag{1}$$

---

*I follow closely some lectures given by Victor Weisskopf at CERN (Lectures in the CERN Summer Vacation Programme, 1969, CERN report number 70-08).

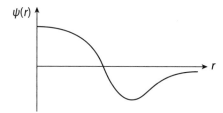

FIGURE 1. The $n=2$ wave function has to have a zero.

Note the dimensions: $[A]=EL^2$, $[B]=EL$. Recall that we have set $p\simeq\hbar/r$ by appealing to the uncertainty principle. You will soon see the advantage of introducing A and B.

To minimize E, the first term wants to drive $r\to\infty$, the second, $r\to 0$. Differentiating, we obtain $A/r^3 = B/r^2$ and thus

$$r = A/B \qquad (2)$$

The energy equals

$$E = \left(-1+\frac{1}{2}\right)B^2/A = -B^2/2A \qquad (3)$$

Keep in mind these two results, as they will be used repeatedly in this chapter.

Since $A=\hbar^2/m$ and $B=e^2$, we have

$$r = \frac{\hbar^2}{me} \equiv a = 1 \text{ Bohr radius} \qquad (4)$$

and

$$E = -\frac{me^4}{2\hbar^2} \equiv -1 \text{ Rydberg} = -13.6 \text{ eV} \qquad (5)$$

Excited states

Our ambition overwhelms us, and we want to go to the first excited state (with principal quantum number $n=2$ but still with zero angular momentum).

As remarked in chapter III.1, by orthogonality, the first excited state has to have one more zero than the ground state has. Since the wave function must have a zero, it has to wriggle twice as fast as the ground state wave function, and so p should be $\simeq 2\hbar/r$, with r yet to be determined. See figure 1. Thus, we should replace $A\propto p^2$ by $4A$. The constant B remains the same.

Referring to (2) and (3), we see that the radius $r=A/B\to 4a$, and the energy $E=B^2/2A\to-\frac{1}{4}$ Rydberg. The radius of the first excited state is 4 times larger, but only 1/4 as deeply bound as the ground state. These happen to be the correct answers.

Thus emboldened, we now guess that $p\simeq n\hbar/r$ for the principal quantum number n s-wave state, since the wave function now has to wriggle n times as fast. So let $A\to n^2A$ with B staying the same. Hence

$$r_n = n^2 \text{ Bohr radius}, \text{ and } E_n = -\frac{1}{n^2} \text{ Rydberg} \qquad (6)$$

We have actually obtained the Balmer[1] series!

The fly by night guess $p \simeq n\hbar/r$ is plausible, but of course cannot be justified, certainly not to the satisfaction of the rigor police. Treat it as a lucky (but highly reasonable) guess.

Helium atom

Moving on to the helium atom, we now picture two electrons, with opposite spins, in the same circular orbit of radius r, around a nucleus consisting of two protons and two neutrons. See figure 2. Note that this picture is essentially classical, with no reference to Schrödinger wave function or the like. We could follow (1) and write down

$$E \simeq 2 \cdot \frac{\hbar^2}{2mr^2} - 2 \cdot 2 \cdot \frac{e^2}{r} + \text{electron-electron repulsion} \qquad (7)$$

I trust the reader to understand the factors of 2 in (7). (Hint: There are two electrons around a nucleus with two protons.) Note that this picture is essentially classical; in "reality," without mutual repulsion, the two electron clouds are superposed on each other, with the spin up and the spin down electrons each occupying the same orbital state.

The hard part is to make a decent fly by night guess for the repulsive energy between the two electrons. We proceed by thinking about the effective or typical distance between the two electrons.

Well, when the two electrons are on opposite sides of the circular orbit, the maximum distance $2r$ is reached. Mathematically, the minimum distance is 0, but hey, we are sensible physicists: the two electrons can't spend much time on top of each other. We guess that the minimum distance is of order r. So let's

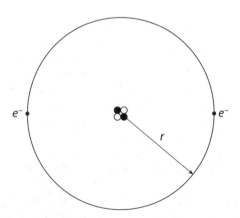

FIGURE 2. A picture of the helium atom; the two electrons are shown as separated by $2r$.

write $r_{\text{effective}} = r/\xi$ with the fudge factor ξ somewhere between $\frac{1}{2}$ and 1, giving a repulsive energy $V_{ee} = e^2/(r/\xi) = \xi e^2/r$.

(While lecturing on this, I took a survey of the class at this point, asking the students whether ξ should be closer to $\frac{1}{2}$ or to 1. They were unanimous in saying "closer to $\frac{1}{2}$," with responses clustering around 0.6 or 0.7.)

We thus arrive at

$$E \simeq \frac{2\hbar^2}{2mr^2} - (4-\xi)\frac{e^2}{r} \tag{8}$$

which in fact has the same form as (1) for the hydrogen problem, but with $A \to 2A$, $B \to (4-\xi)B$. We predict that the helium atom has radius $r = 2a/(4-\xi)$ Bohr radius and energy $E = -(4-\xi)^2/2$ Rydberg. According to Weisskopf, $\xi \sim 0.6$ gives a good fit.

So, $r \simeq 2a/3.4 = a/1.7 < a$. The helium atom is smaller than the hydrogen atom by almost a factor of 2. The doubly charged nucleus pulls the electrons in tighter, but with the electron-electron repulsion inflating the atom a bit.

A remark on the art of the fudge. We could have said that the effect of the electron-electron repulsion would be to modify the second term in (7) to 4η, with η some fudge factor $\lesssim 1$. While this is mathematically identical to (8) of course, there is some kind of pseudopsychological factor at play. A fractional error, say 50%, in ξ would make less of a difference than a fractional error of 50% in η.

Low Z atoms

Talk about being emboldened! Let us now move onto the second row of the periodic table.[2]

Remember that the first row consists of[*] $n = 1$, $l = 0$, with either one or two electrons, corresponding to H and He. The second row consists of $n = 2$, $l = 0$, 1, with one s-wave and three p-wave states, thus accommodating $2(1+3) = 8$ electrons. These correspond to the eight elements with the symbols[3]

$$\text{Li} \quad \text{Be} \quad \text{B} \quad \text{C} \quad \text{N} \quad \text{O} \quad \text{F} \quad \text{Ne}$$

ranging from $Z = 3$ to $Z = 10$.

The inner two electrons see a nucleus with charge Ze. Compared to the hydrogen atom, $A \to A$, $B \to ZB$, and hence $r = A/B \propto 1/Z$ is much smaller than the Bohr radius, particularly for neon. These two electrons are tightly bound and thus the remaining electrons effectively see a nucleus with charge $\tilde{Z} = Z - 2$.

Comparing with (8) for the helium atom, we write down

$$E \simeq (2^2 \tilde{Z})\frac{\hbar^2}{2mr^2} - \left(\tilde{Z} \cdot \tilde{Z} - \frac{1}{2}\xi \tilde{Z}(\tilde{Z}-1)\right)\frac{e^2}{r} \tag{9}$$

---

[*]Here l denotes orbital angular momentum. You need to know that for a given n, l runs from 0 to $(n-1)$, and for each l there are $2l+1$ states, each capable of accommodating up to two electrons.

I suggest that you put this book aside and see whether you can obtain this expression before reading on.

Let us go through the various factors in (9) term by term. First term: 2^2 since $n = 2$ (recall the treatment for the $n = 2$ state of the hydrogen atom), $\tilde{Z} =$ number of electrons (outside the inner core, understood from now on); second term: $\tilde{Z} =$ number of electrons, $\tilde{Z} =$ effective charge of nucleus; and third term: $\frac{1}{2}\tilde{Z}(\tilde{Z} - 1) =$ number of pairs of mutually repelling electrons. Again, ξ is a fudge factor.

Comparing with the hydrogen atom (1), we see that $A \to 4\tilde{Z}A$, and $B \to \tilde{Z}^2 - \frac{1}{2}\xi\tilde{Z}(\tilde{Z} - 1) = \frac{1}{2}\tilde{Z}\big((2 - \xi)\tilde{Z} + \xi\big)$. Thus, we obtain

$$r \simeq A/B = \frac{8}{\big((2 - \xi)\tilde{Z} + \xi\big)} \text{ Bohr radius} \tag{10}$$

and

$$-E \simeq B^2/2A = \frac{1}{16}\tilde{Z}\big((2 - \xi)\tilde{Z} + \xi\big)^2 \text{ Rydberg} \tag{11}$$

You could regard this as a one-parameter (the fudge factor ξ) fit to $2 \times 8 = 16$ experimental numbers. Very impressive already, but Weisskopf actually uses $\xi \simeq 0.6$ from the helium atom. He quotes the result (I show only a few examples) for the binding energy in Rydbergs: for C, 9.6 versus 10.9; for O, 30.5 versus 31.8; for Ne, 69 versus 70. (The number after "versus" is experimental.) The agreement is almost too good to be true.

Inspecting (10) and (11), we see that the radius decreases[4] like $\tilde{Z}$ and the binding energy increases like $\tilde{Z}^3$, for $\tilde{Z}$ largish. Neon,[5] with its completed shell, is tightly bound and the smallest after helium.

Weisskopf goes even further and extends this discussion to general n merely by replacing the 2^2 in the first term in (9) by n^2.

Note that, strictly speaking, the size of an atom is not a sharply defined quantity. Usually, by "size" we mean the expectation value of r, but we could easily consider other possibilities, for example, the inverse of the expectation value of $1/r$.

Here is a critical remark, as promised at the start of this chapter. For hydrogen, the binding energy comes out not to depend on l for a given n. This well known fact, usually derived by brute force in elementary texts, actually depends on some subtle group theory involving a hidden[6] $SO(4)$ symmetry for the Coulomb potential. It is certainly not clear to me how this rather delicate effect would persist with many electrons around. In other words, in (9), the effect of the other electrons on a particular electron is only taken into account energetically.

Exercise

(1) Look up the data on the helium atom, and determine Weisskopf's fudge factor ξ.

Notes

[1] For the story of Johannes Balmer, who published one paper in his life, see, for example, *Group Nut*, p. 496.

[2] For a discussion of the qualitative features of the table inspired and illuminated by group theory, see *Group Nut*, pp. 495–496.

[3] Here I tell my class that in the periodic table, the symbols are composed of either the first letter, or the first letter plus another, of the names in English (which of course may be ultimately of Greek or Latin origin), for example, Li and C, with the following exceptions. The symbols of 10 elements are taken directly from their Greek or Latin names (for example, Na and Fe), while the symbol for one and only one element is taken from German. Which element is it?

[4] Sizes of the 11 smallest atoms (pm = picometer = 10^{-12} m): helium, 31 pm; neon, 38 pm; fluorine, 42 pm; oxygen, 48 pm; hydrogen, 53 pm; nitrogen, 56 pm; carbon, 67 pm; chlorine, 79 pm; boron, 87 pm; beryllium, 112 pm; and lithium, 167 pm. Taken from http://periodictable.com/Properties/A/AtomicRadius.v.html.

[5] Which, by the way, just means "new."

[6] Intriguingly enough, this is related to the mystery of orbits closing in Newtonian gravity. For an explanation, see *GNut*, p. 30, and *Group Nut*, chapter VII.i1.

Black body radiation

Black is not dark

When I was a student, I was puzzled by two aspects of black body radiation. I could follow the math, but what business does a black body have radiating light? And why were the Germans heating up everything in sight and measuring the frequency of the radiation?

Years later, I learned that the words "black" and "dark" have distinct meanings in physics. An ideal black body is a perfect absorber (hence, black) and by time reversal,[1] is also a perfect emitter of electromagnetic waves. In contrast, dark matter does not even interact with the electromagnetic field. Dark will never be the new black.* As for my historical question, I learned[2] decades later that the industrial nations were racing to find the most efficient electric light bulb.

I mentioned in the preface that I would have to cut corners once in a while. In this chapter, I will have to do a bit of that.

So, let us review the story of how Planck, by trying to understand cavities filled with electromagnetic waves, found the quantum!

Maxwell and Boltzmann

To begin, we have to go back to the Maxwell-Boltzmann distribution. Once physicists suspected that gases consist of atoms or molecules moving around and that temperature T corresponds to the typical energy of the atom, they wanted to know the probability distribution of the momenta $\vec{p}$ of the atoms. Evidently, at any given instant, some atoms are moving fast, others more slowly.

---

*English is particularly obscure in this connection. The distinction is clearer in other languages.

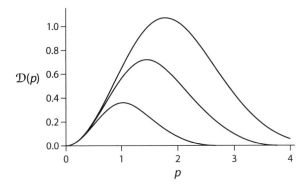

FIGURE 1. Maxwell-Boltzmann distribution plotted for $mT = 0.5, 1, 1.5$ (in arbitrary units); note that the location of the peak of the distribution as a function of p increases as $\sqrt{2mT}$.

A central result of statistical mechanics[3] states that the probability of an atom having energy E is proportional to $e^{-E/T}$. Applied to our gas, this implies that the probability for an atom to have momentum $\vec{p}$ is proportional to $e^{-\varepsilon(p)/T}$. Here $\varepsilon(p) = p^2/2m$, and p denotes the magnitude of $\vec{p}$.

To find the statistical average of various physical quantities $X(\vec{p})$, we need to integrate over all momenta with the measure d^3p. For quantities independent of the direction of $\vec{p}$, the measure or geometric factor d^3p may be replaced effectively by $4\pi\, dp\, p^2$, and thus we encounter integrals of the form $\int_0^\infty dp\, p^2 e^{-p^2/2mT} X(p)$.

Figure 1 shows the Maxwell-Boltzmann distribution $\mathcal{D}(p) \equiv p^2 e^{-p^2/2mT}$ for various values of mT.

An act of desperation

Now let us join Planck puzzling over the black body data. Even better, since this is a totally ahistorical account,[4] imagine yourself as a young Planck in a civilization far far away in another galaxy.[5] You understand that the cavity is full of electromagnetic waves zipping this way and that,* with wave vector $\vec{k}$ and frequency $\omega = c|\vec{k}| = ck$. Again, the geometric factor d^3k can (often) be replaced effectively by $4\pi\, dk\, k^2 \propto d\omega\, \omega^2$. In analogy with the Maxwell-Boltzmann distribution, denote the frequency distribution by $\mathcal{D}(w)$.

With $\mathcal{D}(\omega)$ plotted as a function of ω, the data look very similar to that shown in figure 1. So you try to mimic Maxwell and Boltzmann, and write something like $\mathcal{D}(\omega) \equiv \omega^2 e^{-\omega/T}$ for the frequency distribution.

---

*Note that even in Maxwell's electromagnetism, an electromagnetic wave of frequency ω carries momentum, according to the Poynting vector, proportional to $\omega/c = k$ in the direction of $\vec{k}$.

But this fails resoundingly! Dimensions do not match: The frequency ω has dimension of an inverse time so that $[\omega] = 1/T$, while the temperature* T has dimension of energy $[T] = [E] = ML^2/T^2$.

In physics, revolutions are marked by the introduction of a hitherto unknown fundamental constant: Newton with his G, and Einstein with his[6] c. Only the brave, or at least the reckless, can drive physics through perceived impasses.

You (or Planck) boldly invent a brand new fundamental constant $\hbar$ just so that $\hbar\omega$ has the same dimension as temperature T, namely, that of energy: $[\hbar\omega] = E$. Thus, the new constant must have dimension of energy times[7] time:

$$[\hbar] = ET \tag{1}$$

But what is the energy $\hbar\omega$? As you know, Planck made an inspired leap of faith,[†] guessing that an electromagnetic wave of frequency ω actually consists of tiny packets[8] of energy $\hbar\omega$, later named photons. Surely one of the most amazing guesses in the history[9] of physics!

Distribution of frequencies in a cavity filled with electromagnetic waves

With the new constant $\hbar$, you are tempted to propose the distribution $\mathcal{D}(\omega) \equiv \omega^2 e^{-\hbar\omega/T}$, but this only fits the data at high ω. So, let us retreat and write $\mathcal{D}(\omega) \equiv \omega^2 f(\hbar\omega/T)$ without specifying what $f(\hbar\omega/T)$ is.

Well, we already "know" that $f(x) \to e^{-x}$ for x large.

How about the behavior of $f(x)$ as $x \to 0$?

First, note that the energy density in the cavity is given by $\int_0^\infty d\omega\, \omega^2 f(\hbar\omega/T)\hbar\omega$. Thus, the integral converges exponentially at the high-frequency end. What we are asking for is the behavior at the low-frequency end.

Remarkably, without knowing what f is, we can still do the integral by scaling. Write $\hbar\omega = xT$. The integral scales like the 4th power of ω, and so instantly we obtain

$$\frac{E}{V} \propto T^4 \tag{2}$$

Historically, this corresponds to the Stefan-Boltzmann law. Later, at our leisure, we will fill in $\hbar$ and c by dimensional analysis.

Let us now see whether we can extract more information about the function $f(\hbar\omega/T)$, which determines the spectrum (namely, the energy $d\omega\, \omega^2 f(\hbar\omega/T)\hbar\omega$ contained in waves with frequency between ω and $\omega + d\omega$).

One clue is that at the low-frequency limit the energy spectrum should go over to the classical limit, known as the Rayleigh-Jeans law[10] $d\omega\, \omega^2 T$.

---

*Again I plead the poverty of the alphabet: The two Ts mean different things, but surely you get it.

†Planck later described his move as "an act of desperation."

(My understanding is that this law was "derived" invoking the equipartition theorem, assigning T to each degree of freedom, and mixing it with some hocus-pocus about oscillators, but it is all history now, given the Planck law to be presented presently.) In other words, we want $f(\hbar\omega/T)\hbar\omega \to T$ for $\omega \to 0$ or equivalently, for $T \to \infty$.

Note that the mysterious Mr. H-bar goes away in this limit, as he should. High temperature should correspond to the classical limit.

Thus, $f(x) \to 1/x$ as $x \to 0$.

The Planck distribution

I will pause and let you think of a function that has this small-x behavior, as well as $f(x) \to e^{-x}$ as $x \to \infty$.

The brain of the fly by night physicist finds $1/x$ harder to grasp than x, so the first step[11] is to define $g(x) = 1/f(x)$. We want $g(x) \to e^x$ for $x \to \infty$ and $g(x) \to x$ for $x \to 0$. The reader with a good memory might recall that I already posed this question back in chapter I.4.

A plausible interpolation would be[12] $g(x) = e^x - 1$. So, did you guess $f(x) = \frac{1}{e^x - 1}$?

Thus, the Planck distribution is given by, in the notation used here,

$$\mathcal{D}(\omega) = \omega^2 \frac{1}{e^{\frac{\hbar\omega}{T}} - 1} \tag{3}$$

For use below, let us define $n(\omega) = 1/(e^{\frac{\hbar\omega}{T}} - 1)$, namely, $\mathcal{D}(\omega)$ shorn of the kinematic ω^2.

Of course, all is clear in hindsight, but still, that shouldn't have been too hard after the act of faith of introducing the new constant $\hbar$. Historically, Planck had more to guide him. In particular, Wien's displacement law[13] states how the peak frequency ω_{max} of the distribution increases linearly with T.

An important point and the moral of the story. Derive? I don't see Planck's result being derived here.[14] Only an inspired guess backed by deep insight.[15]

Phase space

We can now put $\hbar$ and c into (2), as promised. Since $[\hbar c] = ET(L/T) = EL$, we note that $[T/\hbar c] = E/EL = 1/L$ is an inverse length, and so $(T/\hbar c)^3$ is an inverse volume. Hence, the energy per unit volume $E/V \sim T(T/\hbar c)^3 \sim T^4/(\hbar c)^3$.

Indeed, you could almost write down the exact energy spectrum, if you know[16] that phase space in statistical mechanics is given by $V d^3 p/(2\pi\hbar)^3 = V d^3 p/h^3$.

Trading the angular integration for 4π as usual, writing $p = \hbar k = \hbar\omega/c$ so that $d^3p \to 4\pi(\hbar/c)^3 d\omega\omega^2$, we obtain[17] the energy density

$$\varepsilon \equiv \frac{E}{V} = \frac{1}{\pi^2 c^3} \int_0^\infty d\omega\omega^2 \frac{\hbar\omega}{e^{\frac{\hbar\omega}{T}} - 1} = \frac{T^4}{\pi^2(\hbar c)^3} \int_0^\infty dx \frac{x^3}{e^x - 1} \qquad (4)$$

In the last step, we scaled by writing $\hbar\omega = xT$.[18]

I will leave it to you to fly by day[19] and show off your prowess in doing the integral in (4), if you wish.

That amazing Einstein

Any thinking theoretical physicist should become more and more in awe of that amazing Einstein as the years go by. Looking at (3), Einstein discovered yet another interesting phenomenon: stimulated emission.

Caution: Our treatment of black body radiation is necessarily sketchy and schematic, and that of Einstein's stimulated emission, even more so. The interested reader is referred to more specialized textbooks for more thorough discussions.[20]

Consider a bunch of two-state atoms in equilibrium with a gas of photons. Let the difference in energy between the excited and the ground states be $E_1 - E_0 = \hbar\omega$. Boltzmann tells us that the ratio of excited atoms to grounded atoms is equal to* $\frac{N_1}{N_0} = e^{-\hbar\omega/T}$. On absorbing a photon of frequency ω, an atom in the ground state jumps to the excited state, while an atom in the excited state relaxes to the ground state by emitting a photon of frequency ω.

Einstein now exploits the fact that, in equilibrium, the two rates must be equal.

From (3) and (4), we know that the number of photons with frequency in the interval $(\omega, \omega + d\omega)$ is equal to $\frac{V}{\pi^2 c^3} d\omega\omega^2 n(\omega)$, with $n(\omega) = 1/(e^{\frac{\hbar\omega}{T}} - 1)$. But to keep things as clear as possible, let us strip the inessential factor $\frac{V}{\pi^2 c^3} d\omega\omega^2$.

The rate at which photons of frequency ω are being absorbed is evidently proportional to N_0 and $n(\omega)$: hence, $n(\omega)N_0$. The corresponding rate at which photons of frequency ω are being emitted by the excited atoms is similarly proportional to $e(\omega)N_1$ (which you can regard as the definition of $e(\omega)$ if you like).

If I have ever seen an elegant back of the envelope calculation, then this is it. Equating the two rates, we obtain, following Einstein of course,

$$e(\omega) = n(\omega)\frac{N_0}{N_1} = n(\omega)e^{\frac{\hbar\omega}{T}} = \frac{e^{\frac{\hbar\omega}{T}}}{e^{\frac{\hbar\omega}{T}} - 1} = \left(1 + \frac{1}{e^{\frac{\hbar\omega}{T}} - 1}\right) = 1 + n(\omega) \qquad (5)$$

I trust that you agree with me that the way $n(\omega)$ is defined, with ω^2, π, and whatnot suppressed, enables us to see more clearly the forest for the trees.

---

*This and the relations we obtain below are to be understood probabilistically.

So amazingly, $e(\omega) = 1 + n(\omega)$.

The "1" represents the fact that the excited atoms would still come down to ground even when a photon gas is not present, that is, when $n(\omega) = 0$. What is amazing about the result $e(\omega) = 1 + n(\omega)$ is the $n(\omega)$ term: The presence of a photon gas enhances the tendency of the excited atoms to emit.

The inclination of photons to hang out with each other[21] also leads them to encourage the atoms to create more photons for them to hang out with. We all know that the laser and much else in our so-called civilization depend on stimulated emission.

Note that the derivation in (5) depends explicitly on the actual form of $n(\omega)$.

Creation and annihilation operators

For readers who know about raising and lowering operators for the harmonic oscillator and how they morph into creation ($a^\dagger$) and annihilation (a) operators in quantum field theory, I sketch how contemporary textbooks derive the Planck distribution starting with the commutation relation $[a, a^\dagger] = 1$.

First, a few words about quantum field theory. After being Fourier transformed, the electromagnetic potential $A_\mu(\vec{x}, t)$ can be seen[22] to behave like harmonic oscillators,[23] one at each point $\vec{x}$ in space.

Let us now quantize A_μ, but stay inside a large box of volume V, so that the modes of the electromagnetic field, labeled by $\vec{k}$ (with $k = \omega/c$), are discrete and countable. Focus on a specific mode. Then it makes sense to talk about the number n of photons in that mode. (Again, apologies for the finite number of letters in the alphabet. In this section, n will denote an integer, not to be confused with $n(\omega)$.)

In standard textbooks on quantum mechanics, it is explained that the states of the harmonic oscillator are labeled by an integer n (we derived in chapter III.2 the equal spacing law of the energy levels, which fact is crucial for the formulation of quantum field theory, as was also mentioned), written as $|n\rangle$ in Dirac's notation. Then the commutation relation $[a, a^\dagger] = 1$ is solved* by

$$a|n\rangle = \sqrt{n}|n - 1\rangle \tag{6}$$

and[24]

$$a^\dagger|n\rangle = \sqrt{n+1}|n + 1\rangle \tag{7}$$

Denote by $|n, 0\rangle$ the quantum state consisting of an atom in the ground state $|0\rangle$ in the presence of n photons of the appropriate frequency. Similarly, $|n, 1\rangle$ denotes the state with the atom in the excited state $|1\rangle$ in the presence of n photons. The amplitude for the absorption process discussed in the preceding section is then given by $\langle n - 1, 1|a\mathcal{O}|n, 0\rangle = \langle n - 1|a|n\rangle\langle 1|\mathcal{O}|0\rangle = \sqrt{n}\langle 1|\mathcal{O}|0\rangle$, where $\mathcal{O}$ represents an operator acting on the atom that turns

---

*$a^\dagger a|n\rangle = \sqrt{n}a^\dagger|n - 1\rangle = n|n\rangle$, and $aa^\dagger|n\rangle = \sqrt{n+1}a|n + 1\rangle = (n + 1)|n\rangle$, so that $[a, a^\dagger]|n\rangle = 1|n\rangle$.

$|0\rangle$ into $|1\rangle$. Similarly, the amplitude for the emission process is given by $\langle n+1, 0|a^\dagger \mathcal{O}^\dagger |n, 1\rangle = \langle n+1|a^\dagger |n\rangle \langle 0|\mathcal{O}^\dagger |1\rangle = \sqrt{n+1}\langle 0|\mathcal{O}^\dagger |1\rangle$. Squaring these amplitudes to obtain the transition probabilities, we see that they are in the ratio n to $n+1$, in accordance with Einstein.

Indeed, including the relative probability $e^{-\frac{\hbar\omega}{T}}$ of finding an atom in the excited state versus the ground state and playing a bit fast and loose, we could derive the Planck distribution: $(n+1)e^{-\frac{\hbar\omega}{T}} = n$ implies $n = e^{-\frac{\hbar\omega}{T}}/(1 - e^{-\frac{\hbar\omega}{T}}) = 1/(e^{\frac{\hbar\omega}{T}} - 1)$.

Time reversing the invention of integral calculus

If the probability of a system (here we have in mind the electromagnetic field, but the discussion to follow is quite general) at temperature T having energy E is given by $e^{-E/T}$, and if E can take on a continuum of values, then the expected value of E equals $\int_0^\infty dE E e^{-E/T} / \int_0^\infty dE e^{-E/T} = T$. We all learned how Newton and Leibniz invented integral calculus.

Suppose we time reverse this amazing creation, and replace the integrals just written down by sums. Allow E to take on only the discrete values $E_n = n\varepsilon$ with $n = 0, 1, \ldots, \infty$. The expected value of E is now given by

$$\sum_{n=0}^{\infty} n\varepsilon e^{-n\varepsilon/T} / \sum_{n=0}^{\infty} e^{-n\varepsilon/T} = \frac{\varepsilon}{(e^{\varepsilon/T} - 1)} \tag{8}$$

We obtain[25] the Planck distribution, with the benefit of hindsight (and of Planck's and Einstein's profound insights). In the limit $\varepsilon \to 0$ (or equivalently, $T \to \infty$), this expression $\to T$, and we recover the continuum result, as Newton and Leibniz discovered.

This sum is of course done in any statistical mechanics textbook, but I feel that, for many students, the physics is often lost amid all the πs, k_B, and other distracting stuff zinging around.

The universe and stars as black bodies

Going back to the confusion that I suffered from when I first learned about black bodies (which the professor was unable to explain), I show you in figure 2 the actual spectrum[26] of different types of stars, with temperature ranging from 3,500 K to 40,000 K, being fitted to Planck's distribution.

Admit it, you don't normally think of stars as black bodies!

We will also see (in chapter VII.3) that the early universe, amazingly enough, can be considered as a box filled with black body radiation. Furthermore, the cosmic microwave background of the present universe fits Planck's distribution almost perfectly. Indeed, it provides a "history book" of the universe!

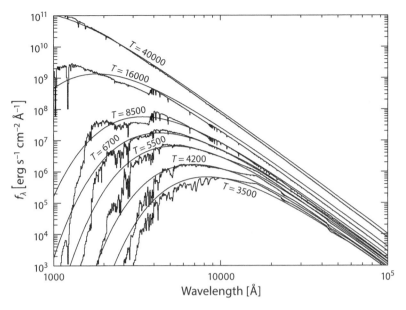

FIGURE 2. Stars are almost black bodies! Flux per wavelength interval emitted by different types of stars. This is fitted to the black body spectrum at various temperatures ranging from $T = 3{,}500$ K to $T = 40{,}000$ K. Note that the Planck distribution is plotted here as a function of wavelength, not frequency. Maoz, D. *Astrophysics in a Nutshell*. Princeton University Press, 2016.

Exercise

(1) Wien's displacement law, which played such a crucial role in the eventual discovery of quantum mechanics, is actually quite general. Show that if the distribution of a physical quantity ζ has the form $\zeta^a F(\zeta/T)$, with a an arbitrary constant and $F(x)$ an arbitrary function, the peak value ζ_{max} of the distribution increases linearly with T. Indeed, see also the caption to figure 1.

Notes

[1] See chapter VI.1.

[2] From B. Brown's biography, *Planck: Driven by Vision, Broken by War*, Oxford University Press, 2015.

[3] Strictly speaking, only in the quantum world could we count and make sense of the statistics implied by the term "statistical mechanics." Nevertheless, we can talk about probability distributions, even in a classical gas.

[4] For historical accounts, see, for example, B. Brown, *Planck*; D. Stone, *Einstein and the Quantum*, Princeton University Press, 2013; A. Pais, *Subtle Is the Lord*, Oxford University Press, 1982, chapter 19.

[5] For more intergalactic fables, see *GNut*.

[6] The speed of light was measured and known by the time Einstein came along, but was not thought of as fundamental somehow and was instead assumed to have different values in different frames. Of course, things tend to be obvious in hindsight.

[7] Oops, multiplied by. English is like that.

[8] According to Stone, *Einstein and the Quantum*, p. 118, as late as 1908, when Planck was nominated for the Nobel Prize (but didn't get it), most physicists, with a few exceptions such as Einstein and Lorentz, didn't understand the significance of what Planck had done. The nomination was made in ignorance rather than in appreciation.

[9] For the actual history, see for example the biographies by Brown and Stone mentioned in note 4.

[10] I was surprised to learn that the Rayleigh-Jeans law was actually proposed after the Planck law. (When I was an undergrad, my friends and I thought that "ultraviolet catastrophe" would make a great name for a rock group. "Catastrophe" refers to what would happen if you integrated the Rayleigh-Jeans expression all the way out to $\omega = \infty$.)

[11] Evolution did not equip us to invert, least of all matrices.

[12] I already noted in chapter I.4 that killjoys would immediately object that there are an infinite number of such functions, for example, $g(x) = e^x x/(1 + x)$. Or how about $g(x) = e^x x(1 - x)^{2n}/(1 + x)^{2n+1}$ for n any positive integer? To all such people, I say, fine, be my guest. Nature is kind.

[13] I chose not to discuss this law and subsumed it as being in the data available to Planck. In fact, the derivation is an elegant piece of fly by night physics involving scaling. See Cheng, *Einstein's Physics*, pp. 36–37.

[14] A frustrated Lorentz tried hard to derive the Planck law but kept on getting the Rayleigh-Jeans law instead. It is clear in hindsight: There is no way you can get it without using quantum mechanics. Of course, eventually, Bose derived (3) by proposing the correct quantum statistics for the photon.

[15] See Pais, *Subtle Is the Lord*, p. 371: "[Planck's] reasoning was mad, but his madness has that divine quality that only the greatest transitional figures can bring to science."

[16] A quick derivation of this basic fact if you don't know it. The factor of $\hbar^3$ is needed by dimensional analysis. The precise result stated here follows in a few steps from the basic commutation relation, $[x, p] = i\hbar$ of quantum mechanics. One simply counts

plane wave states, as explained in every statistical mechanics textbook. OK, it is so easy that I will just show you with a minimum of words. Commutation $[x, p] = i\hbar \implies p = -i\hbar \frac{\partial}{\partial x} \implies e^{ipx/\hbar}$ in one dimension, periodic boundary condition $pL/\hbar = 2\pi j \implies \Delta j = \Delta pL/(2\pi\hbar)$ with j an integer. Cubing this to go to three dimensions gives the number of states in $(dp)^3$ as $d^3pL^3/(2\pi\hbar)^3$ in the continuum limit, as was to be shown. An alternative (but essentially the same) argument could be given by using the de Broglie relation. From Planck's relation $E = \hbar\omega$, we have $p = \hbar k = \hbar(2\pi/\lambda) = h/\lambda$, where we have defined $h = 2\pi\hbar$. Imposing a periodic boundary condition (or equivalently, fitting in integer multiples of de Broglie waves) gives us the same result: The number of states in $(dp)^3$ equals d^3pL^3/h^3. Keep in mind that the volume of a unit cell in phase space is given by h^3, not $\hbar^3$.

[17] Note that $2d^3p/(2\pi)^3 \to 2(4\pi)dkk^2/8\pi^3 \to d\omega\omega^2/\pi^2c^3$.

[18] Remembering that the photon has two spin degrees of freedom, we have multiplied the first integral in (4) by 2.

[19] Actual value: $\varepsilon = \pi^2 T^4/45(\hbar c)^3$.

[20] For example, see Cheng, *Einstein's Physics*, pp. 82–89, or any decent quantum mechanics book, such as Sakurai and Napolitano, *Modern Quantum Mechanics*.

[21] See chapters V.3 and V.5, where we discuss Bose statistics.

[22] I am being necessarily vague here. To truly understand this statement, one would have to read a book on quantum field theory, such as *QFT Nut*.

[23] This is because in Maxwell's electromagnetism, the Lagrangian and the Hamiltonian are both quadratic in the electromagnetic fields. See chapter IX.2.

[24] A common confusion: Equations (6) and (7) are not hermitian conjugates of each other. The hermitian conjugate of $a|n\rangle = \sqrt{n}|n-1\rangle$ is $\langle n|a^\dagger = \sqrt{n}\langle n-1|$.

[25] For the abecedarian, the sum is most easily done by defining $\beta = 1/T$, so that the ratio of the two sums equals $-\frac{\partial}{\partial\beta} \log \sum_{n=0}^{\infty} e^{-n\beta\varepsilon} = \frac{\partial}{\partial\beta} \log(1 - e^{-\beta\varepsilon})$.

[26] Taken from D. Maoz, *Astrophysics in a Nutshell*, p. 11, Princeton University Press.

A word of encouragement to the reader
on getting confused by physics

If you have gotten this far without getting confused at any point, good for you! If you did get confused, I offer you what Feynman said about getting confused:

> Whenever Fermi lectured about any subject whatever, and I've heard many, or talked about any subject whatever that he'd thought about before, the clarity of the exposition, the perfection with which everything was put together to make everything look so obvious and beautifully simple, gave me the impression that he did not suffer from a disease of the mind which I suffer from, which is CONFUSION! When I think about something, I go along in a certain way, and then I get balled up, and then I go back, and I think—I get mixed up easily. I easily get into confusion, which is the horror of the whole business when you're thinking.... But when he'd give a lecture, or when he'd talk about anything, up to that time—I had heard him, ... and he comes out clear—even when I was arguing with him about the piles, in what he was saying he was not confused. He was just not understanding what I was talking about. There was never any sign of confusion. And so I asked him about this afterwards. I said, 'I'd got the impression that you don't get confused.' He said, 'That's impossible! I always get—' It's just that I hadn't realized. I thought he was so perfect that he didn't have this difficulty of getting confused as he went along. But apparently everybody does. That's what stops you from proceeding. You get balled up and forget and get mixed up. So anyway, I found him mixed up, just like me, at this same time. That's another story—that

the great Fermi can make a mistake, or can be confused about a simple idea.*

To be fair to these two greats, though, keep in mind that they are talking about doing research, not reading a textbook.

---

*See https://www.aip.org/history-programs/niels-bohr-library/oral-histories/5020-4.

Part IV

Planck gave us units: black hole radiation
and Einstein gravity

Planck gave us God-given units ▬▬▬▬

To understand the universe at a fundamental level

Once upon a time, boys and girls, we used some English king's feet to measure lengths with. You laugh, but a metal bar preserved in Paris and decreed by a bunch of French revolutionaries is not much more intrinsic. To understand the universe at a fundamental level, we ought not to have to use some absurdly human invention, such as the imperial or metric system.*

Einstein recognized that with the universal speed of light c, we no longer need separate units for length and time. Even the proverbial guy and gal in the street understand that henceforth we could measure length in light years.

We and another civilization, be they in some other galaxy, would now be able to agree on a unit of distance, if we could only communicate to them what we mean by one year or one day. Therein lies the rub: our unit for measuring time derives from how fast our home planet spins and revolves around its star. Only homeboys would know. How could we possibly communicate to a distant civilization this period of rotation we call a "day," which is merely an accident of how some interstellar debris came together to form the rock we call home?

---

*Notions we take for granted today still had to be thought up by someone. Maxwell, in his magnum opus on electromagnetism, proposed that the meter be tied to the wavelength of light emitted by some particular substance, adding that such a standard "would be independent of any changes in the dimensions of the earth, and should be adopted by those who expect their writings to be more permanent than that body." The various eminences of our subject could be quite sarcastic.

Two out of three taken care of

Newton's discovery of the law of gravity brought the first universal constant G into physics. The emphasis here is on the first, and the one and only one at that time.

Next, Maxwell and Einstein brought the second fundamental constant, c, into physics. Henceforth, physicists proudly possessed two universal constants, G and c.

Comparing the kinetic energy $\frac{1}{2}mv^2$ of a particle of mass m in a gravitational potential with its potential energy $-GMm/r$ and canceling off m, we see that the combination[1] GM/c^2 has dimension of length.

Now we are empowered to measure masses in terms of our unit for length (or equivalently, time), or lengths in terms of our unit for mass.

Some readers might argue that we could also use something like the mass of a particle such as the proton, or the electron, or perhaps even the quark. That would indeed suffice for communicating with another civilization, but particle theorists generally believe that these masses were generated.[2] Specifically, in the early universe, these particles are thought to have been massless, or nonexistent (the proton would have fallen apart into three quarks).

We would prefer to base our units on the fundamental laws of physics.

You realize that to do physics at the fundamental level, without recourse to somebody's foot, and not even to a possibly ephemeral particle (such as the proton), we need another constant on the same level as G and c. What could that be? Can you guess?

Planck's great contribution to physics: for all civilizations, extraterrestrial and nonhuman

> The two[3] constants ... which occur in the equation for radiative entropy offer the possibility of establishing a system of units for length, mass, time and temperature which are independent of specific bodies or materials and which necessarily maintain their meaning for all time and for all civilizations, even those which are extraterrestrial and non-human. [Max Planck]

Surely you guessed! Max Planck[4] is properly revered for his introduction of the fundamental constant $\hbar$ into physics. With this far-reaching and magnanimous gesture, he gave us a natural system of units, sometimes known as God-given units.

In a tremendously insightful paper, Planck pointed out that with the three fundamental constants[5] G, c, and $\hbar$, in order of their entrance into the grand drama of physics, we finally have a universal set of units for mass M, length L, and time T, the three basic concepts we need to do physics with.

Three big names, three basic principles, three natural units

To see how these units are defined, note that Heisenberg's uncertainty principle tells us that $\hbar$ over the momentum Mc is a length. Equating the two lengths GM/c^2 and $\hbar/Mc$, we see that the combination $\hbar c/G$ has dimension of mass squared. In other words, the three fundamental constants G, c, and $\hbar$ allow us to define a mass,[6] known rightfully as the Planck mass:

$$M_P = \sqrt{\frac{\hbar c}{G}} \tag{1}$$

We can then immediately define, with Heisenberg's help, a Planck length:

$$l_P = \frac{\hbar}{M_P c} = \sqrt{\frac{\hbar G}{c^3}} \tag{2}$$

and, with Einstein's help, a Planck time:

$$t_P = \frac{l_P}{c} = \sqrt{\frac{\hbar G}{c^5}} \tag{3}$$

Newton, Einstein, Heisenberg, three big* names, three basic principles, three natural units to measure space, time, and energy. We have reduced the MLT system to "nothing"! We no longer have to invent or find some unit, such as the transition frequency of some agreed-on atom,[7] to measure the universe with. We measure mass in units of M_P, length in units of l_P, and time in units of t_P.

Another way of saying this is that in these natural units, $c = 1$, $G = 1$, and $\hbar = 1$. The natural system of units would be understood no matter where your travels might take you, within this galaxy or far beyond.

Newton small, so Planck huge, and the Mother of All Headaches

The Planck mass works out to be about 10^{19} times the proton mass m_p. That humongous number 10^{19} is responsible for the Mother of All Headaches plaguing fundamental physics today.[8] That M_P is so gigantic compared to the known particles can be traced back to the extreme feebleness (as was mentioned in chapter I.2) of gravity: G tiny, so M_P enormous.

As the Planck mass is huge, the Planck length and time are tiny. If you insist on contaminating the purity of natural units by human-made units, t_P comes out to be $\simeq 5.4 \times 10^{-44}$ second, the Planck length $l_P \simeq 1.6 \times 10^{-33}$ centimeter, and the Planck mass[9] $M_P \simeq 2.2 \times 10^{-5}$ gram!

---

*As big as they come!

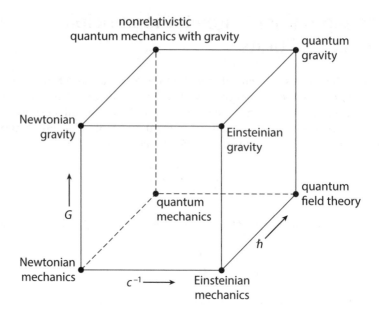

FIGURE 1. The cube of physics. From Zee, A. *Einstein Gravity in a Nutshell*. Princeton University Press, 2013.

It is important to realize[10] how significant Planck's insight was. Nature Herself, far transcending any silly English king or some self-important French revolutionary committee, gives us a set of units to measure by. We have managed to get rid of all human-made units. We needed three fundamental constants, each associated with a fundamental principle, and we have precisely three!

This suggests that we have discovered all[11] the fundamental principles that there are. Had we not known about the quantum, then we would have to use one human-made unit to describe the universe, which would be weird. From that fact alone, in my opinion, we would have to go looking for quantum physics.

The cube of physics

That we need three fundamental constants associated with three fundamental principles suggests that we could neatly summarize all of physics as a cube. See figure 1.

Physics started with Newtonian mechanics at one corner of the cube, and is now desperately trying to get to the opposite corner, where sits the alleged "Holy Grail." The three fundamental constants, c^{-1}, $\hbar$, and G, characterizing Einstein, Planck[12] or Heisenberg, and Newton, label the three axes. As we "turned on" one or the other of three constants, in other words, as each of these constants came into physics, we took off from the home base of Newtonian mechanics.[13]

Much of 20th-century physics consisted of getting from one corner of the cube to another. Consider the lower face[14] of the cube. When we turned on c^{-1}, we went from Newtonian mechanics to special relativity. When we turned on $\hbar$, we went from Newtonian mechanics to quantum mechanics. When we turned on both c^{-1} and $\hbar$, we arrived at quantum field theory, in my opinion the greatest monument of 20th-century physics.

Newton himself already moved up the vertical axis from Newtonian mechanics to Newtonian gravity by turning on G. Turning on c^{-1}, Einstein took us from that corner to general relativity or Einsteinian gravity.

All the Stürm und Drang of the past few decades is the attempt to cross from that corner to the Holy Grail of quantum gravity, when (glory glory hallelujah) all three fundamental constants are turned on.[15]

You might be wondering about the corner with $c^{-1} = 0$ but $\hbar \neq 0$ and $G \neq 0$. That corner, relatively unpublicized and generally neglected, covers phenomena described adequately by nonrelativistic quantum mechanics in the presence of a gravitational field.[16]

In our everyday existence, we are aware of only two corners of this cube, because these three fundamental constants are either absurdly small or absurdly large compared to what humans experience.[17]

Setting $\hbar$, c, and G to unity, or not

The Planckian system of units amounts to setting $\hbar$, c, and G to unity, but often, depending on which area of physics you work in (or which corner of the cube of physics you are on), it would be inconvenient, or even inappropriate, to do so. For instance, in electromagnetism, $\hbar$ and G do not even enter. Setting $c = 1$ would be quite appropriate, but occasionally it would be useful to keep it around, for example, to show the weakness of the magnetic force compared to the electric force.

Particle physicists deal with relativistic quantum phenomena, and so routinely set $\hbar$ and c to 1, but not G, which does not enter until one starts discussing quantum gravity. In theories of quantum gravity (for example, string theory), G is routinely also set to 1.

Did we need to include k?

And now I come to my pet peeve. No doubt that Boltzmann's constant[18] k played a pivotal role in physicists' struggle to understand the discreteness of matter. But now that the reality of the atom has long been established, k should be retired. Temperature is an energy, period. Boltzmann's constant k is just a conversion factor between energy units and some quaint markings on a tube of mercury.

Of course I do not object to the fact that temperature is often measured in degrees, but then degree should be considered a unit for energy, just like the erg or the British Thermal Unit, albeit a rather peculiar unit. The constant k could then be suppressed. Otherwise, why not introduce a fundamental constant called $\kappa = 2.54$ cm/in, measure in inches, and pepper formulas in physics with expressions like $Gm_1 m_2/(\kappa r)^2$? The appearance of k similarly stings my eyes.

We can imagine a world with $\hbar = 0$. Indeed, physicists lived in that world until Planck came along. Similarly, we can imagine worlds with $G = 0$, or $c^{-1} = 0$. But what would it mean to have a world with $k = 0$? Instead of filling glass tubes with mercury, we fill them with a liquid whose coefficient of thermal expansion is infinite?

I have been surprised that distinguished physicists continue to write kT, where T would have sufficed. Perhaps they are so used to it that they think of kT as a new letter[19] in some exotic alphabet.

Postscript: Some colleagues who read this chapter in manuscript urge me to strengthen my rant and rave[20] against k even more. Why keep on writing k after an entire century has passed and after our long sojourn in the quantum world? Is it merely to confuse some of the weaker students into thinking that k has the same status as the three fundamental constants G, c, and $\hbar$?

Notes

[1] You already learned what this combination means physically back in chapter I.2.

[2] By the Higgs mechanism or some other mechanism yet unknown.

[3] He included k.

[4] In his personal life, Planck suffered terribly. He lost his first wife, then his son in action in World War I, then both daughters in childbirth. In World War II, bombs totally demolished his house, while the Gestapo tortured his other son to death for trying to assassinate Hitler.

[5] There are those who speculate that what we believe to be fundamental constants may actually vary with time as the universe evolves. Meanwhile, I go on living my life. I figure that they will tell me when conclusive evidence for time varying fundamental "constants" is established.

[6] Some readers might wonder why we do not use the mass of the electron m_e. I already addressed this issue, but let me elaborate. In modern particle physics, the electron may not always have had the mass it has now, and in fact could have been massless in the early universe. The masses of elementary particles depend on quantum field theoretic notions known as spontaneous symmetry breaking. The Planck mass is almost surely more fundamental than the electron mass. We should express m_e in terms of M_P, not M_P in terms of m_e. Of course, for convenience, in different areas of physics, different units are used, for example, the size of the hydrogen atom as a length unit.

[7] Experimentalists of course still need a practical set of units to measure stuff with. Some august body of prominent humans has to meet periodically. See *Physics Today*, article 2017.

[8] For one thing, we cannot perform experiments to help us understand quantum gravity. Hence we rely on physics by pure thought. See *GNut*, chapter X.8, for example.

[9] While t_P and l_P are remote from human experience, the Planck mass M_P is almost within human scale. Try giving some examples, such as a bit of hair.

[10] Many observers at the time, and many even now, failed to appreciate this. Ernst Mach said that Planck's "concern for a physics valid for all times and all people, including Martians, seems to me very premature and even almost comic."

[11] These days, fundamental principles are posted on the physics archive with abandon. There might be hundreds by now.

[12] Planck was known for his dry lectures, but smoothly given without referring to notes. According to one account, "There were always many standing around the room. As the lecture room was well heated and rather close, some of the listeners would from time to time drop to the floor, but this did not disturb the lecture."

[13] By this I mean the three laws, $F = ma$ and so on, not including the law of universal gravitation.

[14] This face, regarded as a square, was discussed in the very first section of the first chapter of *QFT Nut*.

[15] This statement carries a slight caveat, which we will come to in chapter IV.2.

[16] Two fascinating experiments in this area are (1) dribbling neutrons like basketballs, and (2) interfering a neutron beam with itself in a gravitational field. See the appendix to chapter X.8 in *GNut*. For more details, see J. J. Sakurai and J. Napolitano, *Modern Quantum Mechanics*, pp. 110, 133.

[17] Other huge numbers, such as the number of nucleons in your body and in the earth, while not fundamental, are important to us, a fact that explains why G was known much earlier than c^{-1} or $\hbar$.

[18] A historical oddity: at the time, many also called Planck's constant "k," to the confusion of the multitude. See Stone, p. 115.

[19] Even worse, some books, such as Kittel and Kroemer, use another letter, τ, to denote kT.

[20] "Rave" is related to "rage," a word with no direct connection with "outrage," which traces back to "ultra" via the French word "outré." See E. Maleska, *A Pleasure in Words*, Simon and Schuster, 1981.

A box of photons and
the power of natural units

Entropy and energy of a photon gas

Consider a box of volume V containing photons characterized by a temperature T. The entropy and energy of a photon gas are worked out in textbooks on statistical mechanics. Here we simply dimensionally analyze using sensible units.

In natural units, temperature T has dimension of energy or inverse length: $[T] = M = L^{-1}$. On the other hand, entropy S is dimensionless and proportional to the volume of the box $V \sim L^3$, with L the characteristic size of the system. So, the entropy can only be

$$S \sim L^3 T^3 \sim VT^3 \tag{1}$$

Similarly, the energy density ε has dimension of $[\varepsilon] = M/L^3 = M^4$ in natural units. It follows immediately, by dimensional analysis, that $\varepsilon \sim T^4$, leading to a total energy of

$$E \sim L^3 T^4 \sim VT^4 \tag{2}$$

Indeed, this is just the Stefan-Boltzmann law mentioned in chapter III.5 on black body radiation.

I trust that you see the power of using natural units. These two important results pop out almost instantaneously. Incidentally, the number of photons in the box is given by $N \sim VT^3$, since as far as dimensional analysis is concerned, N and S are both dimensionless quantities.

Putting $\hbar$ and c back

It is easy to put $\hbar$ and c back (since gravity does not enter here, we should leave G out) using dimensional analysis. In non-Planckian units, T has dimension of energy, and hence $[VT^4] = L^3 E^4$.

Note that I am charging the letter E with two tasks here: to represent the energy of a box of photons and, for our convenience, to substitute for M in dimensional analysis. Also, do not confuse T denoting temperature in this specific problem and time in dimensional analysis. Once again, our recurring lament: only so many letters in the alphabet!

After this cautionary note, let us go back to energy expressed as $L^3 E^4$. To get rid of the L^3, we should divide by c^3, so that $[VT^4/c^3] = T^3 E^4$. But $[\hbar] = ET$. Hence, dividing by $\hbar^3$, we conclude that $E \sim VT^4/(\hbar c)^3$.

Similarly, $S \sim VT^3/(\hbar c)^3 \sim N$. This all makes sense: $E \sim NT$, since T is just the average energy carried by each of the N photons.

Note that for fixed T, as $\hbar \to 0$, E, S, and N all go to infinity. In the classical world, these quantities do not make sense. Photons? What photons?

Of course, if we know Planck's distribution, which started this whole $\hbar$ business, then we could calculate E exactly and determine the numerical coefficient in $E \sim VT^4/(\hbar c)^3$. Indeed, this number was given in the form of an integral in chapter III.5.

Pressure by dimensional analysis, by thermodynamics, and by scale invariance

What about the pressure P exerted by the photon gas? Pressure is force per unit area. In natural units, force has dimension $[F] = ML/T^2 = M^2$, and area has dimension $L^2 = 1/M^2$. Thus, pressure has dimension $[P] = M^2/(1/M^2) = M^4$, the same dimension as energy density. We expect that

$$P \sim \varepsilon \qquad (3)$$

Invoking thermodynamics, we could even fix the numerical coefficient in this relation. As you will see presently, the overall numerical factors in (1) and (2) (call them K and K') do not matter at all.

The first law of thermodynamics states that $dE = TdS - PdV$.

Remark to students: The key to acing a course on thermodynamics is simple. Always remember what you are holding fixed every time you differentiate. Here we have to hold the entropy S fixed, so that $dS = 0$, and hence $dE = -PdV$. In other words, $P = -\frac{\partial E}{\partial V}|_S$.

Thus, we cannot blindly differentiate (2), but must first use (1) to express E in terms of S:

$$E = KVT^4 = KV\left(\frac{S}{K'V}\right)^{\frac{4}{3}} = KV^{-\frac{1}{3}}\left(\frac{S}{K'}\right)^{\frac{4}{3}} \qquad (4)$$

Now differentiate:

$$P = -\frac{\partial E}{\partial V}\bigg|_S = \frac{1}{3}K\left(\frac{S}{K'V}\right)^{\frac{4}{3}} = \frac{1}{3}KT^4 = \frac{1}{3}\frac{E}{V} = \frac{1}{3}\varepsilon \tag{5}$$

As promised, the two constants K and K' drop out; they are not invited to the party.

The pure number $\frac{1}{3}$ in (5) is not accidental. The 3 refers to the 3-dimensional space we live in. Can you see why?

For those readers who know[1] some special relativity, there exists a more sophisticated, yet illuminating, derivation of the important relation $P = \frac{1}{3}\varepsilon$. With the Lorentz transformation unifying time and space into a 4-vector $x^\mu = (x^0, x^1, x^2, x^3) = (t, x, y, z)$, energy density ε and pressure P are also unified and packaged into a 4-tensor, the energy momentum tensor $T^{\mu\nu}$ (with $\mu, \nu = 0, 1, 2, 3$). I introduce the energy momentum tensor in appendix Eg, and explain why, physically, it must be a tensor.

In a local rest frame, this tensor takes the diagonal form[2]

$$T^{\mu\nu} = \begin{pmatrix} \varepsilon & 0 & 0 & 0 \\ 0 & P & 0 & 0 \\ 0 & 0 & P & 0 \\ 0 & 0 & 0 & P \end{pmatrix} \tag{6}$$

Energy density ε is the time-time component T^{00}, while pressure P is the space-space component, and since no special direction exists, $T^{11} = T^{22} = T^{33}$.

In a relativistic theory, scale transformations are generated by the trace of the energy momentum tensor, and thus, in a scale invariant[3] system (such as the photon gas, which has no mass and no length scale), the trace $T = \eta_{\mu\nu}T^{\mu\nu}$ (with $\eta_{\mu\nu}$ the Minkowski metric, namely, $\eta_{00} = 1$, $\eta_{11} = \eta_{22} = \eta_{33} = -1$, with all other components equal to 0) vanishes. This implies that $T = T^{00} - T^{11} - T^{22} - T^{33} = \varepsilon - 3P = 0$, as was to be demonstrated. You see again that the 3 comes from the 3 dimensions of space.

Where can we find a box of photons?

Where can we find a box of photons?

Lots of places. First of all, you might have noticed that the only property of a photon we have used is its lack of a mass. Hence, all our results carry over to a box filled with relativistic matter. By "relativistic matter," I mean matter consisting of particles whose masses are negligible compared to their energies.

The early universe is a box of relativistic matter. In the standard model of particle physics, masses were generated later. In any case, when the temperature far exceeds the mass of a given elementary particle, that particle can be regarded as relativistic matter. Therefore, in the early universe, our results (1) and (2) hold true as long as we multiply each of them by some counting factor $g(t)$, which simply counts how many species are effectively massless, and which therefore depends on the age t of the universe. (Fermions and bosons are counted differently[4] due to the different degrees of freedom they have.) As the universe cools (see below), and as the particles acquire masses, when the temperature drops below the masses of various particles, $g(t)$ decreases. For instance, at present, only the photon, and possibly one or more species of neutrinos,[5] are effectively massless.

The universe cools

Normally, we think of entropy as ever increasing due to various dissipative processes, but on the scale of the universe, entropy was conserved as the early universe smoothly expanded. Thus, the result (1) immediately implies that the temperature T of the early universe dropped according to the inverse of the scale size* $a(t)$ of the universe:

$$T \propto \frac{1}{a(t)} \tag{7}$$

---

*Even in an infinite universe, a time dependent scale size $a(t)$ could be defined. See chapter VII.3.

Notes

[1] For those who don't, consult, for example, *GNut*, p. 226.

[2] See, for example, *GNut*, p. 230.

[3] See, for example, *GNut*, p. 621, if you do not know this basic fact.

[4] This is exactly the kind of stuff we will learn in chapters V.4 and V.5, if we would fly by day and do lots of integrals.

[5] See chapter IX.3.

IV.3
CHAPTER

Black holes have entropy
Hawking radiation

Quantum fluctuations can set you free

Nothing can get out of black holes. Spacetime is curved around a black hole in such a way that, once inside the horizon with $r_S = 2GM/c^2$, even light can never emerge, as was discussed in chapter I.2. A black hole is a warp in spacetime into which bodies can fall but can never come out.[1]

But as you have no doubt heard, that picture, painted exclusively with classical physics, no longer holds true when quantum effects are turned on. Black holes radiate as black bodies, with a temperature characteristic of each black hole.

For something as fundamental as black hole radiation, we politely decline to use ludicrous units such as joules per degree centigrade. Indeed, I will show you the power of using appropriate units, as was mentioned in the preceding chapter. We will start with so called particle physics units, with $\hbar$ and c, but not G, set equal to 1, and proceed to natural or God-given units later.

Hawking radiation

Surely you have heard of Hawking radiation. In an extremely influential series of papers, Stephen Hawking (see figure 1), building on earlier work by Jacob Bekenstein and others, and in collaboration with Gary Gibbons, pointed out that the purely classical picture needs to be amended when quantum effects are included: black holes radiate and hence evaporate.

Let us now use[2] dimensional analysis with natural units to estimate the characteristic energy of the particles emitted, that is, the temperature of the

FIGURE 1. In memory of Stephen Hawking (photo taken August 2011 after a lecture I gave in Cambridge). Coincidentally, the sad news of his death came while I was teaching from this chapter.

radiation, known as the Hawking temperature T_H of the black hole. You may be puzzled,* since there are two masses in the problem, the mass M of the black hole and the Planck mass M_P. With two masses, any function of M/M_P is admissible, and so dimensional analysis appears to be inapplicable.

Indeed, we need one more piece of information, namely, a fact I emphasized way back in chapter I.2. The key is that Newton's constant G is a multiplicative measure of the strength of gravity. In Einstein's theory as well as in Newton's, the gravitational field around an object of mass M can only depend on the combination[3] GM.

With $\hbar = 1$ and $c = 1$, the combination GM is a length and hence an inverse mass.

Boltzmann and the founding fathers of statistical mechanics had long ago revealed to us that temperature, a highly mysterious concept at one time, is merely the average energy[†] of the microscopic constituents of macroscopic matter. Hence, temperature has the dimension of energy, that is, the dimension of a mass in natural units.

It follows instantly that

$$T_H \sim \frac{1}{GM} \sim \frac{M_P^2}{M} \tag{1}$$

---

*A distinguished condensed matter physicist once told me that he was puzzled about precisely this point. So your unspoken question may be widespread.

†The Boltzmann constant k, which (as was emphasized in chapter IV.1) is merely a conversion factor between energy units and the markings on some tubes containing mercury known as "degrees," has been set to 1.

This "sophisticated" dimensional analysis captures an essential piece of physics: The radiation is explosive! As the black hole radiates energy, M goes down, and T_H goes up, and thus the black hole radiates faster. The radiative mass loss accelerates. Certainly not something you want to see in the kitchen: An object that gets hotter as it loses energy.

If you so wish, we could easily write T_H in everyday unnatural units by restoring c and $\hbar$. Way back when, we derived Kepler's law without breaking a sweat by noticing that GM has dimension of L^3/T^2. Since temperature has the dimension of energy, and since $[\hbar] = ET$, we form $\hbar/GM$ with dimension of ET^3/L^3. Hence, the Hawking temperature, which, as has already been remarked, has the dimension of energy, is given by

$$ T_H \sim \frac{\hbar c^3}{GM} \tag{2} $$

Very gratifying to see that, indeed, with $\hbar = 0$ and quantum effects turned off, $T_H = 0$ and the black hole does not radiate. Physics is consistent!

In fact, the overall numerical constant in (2), which turns out to be $\frac{1}{8\pi}$, could be determined, with a touch of sophistication,* in a couple lines of algebra.[4]

A man of deep integrity

Jacob Bekenstein, a student of John Wheeler's, was the first to realize that black holes have entropy. I will let Wheeler tell the story in his trademark style:[5]

> One afternoon in 1970, ... I told [Bekenstein] of the concern I always feel when a hot cup of tea exchanges heat energy with a cold cup of tea. By allowing that transfer of heat ... I increase (the universe's) microscopic disorder, its information loss, its entropy. "The consequences of my crime, Jacob, echo down to the end of time," I noted. "But if a black hole swims by, and I drop the teacups into it, I conceal from all the world the evidence of my crime. How remarkable!" Bekenstein, a man of deep integrity, takes the lawfulness of creation as a matter of the utmost seriousness.[6] Several months later, he came back with a remarkable idea. "You don't destroy entropy when you drop those teacups into the black hole. The black hole already has entropy and you only increase it!"

Black holes have entropy

For our purposes here, once Hawking has told us that the temperature of a black hole is given by (2), we can use thermodynamics to determine the entropy.

---

*The first step is to notice that the spacetime metric near the horizon on a black hole resembles the metric near the origin of the plane in polar coordinates. Mysterious, eh?

Thermodynamics states that entropy S is given by[7] $dE = TdS$. Here E is just the mass M of the black hole, so that $T_H dS = dM$.

Integrating $\frac{dS}{dM} = \frac{1}{T_H} \sim GM$, we obtain

$$S \sim GM^2 \sim \frac{M^2}{M_P^2} \tag{3}$$

Note that, as understood and expected, S is dimensionless. Again, if desired, we could readily restore $\hbar$ and c:

$$S \sim \frac{GM^2}{\hbar c} \tag{4}$$

A black hole in the quantum world

In fact, in classical general relativity, not only does a black hole have entropy, as Bekenstein realized, it actually (since $\hbar = 0$) has an infinite amount of entropy, as (4) shows. This makes sense, since entropy is the logarithm of the number[8] of microstates[9] that correspond to a single equilibrium macrostate, and we could make a black hole of a given mass M in an infinite number of ways, by throwing any amount and any variations of stuff into it, provided that the total mass adds up to M. With $S = \infty$, for $dM = TdS$ to be satisfied, we have to set $T = 0$, consistent with (2). In the classical world, black holes do not radiate.

This suggests a fly by night argument for Hawking radiation, due to Gibbons and Hawking. For the entropy S of a black hole to be finite, quantum physics must somehow limit the number of ways a black hole could be made. Let us focus on the difference in entropy between two black holes of mass M and $M + dM$.

An elementary fact of quantum physics is that the size of particles is characterized by the de Broglie wavelength. A particle whose wavelength is much smaller than the Schwarzschild radius could be regarded as a point particle and could fall in (depending on its velocity and impact parameter, etc.), but a particle whose wavelength is larger than the Schwarzschild radius could simply pass the black hole by. Too big to fit!

Thus, a particle whose wavelength is larger than GM but smaller than $G(M + dM)$ is less likely to fall into the smaller black hole. We thus argue that the change in entropy dS is actually finite when quantum mechanics is turned on. Once you admit that dS is not infinite, then the relation $dM = TdS$ no longer forces T to vanish, and once you admit that $T \neq 0$, we can then run our dimensional analysis argument.

The entropy of a black hole is finite, and so Wheeler was not able to violate the second law of thermodynamics by throwing cups of tea into a passing black hole. As Bekenstein explained to him and to the rest of us, he had merely increased the entropy of the black hole. If Wheeler were right, we could all help to decrease the disorder in the universe by simply dumping our mess into passing black holes.

We don't need quantum gravity for Hawking radiation

At this time, I should clear up a point that puzzles many people. To derive his radiation, Hawking did not need a theory of quantum gravity. We do not have to quantize the general relativity field. What we have to quantize is the field of the particle (for example, an electron) being emitted.

In quantum field theory, fields corresponding to particles are constantly fluctuating out of, and back into, the vacuum, thus producing a particle and anti-particle pair (for example, an electron and a positron), which quickly annihilate each other. Normally, for a particle of mass m, the fluctuation lasts for only a short time of order $\Delta t \sim \hbar/2mc^2$. But near a black hole, the electron could fall into the black hole never to be seen again, while the positron escapes to infinity. Or the roles could be reversed. An observer far away from the black hole would thus see a steady stream of electrons and positrons with characteristic energy T_H.

The radius at which the radiation originates is where the gravitational potential energy GMm/r overwhelms the rest energy $2mc^2$ needed for creating a particle-antiparticle pair. But this is just the Schwarzschild radius r_S mentioned at the beginning of this chapter and discussed in chapter I.2. I need hardly emphasize that only a compact object whose physical size is smaller than its Schwarzschild radius qualifies as a black hole and can Hawking radiate.

The Boltzmann factor tells us that only black holes with temperature T_H much greater than the electron rest mass (or equivalently, black holes with size much less than the de Broglie wavelength of the electron) could radiate electrons.

Gravity's task is to change the causal structure of spacetime in order to trap the particle (or antiparticle) which falls in, and Einstein's classical theory is entirely up to the job. No quantum gravity is needed.

When do we have to worry about the quantum nature of the gravitational field?

This may be an appropriate occasion to give a handwaving argument[10] regarding when we have to worry about the quantum nature of the gravitational field. Consider an object of mass M: you, for example. As you walk around, you are surrounded by a gravitational field that in reality consists of a swarm of gravitons. Let's estimate N, the number of quanta in the swarm. If the number of quanta in the field is of order 1, then we would certainly have to deal with the quantum nature of the field. But if $N \gg 1$, then the field can be treated classically. To estimate N, let the object be spherical,[11] and imagine the swarm of gravitons spread out in a spherical distribution with a characteristic size L. By the uncertainty principle, the characteristic energy of a graviton is then of order $\varepsilon \sim 1/L$. The total energy contained in the gravitational potential $\phi = -GM/r$ is, according to Newton (or Einstein), given by $E \sim G^{-1} \int d^3x \ (\nabla\phi)^2 \sim G^{-1} \int_L^\infty dr \ r^2 \ (GM/r^2)^2 \sim GM^2/L$. Thus, the number of quanta equals $N \sim (GM^2/L)/(1/L) \sim GM^2 \sim (M/M_P)^2$.

This is a pleasing result and presumably accords with your intuition: Unless the mass M is comparable to the Planck mass M_P, you don't need to lose any sleep over quantum gravity at all. You certainly did not expect that the field surrounding you could not be treated classically, did you? This heuristic argument applies to all masses, including black holes. Thus, you only have to worry when the mass of the black hole drops to $\sim M_P$ as it approaches its explosive end.[12] Astrophysicists studying massive black holes at the centers of galaxies are not concerned about Hawking radiation.

Many discussions about quantum gravity, since we do not yet have a definitive theory, involve this type of loose fly by night arguments.[13]

Full Planckian and a shock

We are finally ready to go full Planckian and use natural units. As noted in (3), the entropy S of a black hole is just the square of its mass measured in Planckian units. Using the fact that the black hole has radius $R \sim GM$ and hence surface area $A \sim R^2$, we can also write

$$S \sim GM^2 \sim \frac{M^2}{M_P^2} \sim \frac{R^2}{G} \sim \frac{A}{l_P^2} \tag{5}$$

You should be shocked, shocked, shocked. Most theoretical physicists were, and are.

Not shocked?

Normally, the entropy of a system is extensive,[14] that is, proportional to its volume. Think of an everyday container of gas, for example.

Somehow, a black hole has an entropy proportional to its surface area rather than its volume. It is as if the entropy of a black hole resides completely on its surface. Indeed, imagine laying down a grid on the surface of a black hole. Somehow, each Planck-sized cell with area l_P^2 contains one unit of entropy, as indicated by (5). This mysterious property of black holes, which represents one of the deepest puzzles in theoretical physics, led 't Hooft and Susskind separately to formulate the holographic principle.[15] Many fundamental physicists believe that this mysterious property of black holes holds the key to quantum gravity.

On the verge of collapse

An almost universally accepted (but not yet mathematically proven) folk belief is that if the size L of a physical system is smaller than its Schwarzschild radius $r_S \sim M$, then it will collapse into a black hole. In other words, an object too compact for its mass will collapse. (Note that here we have effectively set the Planck mass and the Planck length to 1.)

Recall that the energy and entropy of a box of photons are given by $E \sim L^3 T^4$ and $S \sim L^3 T^3$, respectively. Now consider a box of photons so hot that, if we throw in just a bit more energy, the box would collapse and become a black

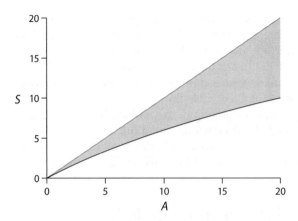

FIGURE 2. The entropy o a black hole S_{BH} and 't Hooft's upper bound on the entropy of a box of radiation $S_{\text{max bo}}$ are plotted schematically as a function of their surface area A in units of the Planck are . For $A > 1$, $S_{BH} > S_{\text{max box}}$. For $A \gg 1$, the entropy of a black hole far exceeds that of a box of radiation. Since nobody understands quantum gravity, we cannot trust this plot for $A < 1$ and perhaps not even for $A \gtrsim 1$.

hole. The condition of being on the verge translates into $E \sim L^3 T^4 \lesssim L$, that is, $T \lesssim 1/L^{\frac{1}{2}}$. The entropy of the box then equals

$$S \sim L^3 T^3 \lesssim L^{\frac{3}{2}} \sim A^{\frac{3}{4}} \tag{6}$$

A box of electromagnetic radiation hot enough to almost be a black hole has an entropy that can grow at most like the $\frac{3}{4}$ power of area, rather than like the area, as is the case for a black hole. Remember that we are using the Planck area to measure area with. Thus, for $A \gg l_P^2$, the entropy of the box of photons is tiny compared to that of a black hole with the same area as the surface area of the box. This bound was obtained by 't Hooft in 1993. See figure 2.

Exercises

(1) The Unruh[16] effect states that an accelerating observer in flat spacetime would perceive a bath of thermal radiation as a result of quantum fluctuations. Just like the Hawking effect, a proper derivation of this effect requires some knowledge of quantum field theory. Nevertheless, argue that the temperature of the Unruh radiation is given in terms of the acceleration a by

$$T_U \sim a$$

in natural units.

(2) Restore the factors of $\hbar$ and c in the expression for the Unruh temperature given in exercise (1). Estimate the acceleration needed to reach room temperature.

(3) Most people think that black holes are invisible, but you know better. Estimate the typical mass of black holes that emit predominantly visible light.

Notes

[1] This is explained in all textbooks on gravity. See, for example, *GNut*, part VII.

[2] The history of black hole entropy illustrates well the dictum that the recognition that some quantity exists in physics is often the hard part. Once the existence of that quantity is understood, then we can try to calculate its value, which may well be far from easy. But dimensional analysis is easy.

[3] Factoid: $GM_\odot$ is known to 8 significant figures, while G is known only to about 4 significant figures.

[4] For example, see *QFT Nut*, pp. 290–291, or *GNut*, pp. 444–446.

[5] J. A. Wheeler, *A Journey into Gravity and Spacetime*, Scientific American Library, W. H. Freeman, 1999, p. 221.

[6] Tragically, Bekenstein died in Helsinki by falling down a flight of stairs, having refused to turn on the lights on a Shabbat (M. Chaichian, private communication).

[7] In elementary texts, for a box of volume V containing some gas, this fundamental law of thermodynamics is stated as $dE = TdS - PdV$, but volume does not enter here.

[8] Strictly speaking, since counting is involved, entropy is a concept of quantum statistical mechanics and does not make complete sense in classical physics.

[9] See, for example, R. Feynman, *Statistical Mechanics*, W. A. Benjamin, 1972.

[10] I heard this argument from G. Dvali.

[11] That is, in the spirit of the famous book *Consider a Spherical Cow*, consider a spherical you.

[12] Indeed, at one point, a contentious subject revolves around what you would expect to see: peculiar remnants or nothing?

[13] More examples can be found in *GNut*, chapter X.8.

[14] This has been proved for systems with short-range interactions.

[15] Which has in turn led to AdS/CFT. See *GNut*.

[16] W. Unruh, *Physical Review* D14 (1976), p. 870; see also S. Fulling, *Physical Review* D7 (1973), p. 2850, and P. Davies, *Journal of Physics* A8 (1975), p. 609.

When Einstein gravity meets the quantum

Newton's constant in the classical and in the quantum world

Before getting to quantum gravity, let's go back to Newtonian gravity, to chapter I.2, where we noted that the combination GM has dimension of $[GM] = L^3/T^2$. (A quick way of remembering this is to note that Newton told us that GM/r^2 is an acceleration with dimension L/T^2.)

Einstein gravity is of course a relativistic theory, and thus with $c = 1$, L and T are of the same dimension. Hence $[GM] = L$ is just a length, so that $[G] = L/M$.

So now what happens when we introduce Einstein gravity to the quantum?

Amazingly, simple dimensional analysis already warns us of unpleasant sparks between the two!

In the quantum world, we could set $\hbar$ to unity, so that $[\hbar] = 1 = ML$. Length has the dimension of an inverse mass (or energy, since the theory is manifestly relativistic). The important observation

$$[G] = \frac{L}{M} = L^2 = \frac{1}{M^2} \tag{1}$$

follows.

Newton's gravitational constant has dimension of inverse mass squared. Indeed, from chapter IV.1, the Planck mass equals $M_P = \sqrt{\hbar c/G}$. So, with $c = 1$ and $\hbar = 1$, $G = 1/M_P^2$.

Collision between two gravitons

When we quantize electromagnetism, the electromagnetic wave is revealed to be a stampede of photons. In precisely the same way, when we quantize gravity, the gravity wave is revealed to be a stampede of gravitons. Newton's G, being a measure of gravity's strength, determines how inclined a graviton is to interact with matter and with other gravitons.

Consider the collision of two gravitons, each with energy E, producing two outgoing gravitons. In quantum physics, scattering processes are described by a probability amplitude. For our purposes here, we do not even need to know how this amplitude, call it $\mathcal{M}$, is computed. Suffice it to know that, since the collision is governed by gravity, $\mathcal{M} \propto G$. Without gravity, the amplitude ought to vanish.

In a classical world, after colliding, these two gravitons would move away far from each other, proceeding to spatial infinity. But in a quantum world, the two gravitons are described by probabilistic wave functions that could interact again. Quantum fluctuations change $\mathcal{M}$. If the two gravitons interact twice, the correction should be $\propto G^2$. In other words, $\mathcal{M} \propto G + KG^2 + \cdots = G(1 + KG + \cdots)$, with some K such that $[K] = M^2$, since $[G] = 1/M^2$.

But E is the only quantity around with dimension of mass. Thus, we conclude by dimensional analysis that the scattering amplitude must have the form

$$\mathcal{M} \propto G\{1 + a\, GE^2 + O(G^2)\} = G\{1 + a\,(E/M_P)^2 + O(G^2)\} \tag{2}$$

with a some dimensionless function of the scattering angles. By definition, the correction to the 1 in the curly brackets is linear in G, and so the correction term has to go like E^2.

As we crank up the energy[1] E past the Planck energy $M_P = \sqrt{1/G}$, the second order term becomes larger than the first order term. Furthermore, in quantum physics, the absolute square of the scattering amplitude determines the probability of the scattering process, and since probability by definition cannot exceed 1, the scattering amplitude cannot be arbitrarily large. You may know this as the unitarity bound. Here we face the potential violation of this bound, suggesting the breakdown of quantum physics at the Planck energy.

We can readily extend this argument to cover the higher order terms. Thus, the next order term in (2) must have, again by dimensional analysis, the form $b\,(E/M_P)^4$, with b some other dimensionless function of the scattering angles.

One possible reaction to this unitarity argument could be "So what? The perturbation expansion fails." It is certainly possible that one day, but a day that theoretical physicists can only dream of now, we will know how to treat quantum gravity nonperturbatively. The series in the curly bracket in (2) could turn out to be the expansion of a function $f((E/M_P)^2)$, which behaves with decency even for $E \gtrsim M_P$. It is also possible that the function is non-analytic and does not admit a perturbative expansion. But these are merely words.

Minimum length

There exist several indications that quantizing gravity would shake up the existing framework of physics. The patient exhibits many symptoms, but apparently all due to one root cause: that G is dimensional, in contrast to the dimensionless couplings governing the strong, electromagnetic, and weak interactions.

Ever since Louis the Duke de Broglie astonished the physics world (and won a Nobel Prize in the process[2]) with his assertion that a particle with momentum p behaves like a wave with wavelength of order $\hbar/p$, particle physicists have been pestering heads of governments (and taxpayers) that they need to build larger and larger accelerators to probe shorter and shorter distances. Given a beam of particles with energy E, they can probe distances of order $l_{dB} \sim \hbar/p \sim \hbar/E$. Allowed enough resources in a world without gravity, they could keep on increasing the energy and happily probe smaller and smaller distances.

But in a world with gravity, we have black holes!

A concentration of mass or energy E in a region smaller than the corresponding Schwarzschild radius $r_S \sim GE$ is expected to collapse into a black hole. Recall chapter I.2. Thus, the colliding beams we use would collapse when $GE \gtrsim l_{dB} \sim \hbar/E$, precisely when $E \gtrsim M_P$. This suggests that the Planck length $l_P \sim 1/M_P$ represents a minimum length, below which we cannot probe.[3]

In the quantum world, physical quantities are constantly fluctuating. The appearance of a fundamental length as soon as gravity is "turned on" suggests that in quantum gravity, spacetime itself is fluctuating on the scale of l_P, thus leading to the picturesque notion of spacetime foam. Does this mean, as was just suggested, that l_P represents the minimum length that we can probe? It seems plausible.

Trying to localize in the presence of gravity

A related argument,[4] not mentioning black holes, was given by Mead[5] in 1964. Again, if we want to localize a particle to within Δx, we need to use a short wavelength, high frequency photon with energy E satisfying $\Delta x \sim \hbar/E$. This is all fine in a world without gravity, but in a world with gravity, the photon will exert a gravitational force on the particle, causing it to accelerate with acceleration $a \sim GE/r^2$. Here r denotes a vaguely defined characteristic distance (r will drop out) describing the interaction between the photon and the particle. The photon traverses this interaction region in time r, during which the particle acquires a velocity $v \sim ar$ and travels a distance $d \sim vr \sim ar^2 \sim GE$. Combining this with Heisenberg's uncertainty principle, we conclude that our knowledge of the position of the particle is limited by what might be called the generalized uncertainty principle:

$$\Delta x \sim \frac{\hbar}{E} + GE \qquad (3)$$

In other words, in addition to the uncertainty imposed by the wavelength of the photon, the particle we are trying to observe has also moved due to its gravitational interaction with the photon. Minimizing this, we see that the best we can do is $\Delta x \sim \sqrt{\hbar G} = l_P$. Again, notice that we have implicitly used special relativity here, equating the general relativity mass with energy.[6]

As soon as you feel that gravity should correct Heisenberg's uncertainty principle to order G, you end up with (3) by dimensional analysis.

Interestingly, string theory also naturally leads to (3). Imagine a graviton, allegedly a closed loop of string. As we pump energy into it, it expands to size GE, thus accounting for the term in (3) additional to Heisenberg's term.

The optimism of two giants and the execution of a young physicist

He who does not believe it owes one dollar. [M. Bronstein]

The reader might recognize that all these arguments amount to essentially different versions of the same argument. Ultimately, they all come down to the fundamental fact that gravity introduces a natural energy or mass scale and a corresponding length scale. This type of argument goes way back, to a little-known paper[7] published in 1935 by the brilliant Russian physicist Matvei Bronstein, who was purged and executed at the age of 31 in 1938. See figure 1.

Historically, Heisenberg and Pauli quantized the electromagnetic field in 1929, concluding rather optimistically that "the quantization of the general relativity field ... may be carried out without any new difficulties by means of a formalism fully analogous to that applied here." Ha! Even quantum electrodynamics was not so easy,[8] let alone quantum gravity.

But the general belief[9] throughout the 1930s was that, once quantum electrodynamics came under control, quantum gravity would follow readily, with perhaps some trivial modifications. With deep insight, the young Bronstein pointed out emphatically[10] that the electromagnetic and the general relativity fields are intrinsically different, because of what was then known as the "general relativity radius" of massive objects.

Black holes are strange, in more ways than one. The founders of quantum physics taught us that the quantum size of a particle of mass m is of order $\hbar/m$: The more massive the particle, the smaller it is in the quantum world. But a black hole of mass M has size $GM = (M/M_P)l_P$. The more massive the black hole, the larger it is, a behavior that is precisely opposite that of all other particles. This peculiar fact underlies the arguments given here.

The presence of the Planck length l_P indicates that the theory of quantum gravity, whatever it turns out to be, cannot possibly be a quantum field theory. For one thing, quantum field theory is based on the notion of local observables, described by fields defined at points in spacetime. But with spacetime itself fluctuating wildly according to the "dance of the quantum," we cannot even locate precisely where we are. In other words, to formulate quantum field theory, we need slices of spacelike surfaces to succeed each other in an orderly progression along a timelike coordinate axis. In his 1935 paper, Bronstein advocated

FIGURE 1. The physicist Matvei Bronstein, a pioneer of quantum gravity.

"a radical reconstruction of the theory ... and the rejection of [a] Riemannian geometry ..., and perhaps also the rejection of our ordinary concepts of space and time, replacing them by some much deeper and nonevident concepts." In the early 21st century, string theorists are saying pretty much the same thing. Indeed, the reader can readily understand that with a fluctuating metric, fundamental concepts that we take for granted in doing physics, such as the arrow of time, the signature of the metric, the topology of spacetime, and so on, would all become problematic.

Notes

[1] Historically, this argument is confusingly phrased in terms of infinities. In more modern treatments of quantum field theory, there are no infinities in physics, only cutoffs. I will not explain this point here. See, for example, *QFT Nut*, chapters III.1–III.2.

[2] In 1929, a year after the death of his mother, who thought that her youngest son would never amount to anything. Born in 1892 and dead in 1987, de Broglie was one of the longest-lived theoretical physicists of his generation. He exerted a negative influence on the development of physics in France.

[3] As is appropriate for a textbook, I am presenting the standard mainstream view here. The statement that we cannot go smaller than l_P is far from settled and has a controversial literature. For a small sampling, see H. Salecker and E. P. Wigner, *Physical Review* 109 (1958), pp. 571–577; R. Gambini and R. Porto, http://arXiv.org/pdf/gr-qc/0603090.pdf sec. II; Y. J. Ng, *Annals of the New York Academy of Sciences* (1995), pp. 579–584; R. Gambini, J. Pullin, and R. Porto, http://arXiv.org/abs/hep-th/0406260; and G. Amelino-Camelia and L. Doplicher, *Classical and Quantum Gravity* 21 (2004), pp. 4927–4940, hep-th/0312313.

[4] For a sampling of arguments about the incompatibility of gravity with quantum physics, see *GNut*, chapter X.8.

[5] C. Alden Mead, *Physical Review* 135 (1964), p. B849.

[6] To me, this kind of handwaving argument is rather fast and loose and should (and could) be refined. Indeed, Mead did refine his argument, first taking into account momentum conservation and then replacing Newtonian gravity by Einsteinian gravity.

[7] See G. Gorelik, *Physics-Uspekhi* 48 (2005), p. 1039.

[8] As the reader probably knows, this early attempt at quantum electrodynamics was afflicted by infinities and various inconsistencies, difficulties that were not cleared up until the late 1940s by the generation consisting of Schwinger, Feynman, Tomonaga, Dyson, Ward, and others.

[9] Keep in mind the enormous confusion at the time, such as Bohr's proposal that energy is not conserved, and the issue of whether the uncertainty principle could be applied to fields. L. Rosenfeld was apparently the first to show that quantum field theory and classical relativity are not consistent together: http://www.sciencedirect.com/science/article/pii/0029558263902797.

[10] He ended his paper with the statement "Wer's nicht glaubt, bezahlt einen Thaler." (He who does not believe it owes one dollar.) Cf. J. Grimm and W. Grimm.

Interlude

Math medley 2

Feynman's integrals

Can you evaluate the integral $\int_0^1 dx \frac{1}{(px+q)^2}$?

Of course you can![1] The denominator is just the square of a linear function of x. But can you calculate it without calculating, in the spirit of math medley 1?

The first step is to rewrite the integral as

$$I = \int_0^1 dx \frac{1}{[ax + b(1-x)]^2} \tag{1}$$

by trivially redefining the constants.

Next, observe that the integral is symmetric under $a \leftrightarrow b$. (Simply change the integration variable $x \to 1 - x$.)

Let a and b have dimension of something; it doesn't matter what (length, mass, whatever), as long as they have the same dimension as evidenced by the denominator (since x is dimensionless, as indicated by the limits on the integral). Then the integral has dimension given by the inverse of this whatever dimension squared.

Combining dimensional analysis and interchange symmetry, we obtain immediately that $I \propto \frac{1}{ab}$. The overall coefficient can be fixed instantly by setting $a = b$ and evaluating the resultant trivial integral $\frac{1}{a^2} \int_0^1 dx$. Thus, $I = \frac{1}{ab}$.

By the way, the symmetry between a and b in (1) can be made manifest by rewriting the integral using the delta function (see appendix Del):

$$I = \int_0^1 dx \int_0^1 dy \frac{\delta(1 - x - y)}{[ax + by]^2} \tag{2}$$

You can also differentiate to obtain less symmetric and hence less obvious integrals, for example, $\int_0^1 dx \frac{x}{[ax+b(1-x)]^3} = \frac{1}{2a^2b}$.

Next, try evaluating this integral:

$$I = \int_0^1 dx \int_0^{1-x} dy \frac{1}{[px+qy+r]^3} \tag{3}$$

Got it! The trick is to write (3) as

$$I = \int_0^1 dx \int_0^1 dy \int_0^1 dz \frac{\delta(1-x-y-z)}{[ax+by+cz]^3} \tag{4}$$

Indeed, doing the z integral amounts to setting $z = 1 - x - y$, so that the square bracket becomes $[ax + by + c(1-x-y)] = [(a-c)x + (b-c)y + c]$. The integral is symmetric under permutations of a, b, c. (Note that $abc = (p+r)$ $(q+r)r$, so that this symmetry is partially hidden in (3).)

Following the same reasoning as above (that is, invoking dimensional analysis and permutation symmetry), we obtain immediately $I = \frac{1}{2abc}$, with the factor of 2 coming from the area of a right triangle with two equal sides. (I am being intentionally cryptic.)

You can obviously generalize these integrals to contain n parameters $a_1, \ldots, a_n$:

$$\frac{1}{a_1 a_2 \cdots a_n} = (n-1)! \int_0^1 \int_0^1 \cdots \int_0^1 dx_1 dx_2$$

$$\cdots dx_n \delta(1 - \sum_j^n x_j) \frac{1}{(a_1 x_1 + a_2 x_2 + \cdots + a_n x_n)^n} \tag{5}$$

Dimensional analysis, symmetry, and evaluating a simple case

My purpose here is to show the combined power of dimensional analysis, permutation symmetry, and going to a simple case (for example, by setting all the a_js equal in (5)) in evaluating a certain class of integrals. In fact, this class of integrals played an important role in the development of quantum field theory, and the identity (5), popularized[2] by Feynman, was essential in evaluating Feynman diagrams, which I mentioned in chapter II.3 and will chat about briefly in part IX. A typical (4-dimensional) integral encountered has the form $\int d^4 q / \{(q^2 + m^2)((q+p)^2 + m^2)((q+k)^2 + m^2)\}$, which becomes doable upon applying (5) to the integrand.[3] By the way, Schwinger also had an identity. Otherwise, how could he have beaten Feynman to the calculation of the magnetic moment of the electron?

Dimensional analysis may be helpful even if you are dealing with pure numbers

Some undergraduates are surprised that dimensional analysis is useful for evaluating integrals. "These are pure numbers," they might protest.

But no, we can assign dimensions to variables, and as long as we do it consistently, dimensions still have to match, for example, in (5) if the a_js are thought of as masses. Take another example. Consider the celebrated Gaussian integral $I = \int_{-\infty}^{\infty} dx e^{-ax^2}$. Assign the dimension of length to x. Then $[a] = 1/L^2$ and $[I] = L$, which implies immediately that $I = C/a^{\frac{1}{2}}$ with some unknown constant[4] C. Differentiating with respect to a, we obtain $\int_{-\infty}^{\infty} dx x^2 e^{-ax^2} = C/(2a^{\frac{3}{2}})$ and so forth.

Exercise

(1) "Guess" what the integral $I = \int_{-\infty}^{\infty} dx/(x^2 + a^2)$ is by using dimensional analysis. Hint: Be careful!

Notes

[1] Otherwise you wouldn't be reading this book.

[2] I use the word "popularize," because I have not verified that Feynman was the first to use these identities on these integrals.

[3] See, for example, (10) on p. 197 of *QFT Nut* on the calculation of the anomalous magnetic moment of the electron. Relentless exploitation (for example, (13) on the same page) of the interchange symmetry evident in (5) and other tricks render this celebrated calculation now doable as a homework problem by students far less gifted than Schwinger and Feynman.

[4] Evidently, dimensional analysis can never give you C; we need to go nonlinear. One nifty trick is to evaluate I^2 by going to polar coordinates. See *QFT Nut*, p. 14.

Part V

From ideal gas to Einstein condensation

Ideal Boltzmann gas

Forgiving operations

Some religions tell us that to forgive is a virtue. The fly by night physicist also loves forgiving operations. With a forgiving operation, we can afford to be sloppy and yet obtain the correct answer. Forgiving operations are in stark contrast to draconian operations, in which any mistake is fatal.

An example of a forgiving operation is $\frac{d}{dx} \log f(x)$. The log turns any constant multiplying $f(x)$ (constant meaning anything not depending on x) into an additive constant, which is then annihilated by the derivative. This example appears, in particular, in the theory of gases.

Ideal gas

The formula $S = \log W$, with the infamous k, is carved on Boltzmann's tomb. Here W denotes the number of different microstates that correspond to the macrostate with energy E of the system that we are studying, and S the entropy of the system. We fly by night physicists are not going to worry about all the ifs and buts in the following discussion, leaving all those caveats and precise definitions to the serious scholars.

An ideal gas is by definition a gas in which the interaction between the atoms (not molecules, just to be definite) crucially brings about the collisions that cause the gas to reach equilibrium but can otherwise be neglected. The microstate is characterized by the momenta $\vec{p}_1, \vec{p}_2, \cdots, \vec{p}_N$ of the N atoms, with N of order 10^{23} for an everyday macroscopic sample of gas. This corresponds to a point in some humongous $3N$-dimensional space defined by the $3N$ momentum components $\{p_a^i, a = 1, \cdots, N, i = 1, 2, 3\}$ of the atoms. The energy of the gas is given by $E = \frac{1}{2m}(\vec{p}_1^2 + \vec{p}_2^2 + \cdots + \vec{p}_N^2)$. Hence, a macrostate with a given energy E describes a sphere of radius $\sqrt{2mE}$, namely, the set specified by

$$\vec{p}_1^2 + \vec{p}_2^2 + \cdots + \vec{p}_N^2 = \sum_{a=1}^{N}\sum_{i=1}^{3}(p_a^i)^2 = 2mE \tag{1}$$

in a 3N-dimensional space. By high school geometry, the area (in the general sense of the word) of a sphere goes like the radius of the sphere raised to a power[1] equal to the dimension of the space the sphere lives in minus one. (Think about this for circles and everyday spheres, and if you don't get it, read the endnote.) Thus, the surface area of our humongous sphere is proportional to $(\sqrt{E})^{3N-1} \sim E^{\frac{3}{2}N}$. Between us friends, what is 1 compared to a number of order 10^{23}?

A microstate specified by $\vec{p}_1, \vec{p}_2, \cdots, \vec{p}_N$ is a point on this enormous sphere. Boltzmann's W is evidently (more below) proportional to the area $\sim E^{\frac{3}{2}N}$ of the sphere. Thus, Boltzmann's profound formula gives the entropy as

$$S = \log W = \frac{3}{2}N \log E + \text{irrelevant garbage} \tag{2}$$

The fundamental law of thermodynamics states that

$$dE = TdS - PdV \tag{3}$$

Since we are keeping the volume V fixed, $dE = TdS$, that is, $\frac{dS}{dE} = \frac{1}{T}$. Differentiating (2), we obtain immediately $\frac{3}{2}\frac{N}{E} = \frac{1}{T}$, that is,

$$E = \frac{3}{2}NT \tag{4}$$

the familiar result for the ideal gas. Note that the 3 comes from the dimension of space.

Boyle's law

If we want to go further and determine the pressure, we have to dumpster dive and examine the garbage that we dismissed so cavalierly earlier. According to (3), we have to track down the dependence on V. Let the gas be contained in a cube with side L. In quantum physics, the atoms of the gas are actually waves satisfying boundary conditions, fixing something like pL to be proportional to some integer times π. Hence the spacing Δp between the allowed values of the momentum p goes like $1/L$, which vanishes (as it should) in the thermodynamic limit $L \to \infty$.

Thus, Boltzmann's W is actually proportional to the area of the 3N-dimensional sphere divided by the area between the points, that is, $\sim (\sqrt{E}/$

$(1/L))^{3N} \sim (E^{\frac{3}{2}} L^3)^N \sim (E^{\frac{3}{2}} V)^N$. So now the entropy equals $S = \frac{3}{2} N \log E + N \log V +$ irrelevant garbage, a more refined version of (2).

In applying (3) to determine the pressure, we are instructed to keep S fixed. Thus,

$$0 = dS = \frac{3}{2} N \frac{dE}{E} + N \frac{dV}{V} \tag{5}$$

giving $P = -\frac{dE}{dV}|_S = \frac{3}{2} \frac{E}{V} = \frac{NT}{V}$. (We used (4) in the last step.) Behold, Boyle's law[2] for the classical gas[3]

$$PV = NT \tag{6}$$

pops out.

Note that while we were sloppy in the intermediate steps and used only the twiddle sign, we are licensed to use equal signs in our final results (4) and (6).

Counting

Some readers may know that here we are using the microcanonical ensemble, whose crucial starting point is to obtain a decent expression for W by counting microstates. This brings me to another pet peeve: The term "classical statistical mechanics" makes no sense. The word "statistical" implies counting, but in classical physics, state variables, such as momentum here, take on continuous values. Textbooks discussing classical statistical mechanics typically devote considerable verbiage to phase space and how counting could make sense in classical physics.[*]

But no matter what, the words "classical" and "statistical" clash. Perhaps this fact contributes to the harsh criticism Boltzmann's fellow physicists bullied him with, which in part drove the leading tragic figure of physics to suicide.

In our example, when we had to count, we invoked quantum mechanics to quantize the momentum eigenstates. Intriguingly, in the calculation, $\hbar$ appears and then disappears.

Free to be sloppy

In physics as in life, it pays to know when one can afford to be sloppy.

---

[*]Strictly speaking, in classical statistical mechanics, one cannot define entropy, only the change in entropy, but for most applications, this suffices.

Notes

[1] Thus, the circumference of the circle of radius r goes like $r^{2-1} = r$, the area of the sphere of radius r like $r^{3-1} = r^2$, the area of the hypersphere of radius r living in 4-dimensional space like $r^{4-1} = r^3$, etc. Care to work out the precise overall constant? The answer is given in *QFT Nut*, p. 539 (just to show you that I can fly by day also).

[2] Discovered by Henry Power. Incidentally, Boyle's law represents an early example of "grand unification" of several previously known laws. For instance, Charles's law states that gases expand when heated.

[3] https://www.youtube.com/watch?v=8FvCviUW-58.

Van der Waals

master of the envelope

Modifying Boyle's law

I like to call van der Waals[1] the master of the envelope.

Start with Boyle's law

$$PV = NT \qquad (1)$$

relating the pressure P and temperature T of an ideal gas containing N molecules (let's say monoatomic for simplicity) confined to a volume V. While this law is good for a first pass at the theory of gases, it fails to describe many interesting phenomena, such as the possibility of a gas going to a liquid phase.

For our discussion here, it is important to distinguish between extensive and intensive variables. An extensive variable, such as N or V, halves when we halve the gas sample being studied. In contrast, an intensive variable, such as P and T, stays the same when we halve the gas sample. For example, you could mentally partition the room you are in into two halves. Clearly, an intensive variable can only be equal to an intensive variable, not to an extensive variable, and vice versa.

Including interactions between gas molecules

By a couple of remarkably simple moves, van der Waals was able to render Boyle's law (1) considerably more realistic.

First, van der Waals says, molecules have a finite size and thus each occupies a volume b. Hence, the actual volume available to the gas is not V, but the smaller volume $V - Nb$.

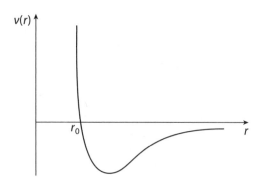

FIGURE 1. Sketch of the potential between two gas molecules.

Second, the transition from gas to liquid is driven by an attractive force between molecules that wants to pull them together. This reduces the pressure P. The amount of reduction should be proportional to the number of attracting pairs $= N(N-1)/2 \sim N^2/2$. Since P is intensive, it would make sense only if the intensive combination N^2/V^2 appears. Thus, we expect the reduction to equal $a(N/V)^2$, with a some constant.

The parameters a and b, both defined to be positive, characterize a gas.

Substituting $V \to V - Nb$ and $P \to P + a(N/V)^2$ into (1), we (or rather van der Waals) obtain

$$\left(P + a\left(\frac{N}{V}\right)^2\right)(V - Nb) = NT \tag{2}$$

To see that the signs are correct in (2), rewrite it as (with the number density $n = N/V$)

$$P = \frac{NT}{V - Nb} - a\left(\frac{N}{V}\right)^2 = \frac{nT}{1 - nb} - an^2 \simeq nT\left(1 + nb - \frac{na}{T} + O(n^2)\right) \tag{3}$$

Indeed, b increases the pressure, while a reduces it. The van der Waals equation of state is evidently a low density expansion to leading order in n.

Microscopic origin of a and b

In the molecular theory of gases, one imagines the interaction potential $v(r)$ between two molecules to have the form sketched in figure 1. The potential contains a hardcore repulsion of range r_0 and a longer ranged attraction that vanishes rapidly with increasing r.

Recalling that pressure has the dimension of an energy density, we obtain the dimension

$$[a] = \left(\frac{E}{L^3}\right)\frac{1}{(1/L^3)^2} = EL^3 \tag{4}$$

An educated guess would be that a is given by the depth of $v(r)$ times the range of $v(r)$ cubed. Similarly, we expect b to equal the range of $v(r)$ cubed. As is shown in appendix VdW, these guesses are correct.

Not merely some dumb garden variety fit to data

Textbooks tend to minimize van der Waals's contribution, and are thus guilty of the bias of hindsight. When van der Waals proposed his equation in 1873, most physicists did not accept the reality of molecules. It is also important to realize that (2) is not merely some dumb garden variety two-parameter fit to data. The two parameters tell us about the size of molecules and the attraction between them. Furthermore, (2) indicates to experimentalists the range of temperature and pressure at which helium, for example, could be liquefied. The impact of van der Waals's work, on theoretical and experimental physics and on technology, was huge.

   Indeed, Maxwell was so impressed by van der Waals's equation that he thought it worthwhile to learn Dutch so as to be able to read the paper in the original. As the reader may also know, Maxwell managed to make an important contribution to the theory of phase transitions with what is now called "Maxwell's construction." See below.

Stability and phase transitions

Let us plot pressure as a function of the specific volume $v \equiv V/N$ using (3) in the form

$$P = \frac{T}{v-b} - \frac{a}{v^2} \tag{5}$$

See figure 2.

   A fundamental stability requirement from thermodynamics is that $\left.\frac{\partial P}{\partial v}\right|_T < 0$: As pressure increases at fixed temperature, the specific volume should decrease. Boyle's law manifestly satisfies this. In contrast, van der Waals's law gives $\left.\frac{\partial P}{\partial v}\right|_T = -\frac{T}{(v-b)^2} + \frac{2a}{v^3}$, which at sufficiently low T could be positive or negative, depending on v, as we can verify by looking at figure 2 with our very own eyeballs. That $\left.\frac{\partial P}{\partial v}\right|_T$ can be positive for some v signals an instability. Note that for $a = 0$ with no attraction between molecules, $\left.\frac{\partial P}{\partial v}\right|_T$ is always negative, and Boyle's gas is always stable, which makes good physical sense.

   The instability informs us of a phase transition. Since at sufficiently low temperatures, $\left.\frac{\partial P}{\partial v}\right|_T$ goes from negative to positive and back to negative as v increases, $P(v)$ has a minimum and a maximum, as we can also see in figure 2. Imagine now raising the temperature. The two extrema move closer to each

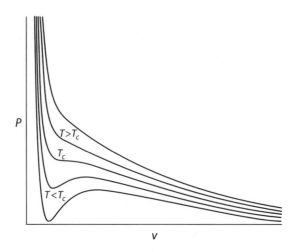

FIGURE 2. A plot of the function $P(v)$ for various values of T.

other and merge at some critical temperature T_c. The values of v and P at which the merging occurs are denoted by P_c and v_c. More discussion of the critical point may be found in appendix Cp.

Since thermodynamics forbids positive $\left.\frac{\partial P}{\partial v}\right|_T$, Maxwell proposed constructing a monotonically nonincreasing P as a function of v. He replaced the offending section of $P(v)$ by a horizontal line defined by $P(v)$ equals some constant. Further discussion is beyond the scope of this book.

Using units the gas actually knows about

From the discussion leading up to the van der Waals equation, we know that it would be sensible to measure the intensive variables $P, T, v = V/N$ not in terms of some arbitrary humanmade units, but in terms of units the gas actually knows about. Well, the gas knows about P_c, T_c, and v_c, respectively the values of P, T, and v at the critical point. (This general philosophy is reminiscent of the God-given units introduced in chapter IV.1: We should use units that the fundamental laws of physics actually know about. I will spare you another rant here.)

The first step is to determine P_c, T_c, and v_c, then write $P = \mathscr{P} P_c, T = \mathscr{T} T_c$, and $v = v v_c$, and plug into (5).

The result, known as the law of corresponding states, reads

$$\left(\mathscr{P} + \frac{3}{v^2}\right)\left(v - \frac{1}{3}\right) = \frac{8}{3}\mathscr{T} \tag{6}$$

and offers a universal description of those gases obeying the van der Waals law. The calculation of these three "interesting" pure numbers is contrary to the spirit of this book, but I will at least sketch how P_c, T_c, and v_c are determined in appendix Cp.

Exercises

(1) Give more examples of intensive and extensive variables.

(2) Compare the calculated pressure of 1 mole of water molecules contained in a 1000 cm^3 box at 20° C when using the ideal gas equation versus the van der Waals equation.

(3) Compare the pressure of 1 mole of helium molecules contained in a 1000 cm^3 box at 20° C when using the ideal gas equation versus the van der Waals equation. How does the discrepancy compare to that found in exercise (2)?

Note

[1] Johannes Diderik van der Waals (1837–1923), Nobel Prize, 1910.

Quantum gases

Quantum particles are indistinguishable

Quantum particles famously come in two[1] types: fermions and bosons. Fermions are loners and cannot share with other fermions. Bosons, in contrast, are gregarious and strongly prefer to hang out with other bosons. Fermions (such as the electron) and bosons (such as the photon or a pair of electrons) obey, respectively, the mysterious[2] Fermi-Dirac and Bose-Einstein statistics.

Quantum particles are indistinguishable.[3] In sharp contrast, classical particles could have been manufactured in a factory somewhere, with minor defects inevitable, and stamped with a serial number.

From the microscopic structure of atoms to the macroscopic structure of neutron stars, a dazzling wealth of physical phenomena would be incomprehensible without quantum statistics. Many features of condensed matter physics (for instance, band structure, Fermi liquid theory, superfluidity, superconductivity, lasers, and the quantum Hall effect) all follow from these two extreme behaviors of quantum particles. It has been remarked[4] that "no one fact in the physical world . . . has a greater impact on the way things are" than quantum statistics.

Ideal quantum gases

My aim in this chapter is extremely limited. I merely want to provide you with some fly by night understanding of ideal quantum gases. Recall from our study of the ideal classical gas that "ideal" means that interactions between the particles in the gas are negligible but are "infinitesimally" present to ensure that the gas reaches statistical equilibrium after a sufficiently long time.

Start with an ideal classical gas obeying* Boyle's law, $P = nT$, relating pressure P to the number density n and the temperature T. When we switch on quantum mechanics, we expect that this law gets changed to $P = nTG(\hbar, n, T, m)$ with an unknown dimensionless function G dependent on Planck's constant, n, T, and the mass of the particle m.

Our first appeal is to dimensional analysis. Let us use mass M, length L, and energy E instead of M, L, and time T. (This is a triviality; I am simply avoiding potential confusion between using the letter T for temperature and for the generic time that appears in dimensional analysis.) So, list $[\hbar] = E/T$, $[n] = 1/L^3$, $[T] = E$, and $[m] = M$. Oops, we should have noticed right away that the unknown function $G(\hbar, n, T, m)$ depends on four variables, and so dimensional analysis alone cannot solve our problem. We could try to form a dimensionless quantity $\hbar^\alpha n^\beta E^\gamma m^\delta$ and go through the somewhat tedious task of trying to determine these four unknown exponents. Boring! But no matter how you slice it, we have only three equations.

Interjecting some physics into dimensional analysis

What to do? Well, the only possibility is to put in some physics! We observe that $\hbar$, m, and T (by controlling the average energy of the particle) all pertain to the particle, but n controls the average separation between the particles. Prince de Broglie taught us that in the quantum world, the particle is actually a wave packet with a typical wavelength $\lambda = \hbar/p \sim \hbar/(mT)^{\frac{1}{2}}$ (since $p^2/2m \simeq T$).

But what do we compare λ to? The only other length available is the average separation between the particles, namely, $(V/N)^{\frac{1}{3}} = 1/n^{\frac{1}{3}}$.

We conclude that the physically relevant dimensionless variable is $\lambda/(1/n^{\frac{1}{3}}) = n^{\frac{1}{3}}\lambda$, namely, the de Broglie wavelength measured in units of interparticle separation. Equivalently, we could use the dimensionless variable $n\lambda^3 \simeq n\hbar^3/(mT)^{\frac{3}{2}}$, namely, the number of particles inside a de Broglie volume. Hence we write

$$P = nT\left\{1 + f\left(\frac{\hbar^3 n}{(mT)^{\frac{3}{2}}}\right)\right\} \tag{1}$$

with f an unknown function measuring the deviation from the classical result.

---

*Recall that in chapter V.2, we determined the corrections to Boyle's law due to interactions.

Fermi gas versus Bose gas at low density or high temperature

After trudging through a course on statistical mechanics, you can readily determine f with a bit of work. But we fly by night physicists want to obtain the leading quantum correction without breaking a sweat.

First, observe that if we take $f(x)$ to be a linear function $\propto x = \hbar^3 n/(mT)^{\frac{3}{2}}$, then quantum corrections would be of order $\hbar^3$. Let us think of a physical reason why quantum effects would only enter in third order.

What, you can't either? I certainly can't. Search me!

Instead, reasonable people would expect the leading quantum correction to be of order $\hbar$. So, we write $f(x) = Cx^{\frac{1}{3}}$, with C a numerical constant of order unity. Hence

$$P = nT\left(1 + C\frac{\hbar n^{\frac{1}{3}}}{(mT)^{\frac{1}{2}}} + \cdots\right) \qquad \text{(fermion)} \qquad (2)$$

This evidently gives a low density or high temperature expansion. Again, after mastering statistical mechanics, you could determine C and calculate as many dots in (2) as your personality demands.

Waving our hands, we now argue that for a Fermi gas, C is positive. Since fermions want to be left alone, quantum statistics urges them to get away from one another. This exerts an additional pressure, known as Fermi pressure, above and beyond what thermal agitation would give classically. For a Fermi gas, we should have $P > P_{\text{classical}} = nT$.

By the same token, we expect quantum statistics to tend to bring bosons together and hence to lower the pressure. So, let us write, in contrast to (2),

$$P = nT\left(1 - C\frac{\hbar n^{\frac{1}{3}}}{(mT)^{\frac{1}{2}}} + \cdots\right) \qquad \text{(boson)} \qquad (3)$$

Here we have made a flying guess: We take the numerical constant to be the same C as in (2).

This is indeed just a flying guess, plausible or not, depending on your undefinable physics sense. Well, classical particles want neither to be alone nor to party, and are thus halfway between fermions and bosons. Perhaps it is not implausible that the classical result $P = nT$ should be half way between (2) and (3).

If you prefer, write C_f and C_b, respectively, for the constants in (2) and (3). Our discussion does not depend on their precise value, only on the fact that they are positive.

Fermi gas at zero temperature

Let us now think about a Fermi gas in the opposite limit $T \to 0$. Referring to (1), write $P = nTF\left(\hbar n^{\frac{1}{3}}/(mT)^{\frac{1}{2}}\right)$. (Of course, $F \equiv 1 + f$, but it turns out to be more convenient to think in terms of F with its argument as indicated. Once again, physics is not math!) So, we have to make a reasonable guess about the behavior of $F(x)$ as $x \to \infty$. What do you guess?

Classically, at low temperature, thermal agitation ceases, and the pressure $P = nT \to 0$. But for a Fermi gas, even with no thermal agitation, by requiring the fermions to stay away from one another, quantum statistics still produces a pressure.

But we also don't have a reason for the pressure to blow up. Since we have no reason to expect $P \to \infty$ and are expecting that P does not vanish as $T \to 0$, we guess that P approaches a constant.

So, requiring P to approach a constant, we determine that $F(x) \propto x^2$ as $x \to 0$. We thus obtain, remarkably,

$$P \sim nT\left(\frac{\hbar^2 n^{\frac{2}{3}}}{mT}\right) \sim \frac{\hbar^2 n^{\frac{5}{3}}}{m} \tag{4}$$

Without any labor, just saying that the pressure P neither blows up nor vanishes, we have determined that P grows as density to the power $\frac{5}{3}$.

This is a famous result, stating that even at zero temperature, a Fermi gas exerts a nonvanishing pressure, known as the Fermi degeneracy pressure, a pressure that is responsible for the existence of neutron stars, as we will see in chapter V.4. There we will also give another derivation of (4). In particular, we will see how the peculiar looking $\frac{5}{3}$ power emerges.

Remarkably, our intuition fixes the small x behavior of $F(x)$ for us. We learned that quantum effects start in order $\hbar^2$ at zero temperature, rather than in order $\hbar$ at high temperatures, as in (3).

The Fermi degeneracy pressure also plays an important role in solid state physics. For instance, a metal may be treated in some approximation as a Fermi gas of electrons.

A hint of Bose-Einstein condensation

You might have noticed that in the preceding section, we specified a Fermi gas. What about a Bose gas? Think.

For a Fermi gas, by requiring the fermions to stay away from one another, quantum statistics still produces a pressure even at zero temperature. Well, this statement would certainly not apply to a Bose gas. Indeed, by requiring the bosons to get close to each other, quantum statistics would produce a negative pressure, which makes no sense. See (3). This actually signals the gas's desire to collapse before we even get to zero temperature.

Einstein had the penetrating insight that Bose statistics leads to the striking quantum phenomenon now (misleadingly) known as Bose-Einstein condensation.[5]

Another physics alert to watch out for: When a manifestly positive quantity threatens to go negative, alarm bells should go off.

Indeed, the result in (3), $P = nT\left(1 - C\dfrac{\hbar n^{\frac{1}{3}}}{(mT)^{\frac{1}{2}}} + \cdots \right)$, already warns us that as $T \to 0$, the correction term becomes large and negative, menacing to overcome the 1. Of course, rigorous types are all screaming by now, saying that (3) was meant to be a high temperature expansion, and now we are talking about low temperatures. But still, when you fly by night, a warning bell is a warning that something drastic might happen.

We will come back to Bose-Einstein condensation in chapter V.5.

Notes

[1] That's why physics is simpler than psychology: Humans come in all types, "all" presumably meaning almost infinite.

[2] One of the greatest triumphs of quantum field theory is the correlation of these two opposing quantum statistics with whether the corresponding quantum particle carries half-integer or integer spin: the celebrated spin-statistic theorem. For a brief discussion, see *QFT Nut*, chapter II.4.

[3] The explanation of this fact represents another striking triumph of quantum field theory.

[4] By I. Duck and E. C. G. Sudarshan, *Pauli and the Spin-Statistics Theorem*, World Scientific, 1998.

[5] This is the only example I know of in which the less famous person gets more credit than he or she deserves. Well, Einstein hardly needs to fight for more credit and can afford to be generous.

Guessing the Fermi-Dirac distribution ━━

Pressure from quantum statistics

A normal fly by day textbook on statistical physics, which I absolutely and unequivocally urge you to work through, would give a precise step by step derivation of the energy distribution of the Fermi gas and of the Bose gas. Instead, we will now make a fly by night guess of the Fermi-Dirac distribution, and do the same for the Bose-Einstein distribution in chapter V.5. Needless to say, you should not regard these informed guesses as substitutes for the actual derivations.

Our starting point is our guesses for how Boyle's law is modified for the Fermi gas and for the Bose gas, and our understanding of the Fermi degeneracy pressure.

The pressure of a classical gas is generated by the thermal motion of the particles in the gas, and hence it tends to zero as the temperature approaches zero. In sharp contrast, the pressure of a Fermi gas is due, in addition to thermal motion, also to quantum statistics: those fermions want desperately to get away from one another. In chapter V.3, we already deduced that, as $T \to 0$, the pressure of a Fermi gas tends to $P \sim h^2 n^{\frac{5}{3}}/m$, known as the degeneracy pressure.

We will now explore the physical origin of the degeneracy pressure, and in the process obtain the peculiar $\frac{5}{3}$ power.

Fermi sphere and Pauli exclusion

Consider a gas of fermions, say, neutrons to be definite. Neutrons have spin $\frac{1}{2}$. For our purposes here, neutrons with spin up and neutrons with spin down may be treated as two separate species. We will focus on neutrons with spin up. The discussion is simply repeated for neutrons with spin down.

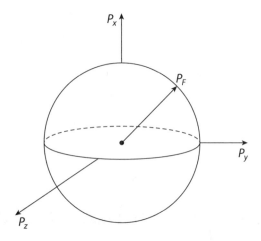

FIGURE 1. Fermi sphere (actually, a ball).

We speak of neutrons here just to be definite, but the discussion here evidently applies to any gas of fermions, for example, the gas of electrons in a metal treated to a first approximation.

We are talking about ideal gases here, in the same spirit as when we discussed classical ideal gases. The interaction between the neutrons is regarded as negligible but is just enough for the gas to reach statistical equilibrium through neutron-neutron collisions. Solving a quantum many body problem is essentially impossible, but Fermi proposes the following approximate treatment, which in practice works extremely well.

For the noninteracting gas, we merely have to solve the elementary problem of a single quantum particle. The single particle energy eigenstates are then simply characterized by momentum $\vec{p}$, with energy $\epsilon(p) = \vec{p}^2/2m$.

Consider the gas confined to a volume V, so that the allowed values of $\vec{p}$ are discrete, as discussed in chapter III.5. Imagine populating the energy eigenstates with neutrons, throwing them in one by one. The Pauli exclusion principle forbids two neutrons (recall that they are all spin up) from going into the same state. The first one goes into the ground state, the second one into the first excited state, and so on, until we have thrown in all N of them. (For applications to macroscopic objects, such as neutron stars or metals, N is a humongous number.)

Imagine drawing a dot in momentum space for every value of $\vec{p}$ with the corresponding state occupied by a neutron. For macroscopic values of V, these dots are microscopically close to one another, so that we can replace the swarm of discrete dots by a continuum. Given the spherical symmetry* enjoyed by the Fermi gas, these dots would form a solid ball in momentum space, known

---

*That is, $\epsilon(p) = \vec{p}^2/2m$ does not depend on the direction of $\vec{p}$, only on its magnitude.

to physicists as a Fermi sphere,* with radius p_F, called the Fermi momentum. See figure 1. By definition, the Fermi momentum p_F is the largest momentum a fermion could have at $T = 0$. We will determine the value of p_F presently. The neutrons are said to fill a Fermi sphere.

Number and energy densities of a Fermi gas

The total number N of neutrons and the total energy E are then given by a straightforward sum over the eigenstates. In the limit $V \to \infty$, the spacing between the eigenstates in momentum space tends to zero, as has already been alluded to, and we can replace the sums by integrals† over the Fermi sphere (again, "ball" for the nitpickers). Thus, evaluating integrals by counting powers of p, we find

$$N = \frac{V}{h^3} \int_{p_F} d^3 p \sim \frac{V p_F^3}{h^3} \tag{1}$$

and

$$E = \frac{V}{h^3} \int_{p_F} d^3 p \, \frac{p^2}{2m} \sim \frac{V p_F^5}{h^3 m}$$

$$\sim \frac{V}{h^3 m} \left(\frac{N h^3}{V} \right)^{\frac{5}{3}}$$

$$\sim \frac{h^2 N^{\frac{5}{3}}}{m V^{\frac{2}{3}}} \tag{2}$$

In the second approximate equality in (2), we use (1) to solve for p_F.

Dividing N and E by V we obtain the number and energy densities of a Fermi gas respectively:

$$n \sim \frac{p_F^3}{h^3} \tag{3}$$

$$\varepsilon \sim \frac{p_F^5}{h^3 m} \sim \frac{h^2 n^{\frac{5}{3}}}{m} \tag{4}$$

(Note that I use ε for the energy density of the entire Fermi gas, and $\epsilon(p)$ for the energy of an individual fermion. One is "curly," while the other is "straight" and evidently depends on p.)

The Fermi momentum (namely, the radius of the Fermi sphere) is determined to be

$$p_F \sim h n^{\frac{1}{3}} \tag{5}$$

---

*Not to be confused with the sphere in chapter V.1, which exists in dimension $\sim 10^{23}$. In contrast, the Fermi sphere, which to be precise, is what a mathematician would call a "Fermi ball," manifestly lives in 3-dimensional space.

†We already showed in chapter III.5 that the number of states in $d^3 p$ is given by $V d^3 p / h^3$. Recall that $h = 2\pi \hbar$.

which we note is the uncertainty momentum associated with the distance separating the particles in the gas. Correspondingly, the Fermi energy is defined by

$$\epsilon_F \equiv \frac{p_F^2}{2m} \sim \frac{\hbar^2 n^{\frac{2}{3}}}{2m} \tag{6}$$

(Do not confuse ε and ϵ_F!)

To find the pressure, use the fundamental law of thermodynamics $dE = TdS - PdV$, which, since $T = 0$, becomes $dE = -PdV$:

$$P = -\frac{\partial E}{\partial V} \sim \frac{\hbar^2 N^{\frac{5}{3}}}{m V^{\frac{5}{3}}} \sim \frac{\hbar^2 n^{\frac{5}{3}}}{m} \tag{7}$$

in agreement with what we had in chapter V.3. Triumph! In particular, from (1) and (2), we see that the power $\frac{5}{3} = \frac{3+2}{3}$ comes from the dimension of space $D = 3$. Note once again that pressure has the same dimension as energy density. In fact, comparing (4) and (7), we have $P \sim \varepsilon$.

Fly by night guess of the Fermi-Dirac distribution

Ready now to guess the energy distribution of the Fermi gas? Recall Planck's black body distribution $n(\omega) = 1/e^{\frac{\hbar\omega}{T}} - 1$ given back in chapter III.5, defined such that the number of photons with frequency in the interval $(\omega, \omega + d\omega)$ is equal to $\frac{V}{\pi^2 c^3} d\omega \omega^2 n(\omega)$. What would be the number of fermions with momentum p in the interval $(p, p + dp)$ in an ideal Fermi gas?

Again, write this as $\frac{4\pi V}{\hbar^3} dp p^2 n(p)$, which serves to define a number distribution $n(p)$ stripped of inessential factors. Incidentally, since $\epsilon = p^2/2m$ and p are directly related, it is often convenient to think of $n(p)$ as a function $n(\epsilon)$ of ϵ. To lessen notational clutter, not only have I confounded $n(p)$ and $n(\epsilon)$, I have also suppressed the dependence of $n(\epsilon)$ on temperature T and on p_F.

Notational alert: You should not confuse either $n(p)$ or $n(\epsilon)$ with the number density n as given in (3). A reminder: $n = \frac{N}{V} = \frac{4\pi}{\hbar^3} \int_0^{p_F} dp p^2 n(p)$.

Our task here is to produce a fly by night guess of the function $n(p)$ as a function of temperature T. How to proceed?

Well, to start with, we know $n(p)$ when $T = 0$. It is specified by the Fermi sphere: States with p less than the Fermi momentum p_F are filled, states with p greater are not. So, $n(p) = 1$ for $\epsilon(p) < \epsilon_F$, and $n(p) = 0$ for $\epsilon(p) > \epsilon_F$ with the Fermi energy defined in (6). But this is just the Heaviside step function

$$\epsilon(p) = \theta(\epsilon_F - \epsilon) \tag{8}$$

See figure 2(a).

Fine.* This is for $T = 0$.

---

*Even though nitpickers may howl: $n(p)$ is a function of p, but we talk as if it were a function of $\epsilon = \epsilon(p)$. I had warned you. Pace.

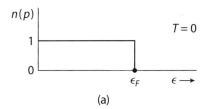

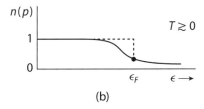

FIGURE 2. The Fermi-Dirac distribution (a) at $T=0$ and (b) at $T \gtrsim 0$.

What happens for $T \gtrsim 0$? By thermal agitation, some fermions with energy just below ϵ_F would acquire energy exceeding ϵ_F, leading to a long Boltzmann tail $\sim e^{-\epsilon(p)/T}$, as shown in figure 2(b). The states with energy just below ϵ_F would be correspondingly depleted, though we expect that $n(p)$ would still be very close to 1 for $\epsilon(p)$ much less than ϵ_F.

Fine again. That is for $T \gtrsim 0$. For high temperature, we expect quantum effects to become unimportant, and the Fermi gas approaches a classical gas, so that $n(\epsilon) \propto e^{-\epsilon/T}$.

Now that we know what $n(\epsilon)$ looks like in three different temperature regimes, we are ready to guess its form. You want to try?

One hint is offered by the distribution of photon energy $n(\epsilon) = 1/(e^{\frac{\epsilon}{T}} - 1)$ with $\epsilon = \hbar\omega$. In chapter V.3, we saw that to go from the pressure of a dilute Fermi gas to the pressure of a Bose gas, we merely have to flip $+1$ into -1, and the expression for the classical Boltzmann gas is just in between, with ±1 replaced by 0.

So, an unjustified fly by night guess might be $n_g(\epsilon) = 1/(e^{\frac{\epsilon}{T}} + 1)$.

But that's wrong. We know what $n(\epsilon)$ is supposed to be for $T=0$. Well, at $T=0$, the exponential $e^{\frac{\epsilon}{T}}$ blows up for $\epsilon > 0$ and vanishes for $\epsilon < 0$. Thus, at $T=0$, our guess is $n_g(\epsilon) = 0$ for $\epsilon > 0$ and $n_g(\epsilon) = 1$ for $\epsilon < 0$. In other words, it is the Heaviside function $\theta(-\epsilon)$.

Wrong wrong guesses are bad, but wrong guesses might be good

In physics (and in mathematics), some results are just wrong wrong, but a result can also be wrong in an interesting way. In this case, the wrong result points us toward the right result.

In fact, we had wanted $n(\epsilon) = \theta(\epsilon_F - \epsilon)$ at $T = 0$, not $\theta(-\epsilon)$. Close, but no cigar.

But this is easy to fix: Simply shift $\epsilon \to \epsilon - \epsilon_F$. Thus, our improved guess is $n_{ig}(\epsilon) = 1/(e^{\frac{(\epsilon - \epsilon_F)}{T}} + 1)$. At $T = 0$, $n_{ig}(\epsilon)$ becomes $\theta(\epsilon_F - \epsilon)$, as desired. (Note: The subscript g is for "guess," ig for "improved guess.")

This is in fact almost the correct answer. While this expression is precisely what we wanted at $T = 0$, as T increases, what we wrote as ϵ_F in n_{ig} has to become a function of T, not fixed as in (6), for a very physical reason, as we will see presently.

A fundamental conservation law

To see what the physical reason is, let us go back to Planck's result for the energy E of a box of radiation of volume V:

$$\frac{E}{V} = \frac{1}{\pi^2 c^3} \int_0^\infty d\omega \omega^2 \frac{\hbar\omega}{e^{\frac{\hbar\omega}{T}} - 1} \tag{9}$$

Since the energy of a photon of frequency ω is $\hbar\omega$, the number of photons N in the box is given by

$$\frac{N}{V} = \frac{1}{\pi^2 c^3} \int_0^\infty d\omega \omega^2 \frac{1}{e^{\frac{\hbar\omega}{T}} - 1} \tag{10}$$

Let us ask what happens when an experimentalist pumps energy into the box. The temperature T increases, of course. Indeed, colloquially, "pumping energy into" is known as "heating up." A teeny bit of mental gymnastics verifies this: As T increases, $\frac{\hbar\omega}{T}$ and hence $e^{\frac{\hbar\omega}{T}}$ decrease, so that the integrand in (9) increases, thus driving E up.

But (10) implies that N also increases as T increases. So what is wrong with that? Nothing. As the experimentalist heats up the box, she is also generating more photons inside the box.

Now that we understand this, let us return to our fermion gas, be it a gas of electrons (in condensed matter physics and in astrophysics) or a gas of neutrons (in astrophysics). The number N of fermions in the box is given by

$$N = \frac{4\pi V}{h^3} \int_0^\infty dp p^2 \frac{1}{e^{\frac{(\epsilon - \epsilon_F)}{T}} + 1} \tag{11}$$

Notice that the integral over p now extends to infinity, since the integrand has an infinitely long Boltzmann tail.

But in sharp contrast to photons, the number of electrons is conserved! If our experimentalist had put in 477 electrons, then no matter what,[1] there will always be 477 electrons in the box, not one more and not one less. Thus, in (11), as T varies, ϵ_F must also vary to keep N fixed.

I want to emphasize that in some sense, this is a trivial notational issue, but a potentially confusing one.[2] So, let ϵ_F be a function of T, some physicists would say. But most physicists, including me, would prefer to continue associating ϵ_F with the Fermi sphere defined at $T = 0$. The Fermi sphere is a ball with a sharply defined radius p_F and the corresponding energy $\epsilon_F = p_F^2/2m$. Okay, no big deal, just define a function $\mu(T)$, known as the chemical potential (for historical reasons), such that $\mu(T = 0) = \epsilon_F$.

The Fermi-Dirac distribution

So finally, at temperature T, the number N of fermions in the box and the energy of the box are given respectively by

$$N = \frac{4\pi V}{h^3} \int_0^\infty dp p^2 \frac{1}{e^{\frac{(\epsilon(p)-\mu)}{T}} + 1} \tag{12}$$

and

$$E = \frac{4\pi V}{h^3} \int_0^\infty dp p^2 \frac{\epsilon(p)}{e^{\frac{(\epsilon(p)-\mu)}{T}} + 1} \tag{13}$$

Generations of students have been confused by these two equations. To avoid joining these people, you should realize, crucially, that μ is most definitely not a constant, but a complicated function of T and N (actually, of the density $n = N/V$, as we can see) determined by (12). In other words, (12) actually defines μ. Once μ is determined (except in various limits, only numerically), we are then supposed to plug it into (13) to determine the energy E for a given N.

A knob we adjust

Think of μ as a knob we adjust so that (12) produces the actual number of electrons our experimentalist put into the box.

At the risk of repeating myself, here is a mnemonic to make sure you understand: Admire (13) as a wonderful result for the energy E of a Fermi gas, but regard (12) as "merely" a definition of μ.

In summary, the distribution, known as the Fermi-Dirac distribution,

$$n(\epsilon) = \frac{1}{e^{\frac{(\epsilon(p)-\mu)}{T}} + 1} \tag{14}$$

has an interesting dependence on temperature T. At $T = 0$, $n(\epsilon) = \theta(\epsilon_F - \epsilon)$, where $\epsilon_F = \mu(T = 0)$. As T increases, the step function "softens," varying from 1 for $\epsilon \ll \mu$ to 0 for $\epsilon \gg \mu$. Meanwhile, μ varies with T to keep N constant.

Factors of 2π

Back in chapter I.1, I had already remarked that if the numerical accuracy of a fly by night result could be improved by including some factors of 2π, there is no reason not to do so. For calculations in statistical physics, it is particularly useful to remember that the volume of a unit cell in phase space is h^3, not $\hbar^3$, as was explained in an endnote in chapter III.5. Thus, in (1), (2), (3), and (4), I specified h^3. Numerically, one easily could be off by $(2\pi)^3$. (This point often vexes particle physicists, who are used to setting $\hbar = 1$. For readers with some familiarity with quantum field theory, recall that the momentum integration in Feynman diagrams involves $d^4p/(2\pi)^4$.) In chapter VII.1, when we discuss white dwarfs and neutron stars, these factors of 2π are important.

Thomas-Fermi model of atoms

Some readers not into atomic physics might wish to skip this brief discussion of the Thomas-Fermi model of atoms with a large number Z of electrons, but I include it as a good example of the fly by night approach. Recall that I have already included an analysis of the resulting equation as an exercise in chapter III.2.

In 1927, Thomas and Fermi independently treated the electrons as a spherically symmetric gas satisfying Fermi statistics at zero temperature. (At this point, some students always question how good this approximation could be. Even for uranium, Z is only 92, a far cry from 10^{23}, but the point is not how accurately this approximation fits the data. Sure, you could fire up the computer and get more precise results.[3] The point is that Thomas and Fermi provided a conceptually interesting starting point for understanding large atoms, and historically, subsequent refinements to this approximation eventually led to the highly successful density functional theory.[4])

We denote the radial distance to the nucleus by r. Let us assume that it makes sense at every point in space to think of a Fermi gas with local number density $n(r)$ and energy density $\epsilon(r)$. I will drop all overall numerical coefficients from now on, and invite you to fill them in.

The electrostatic potential $\phi(r)$ felt by the electrons at r is determined by $\nabla^2\phi \sim en(r)$, with the boundary conditions $\lim_{r\to 0} r\phi = Ze$ (signaling the presence of the nucleus*) and $\lim_{r\to\infty} r\phi = 0$ (indicating that the atom is electrically neutral).

How do we determine $n(r)$? Imagine that at each point in space, a Fermi sphere sits with radius $p_F^3 \sim h^3 n$ (see (3)), which we fill up with electrons. The maximum total energy $\epsilon_{\max}$ of the electrons at this point is given by $\epsilon_{\max} = \frac{p_F^2}{2m} - e\phi(r) = \frac{h^2 n^{\frac{5}{3}}}{2m} - e\phi(r)$. But $\epsilon_{\max}$ must be 0, since otherwise some electrons would become unbound. This condition relates n to ϕ:

$$n(r) \sim \left(\frac{2m}{h^2}e\phi(r)\right)^{\frac{3}{2}} \tag{15}$$

---

*In other words, near the nucleus, $\phi \simeq Ze/r$.

The density $n(r)$ of electrons varies in space as the potential $\phi(r)$ varies. The deeper the electrostatic potential is at that point, the more electrons we can put there. Make sense?

Plugging this into $\nabla^2\phi \sim en(r)$, we obtain a closed equation for $n(r)$, or equivalently, for $\phi(r)$. I will let you have the fun of pursuing this to the end. See chapter III.2 and also the exercise here.

Exercise

(1) Determine the characteristic size of large-Z atoms using a loose fly by night approach. Show that, contrary to what the naive might think, it actually decreases like $\sim Z^{-\frac{1}{3}}$.

Notes

[1] We are always talking about nonrelativistic physics here.

[2] At least I was confused for a while as a student.

[3] One could turn this into an in-joke. What is 92 equal to? A dime a dozen type physicist: 92. A great physicist: ∞.

[4] Speaking as one who has spent much of his life in Kohn Hall.

Einstein condensation

Guessing the Bose-Einstein distribution

For a Fermi gas at temperature T, we learned in chapter V.4 that the number density n of particles and the energy density ε in a box of volume V are given respectively by

$$n = \frac{N}{V} = \int \frac{d^3 p}{(2\pi\hbar)^3} \frac{1}{e^{\frac{\epsilon(p)-\mu}{T}} + 1} \tag{1}$$

and

$$\varepsilon = \frac{E}{V} = \int \frac{d^3 p}{(2\pi\hbar)^3} \frac{\epsilon(p)}{e^{\frac{\epsilon(p)-\mu}{T}} + 1} \tag{2}$$

(While we are not into exact expressions in this book, we have even put in the "phase space" factor we derived in chapter III.5, for the reasons mentioned in chapter V.4. Also, for the sake of uniformity, we have "undone" the angular integration of that chapter and written the integral as over $d^3 p$.)

We arrived at these equations by making an educated guess, starting from the Planck distribution for photons, with the corresponding expressions for the number density and energy density of photons being

$$n = \frac{N}{V} = 2 \int \frac{d^3 p}{(2\pi\hbar)^3} \frac{1}{e^{\frac{\epsilon(p)}{T}} - 1} \tag{3}$$

and

$$\varepsilon = \frac{E}{V} = 2 \int \frac{d^3 p}{(2\pi\hbar)^3} \frac{\epsilon(p)}{e^{\frac{\epsilon(p)}{T}} - 1} \tag{4}$$

where, again for the sake of uniformity in notation, we have written $p = \hbar\omega/c$ and $\epsilon(p) = cp = \hbar\omega$ for the photon momentum and energy, respectively. (Here we even included a factor of 2, which, however, is not relevant for the following discussion, due to the two spin degrees of freedom carried by the photon spin.)

We also learned in chapter V.4 the crucial point that when the number of particles is conserved, we have to introduce a chemical potential μ to ensure that the number of particles is what we say it is. Photons are special, because they can be emitted and absorbed at will by charged particles; hence photons are not associated with a chemical potential. In contrast, in general, a Bose gas (for example, a gas of He^4 atoms) consists of bosons whose number is in fact conserved. We need a chemical potential μ to describe a gas of He^4 atoms.

Given this vast amount of knowledge, you should now try to guess the expressions corresponding to (1) and (2) for a Bose gas. Stare at (3) and (4), as well as (1) and (2), for a while.

Of course, if you have the good sense to take a macho course on statistical physics instead of hanging out with fly by night physicists, you would not have to guess anything. Real men[1] don't guess; they derive and calculate.

Zero denominator alert, and input output

So, did you manage to guess the Bose-Einstein distribution? For a Bose gas, the number density and energy density are given respectively by

$$n = \frac{N}{V} = \int \frac{d^3p}{(2\pi\hbar)^3} \frac{1}{e^{(\epsilon(p)-\mu)/T} - 1} \tag{5}$$

and

$$\varepsilon = \frac{E}{V} = \int \frac{d^3p}{(2\pi\hbar)^3} \frac{\epsilon(p)}{e^{(\epsilon(p)-\mu)/T} - 1} \tag{6}$$

where $\epsilon(p) = p^2/2m$ and μ is determined by n as explained in chapter V.4.

Compare these to the corresponding expressions[2] (1) and (2) for a Fermi gas. We see that a mere sign has been flipped. But what an important sign flip that is!

Immediately, alarm bells start screeching! A basic alert from the day we learned to divide, boys and girls. Could we be dividing by zero here? Yes, when $\epsilon(p) = p^2/2m = \mu$, the denominator vanishes.

After a brief examination, you decide that this is no big deal: since $\epsilon(p) \geq 0$, any possibility of the denominator vanishing is avoided if $\mu < 0$. Fine, we learn that the chemical potential μ for a Bose gas must be negative.

At the risk of interrupting an exciting narrative (and at the risk of belaboring[3] a point, albeit an important point), allow me to say that most students, on first exposure to the Bose gas, miss a crucial difference between (3) and (4) on the one hand, and (5) and (6) on the other. Okay, you say, I see the difference: Generically there is a chemical potential, but not for the photon gas. Well, that implies that n is an output in (3) and an input in (5).

For a gas of, say, He^4 atoms, the experimentalist tells us what the number density n and the energy density[4] ε are, and we solve (5) and (6) for μ and T. When the experimentalist varies n and ε, of course μ and T would vary in response. In sharp contrast, there is no μ to solve for in (3) and (4). For a gas of photons, the experimentalist tells us what the energy density ε is, and we solve

(4) for the temperature T. Then we plug T into (3) and tell the experimentalist how many photons are in the box. The number density of photons is an output, not an input. But the number density of conserved bosons (He^4 atoms, for example) is an input, not an output.

Now a challenge for you. What would happen if the experimentalist keeps on increasing the number density n?

Einstein's insight

> I maintain that, in this case, a number of molecules steadily growing with increasing density goes over in the [ground state] .... A separation is effected; one part condenses, the rest remains a 'saturated' ideal gas. [A. Einstein]

Einstein had the stunning insight that the Bose-Einstein distribution would lead to a dramatic new phenomenon now known as Bose-Einstein condensation.[5] At zero temperature, fermions would all try to get into the ground state but can't, as described in chapter V.4. But bosons, given that they like to hang out with one another, would certainly all want to end up in the ground state. What keeps them from all being in the ground state is of course thermal agitation. But still, you would expect that below some critical temperature T_c, when agitation gives way to serenity, a Bose gas would condense.

In the glare of hindsight, it seems all so clear, no? Is it? It took an Einstein to see this.

An integral does not blow up

This discussion suggests that we should look at what happens to (5) and (6) as we lower the temperature. Conceptually, it turns out to be somewhat easier to crank up the density n. Did you think through the challenge question?

Since $\mu < 0$, it would be clearer to write $\mu = -|\mu|$ and

$$n = \int \frac{d^3 p}{(2\pi\hbar)^3} \frac{1}{e^{(\epsilon(p)+|\mu|)/T} - 1} \tag{7}$$

When $p = 0$, the denominator equals $e^{\frac{|\mu|}{T}} - 1$, which is positive as long as μ stays away from 0, and we are safe.

Now crank up the density n. The integral could follow by increasing the integrand, which it could do by decreasing the denominator $e^{(\epsilon(p)+|\mu|)/T} - 1$, which in turn means decreasing $|\mu|$. Thus, as n increases, the negative quantity μ has to steadily approach 0 from below.

To summarize,

$$n \uparrow \implies |\mu| \downarrow \implies \mu \uparrow \implies \mu \to 0 \tag{8}$$

What happens when μ hits 0? Then the density n reaches its maximum value; we cannot crank it up any further. Setting $\mu = 0$, we obtain the maximum density:

$$n_{\max} = \int \frac{d^3p}{(2\pi\hbar)^3} \frac{1}{e^{p^2/2mT} - 1} \sim \frac{(mT)^{\frac{3}{2}}}{\hbar^3} \int_0^\infty dx\, x^2 \frac{1}{e^x - 1} \sim \frac{(mT)^{\frac{3}{2}}}{\hbar^3} \quad (9)$$

(Scale $p = (2mT)^{\frac{1}{2}}x$.) The integral over x can be done exactly, but in this book,[6] we couldn't care less; all we care about is that the integral is not infinite. Well, it certainly converges at both ends, for large x exponentially, and for $x \simeq 0$, with the integrand going like $\sim x^2/x = x$. The integral is some pure number.

As we crank up the density, at some $n_{\max}$, the integral can't keep up. Usually, in physics, something interesting happens when an integral blows up (that is, diverges). But here, amusingly, something dramatic happens because an integral does not blow up!

Condensing into the ground state

What happens when the density n exceeds $n_{\max}$, which we should perhaps write as n_c for critical density? In statistical mechanics, as the very word "statistical" implies, we start out with a sum over discrete quantum states. By inventing integral calculus, Newton and Leibniz taught us that, under suitable circumstances, the sum could be replaced by an integral. Precisely for $n > n_c$, this is not allowed! A finite fraction $(n - n_c)/n$ of the Bose particles condenses into the ground state and is unaccounted for by the integral.

To summarize, at a given temperature T, Bose-Einstein condensation occurs when n exceeds the critical density

$$n_c(T) \sim \frac{(mT)^{\frac{3}{2}}}{\hbar^3} \quad (10)$$

Equivalently, at a given density n, condensation occurs when the temperature T drops below the critical temperature

$$T_c(n) \sim \frac{\hbar^2 n^{\frac{2}{3}}}{m} \quad (11)$$

The importance of Bose-Einstein condensation can hardly be overstated: A macroscopic number of particles condensing into a single quantum state implies the manifestation of quantum physics on a macroscopic scale. It led to an explanation of superfluidity and superconductivity, which in turn blossomed into all sorts of fascinating physics, such as the Josephson junction and fluids of cold atoms. Perhaps even more importantly, the understanding of superconductivity produced various universal concepts, such as the spontaneous breaking of a gauge symmetry, needed for the foundation of modern particle physics.

The fly by night guess actually works

In chapter V.3 we derived the leading quantum correction to the classical ideal gas law:

$$P = nT\left(1 \pm C\frac{\hbar n^{\frac{1}{3}}}{(mT)^{\frac{1}{2}}} + \cdots\right) \tag{12}$$

with the + sign for the Fermi gas and the − sign for the Bose gas. This result has already signaled to us that, for the Bose gas, the correction could drive the pressure negative, which is manifestly not allowed.

As we can now see (and as is often the case), we should have taken this "warning" seriously, even though the rigorous physicist would surely explain to us that (12) was meant to be a low density and high temperature expansion, from which we could not possibly draw any conclusion about high densities and low temperatures. In fact, the wild fly by night guess, that something would be happening when $\hbar n^{\frac{1}{3}}/(mT)^{\frac{1}{2}} \sim 1$, turned out to be right.

Recalling how we obtained (12) in the first place back in chapter V.3, we now understand the physical origin of Bose-Einstein condensation. The dimensionless quantity $\hbar n^{\frac{1}{3}}/(mT)^{\frac{1}{2}}$ is just the de Broglie wavelength of a particle divided by the average separation between the particles. Clearly, quantum effects rule when this ratio becomes of order unity.

Never be conceptually sloppy

The fly by night physicist could be calculationally sloppy, but he or she would not dream of being conceptually sloppy. I trust that this example drives home the lesson that it is far more important to examine whether an integral diverges or converges than to compute its numerical value. But again, each to his or her taste.

Notes

[1] Real men don't eat quiche, as they used to say long ago.

[2] Note the cute identity $\frac{1}{e^x-1} - \frac{1}{e^x+1} = \frac{2}{e^{2x}-1}$, with which one could play mathematical games with Fermi and Bose gases.

[3] A point already made in chapter V.4 for the Fermi gas.

[4] Of course, since math is often (but not always) a two way street, the experimentalist has the option of specifying the temperature T, rather than energy density ε, which is then given by (6).

[5] A bit of a misnomer, since Bose was not directly involved at all.

[6] Outside this book, of course I care.

Part VI

Symmetry and superb theorems

Prologue to Part VI

Symmetry plays a starring role in modern physics. Entire books could be (and have been) written on this theme. Many theoretical physicists, including me, credit Einstein for pushing symmetry to the fore. We might say that before Einstein, physicists first established the equations of motion (for example, Maxwell's equations) and then extracted the symmetries hidden therein, such as Lorentz invariance. Einstein, in contrast, imposed Lorentz invariance on Newtonian mechanics and deduced marvelous physical results.*

The study of symmetry does not appear to belong in the narrow definition of what is traditionally understood to be back of the envelope physics or guesstimation. But in fact, one could argue that symmetry considerations represent the quintessence of fly by night physics. Indeed, they empower us to determine various equations of motion that either may be unknown (as on the cutting edge of various areas of contemporary physics) or may have temporarily escaped from our puny minds. For instance, you might have forgotten the Navier-Stokes equation for fluid flow (as I have at various stages of my life) but remember that it is an equation for $\frac{\partial \vec{v}}{\partial t}$, with $\vec{v}(\vec{x}, t)$ the fluid's velocity field. By invoking rotational invariance, parity, and time reversal (or lack thereof), you can almost immediately construct the entire equation out of $\vec{\nabla}, \vec{v}$, and so on (as we will see in chapter VI.2).

Another reason for putting an interlude on symmetries here is that I need to invoke some of this stuff later.

Fearful, pp. 96–97. The "fearful" in the title of my symmetry book is variously translated as awe inspiring, magical, vertiginous, timorous, et cetera. Quite a few of my physics friends have taken issue with these adjectives.

Symmetry, fearful or fearless

Symmetries of the laws of physics versus symmetries of a specific physical situation

The reader is surely already familiar with the notion of symmetry and its power. In physics, we are often interested in the symmetries enjoyed by a given physical system. On a deeper and more abstract level, we are interested in the symmetries of the fundamental laws of physics. For example, one of the most revolutionary and astonishing discoveries in the history of physics is that objects do not fall down, but toward the center of the earth. Down is merely an "emergent" concept due to our small size relative to the radius of the earth. Newton's law of gravitation does not pick out a special direction: It is left invariant by rotations.

Starting with Galileo and Newton, the history of theoretical physics has witnessed the discoveries of one unexpected symmetry after another. Physics in the 20th century consisted of the astonishing discovery that as we study Nature at ever deeper levels, Nature displays more and more symmetries.[1] Hopefully the trend will continue.

Rotational invariance

Undergraduates are familiar with rotational invariance, with an understanding bestowed by Newton, as I just said. If the left side of an equation transforms[2] as a vector, the right side must also. Exhibit A: $\vec{F} = m\vec{a}$.

But things are actually a bit more subtle than that. It is not enough to put an arrow on top of a letter; we have to be sure that $\vec{F}$ really is a vector, that is, it transforms properly under rotations.

For instance, with $\vec{g}$ a fixed vector pointing down, $m\vec{a} = \vec{F} = m\vec{g}$, valid near the surface of the earth, is not invariant under rotations in general. This equation is invariant only under rotations around the vertical axis. (Indeed, this will be an issue in chapter VIII.1 when we study water waves.)

Scalars, vectors, and tensors are defined by how they transform under rotation.[3] You should now check that Maxwell's equations and Schrödinger's equation are rotationally invariant. Keep in mind that the gradient operator $\vec{\nabla} = (\frac{\partial}{\partial x}, \frac{\partial}{\partial y}, \frac{\partial}{\partial z})$ also transforms like a vector. For instance, both sides of the equation $\vec{\nabla} \times \vec{E} = -\frac{1}{c}\frac{\partial \vec{B}}{\partial t}$ transform like a vector, thanks to how the vector cross product is defined.

In contrast to rotations, parity and time reversal, which play crucial roles in the fundamental laws of physics, are less emphasized in the undergraduate curriculum. Let's talk about them in turn.

Parity or space reflection

Parity, namely space reflection, $\vec{x} \to -\vec{x}$ (that is, $x \to -x$, $y \to -y$, $z \to -z$), is just a teeny bit subtle in elementary physics. Under parity, denoted by P, time is untouched: $t \to t$. So

$$\text{P}: \quad \vec{v} = \frac{d\vec{x}}{dt} \to \frac{d(-\vec{x})}{dt} \to -\vec{v}, \quad \text{and} \quad \vec{a} = \frac{d\vec{v}}{dt} \to \frac{d(-\vec{v})}{dt} = -\vec{a} \qquad (1)$$

Thus, $\vec{F} = m\vec{a}$ is parity invariant if and only if $\vec{F} \to -\vec{F}$ under P.

Again, the force much studied in introductory physics, $\vec{F} = m\vec{g}$, breaks parity invariance. Things fall down, not up.

However, everyday life involves a narrow view of the truth. Once Newton (him again!) recognizes that terrestrial gravity is actually given by

$$m\vec{a} = -\vec{\nabla}\left(\frac{GMm}{r}\right) \qquad (2)$$

with r the distance to the center of the earth, then parity invariance in fact holds. The crucial point is that $\vec{\nabla}$ flips sign under P. This means that we end up on the other side of the world.

To avoid cluttering up the discussion with jargon, I intentionally do not distinguish between covariance and invariance. Covariance means that $\vec{a}$ and $\vec{\nabla}$ both change sign; invariance means that (2) is unchanged after we cancel off the two minus signs.

In popular books, parity is often explained by asking whether the fundamental laws of physics distinguish between a physical process and its image in a mirror. This is because upon mirror reflection (with the mirror perpendicular to the z-axis, say), we have $x \to x$, $y \to y$, $z \to -z$, and this transformation is identical to space reflection $\vec{x} \to -\vec{x}$ followed by a rotation through π around the z-axis. So if rotational invariance holds, then mirror reflection is equivalent to space reflection.

The issue is whether the laws of physics care about left and right. Another way of defining P invariance is to ask whether the world in the mirror is allowed by physics.[4]

Maxwell's equations (see appendix M for a quick review) are invariant under parity. First, we have to decide how charge and current transform. Under space reflection, $\rho(x, y, z, t) \rightarrow \rho'(x', y', z', t') = \rho(-x, -y, -z, t)$, and $\vec{J}(x, y, z, t) \rightarrow \vec{J}'(x', y', z', t') = -\vec{J}(-x, -y, -z, t)$. In other words, charges just sitting there continue to sit but at the reflected location $\vec{x}'$, while moving charges move in the opposite direction.

Consider first $\vec{\nabla} \cdot \vec{E} = \rho$. We see that it would be invariant if $\vec{E}(x, y, z, t) \rightarrow \vec{E}'(x', y', z', t') = -\vec{E}(-x, -y, -z, t)$. Thus, $\vec{\nabla} \cdot \vec{E}$ is unchanged, just as ρ is unchanged.[5] I am intentionally not spelling everything out in excruciating detail to avoid clutter. You work it out.

Next, the equation $\vec{\nabla} \times \vec{E} = -\frac{1}{c} \frac{\partial \vec{B}}{\partial t}$ informs us that, since $\vec{\nabla} \times \vec{E}$ is left unchanged by space reflection, the magnetic field $\vec{B}(x, y, z, t) \rightarrow \vec{B}'(x', y', z', t') = +\vec{B}(-x, -y, -z, t)$, in contrast to the electric field. Hence, while $\vec{E}$ is a vector, $\vec{B}$ is sometimes referred to as an axial vector or pseudovector.

To summarize,

$$P: \quad \vec{E}(\vec{x}, t) \rightarrow -\vec{E}(-\vec{x}, t), \quad \text{but} \quad \vec{B}(\vec{x}, t) \rightarrow +\vec{B}(-\vec{x}, t) \tag{3}$$

You can now verify that the other two Maxwell's equations are invariant under parity. (All three terms in $\vec{\nabla} \times \vec{B} - \frac{1}{c} \frac{\partial \vec{E}}{\partial t} = \frac{1}{c} \vec{J}$ flip sign, while $\vec{\nabla} \cdot \vec{B} = 0$ is left trivially unchanged.)

The strategy is to determine how $\vec{E}$ and $\vec{B}$ transform under space reflection according to how they are generated, and then to verify that the other two Maxwell's equations are invariant.

Time reversal

Under time reversal (denoted by T), $\vec{x} \rightarrow \vec{x}$, $t \rightarrow -t$, and so

$$T: \quad \vec{v} = \frac{d\vec{x}}{dt} \rightarrow \frac{d\vec{x}}{d(-t)} \rightarrow -\vec{v}, \quad \text{and} \quad \vec{a} = \frac{d\vec{v}}{dt} \rightarrow \frac{d(-\vec{v})}{d(-t)} \rightarrow \vec{a} \tag{4}$$

Compare and contrast with (1).

Thus, if the force $\vec{F}$ is unchanged under time reversal T, Newton's law of motion $\vec{F} = m\vec{a}$ is left unchanged, that is, it is invariant. In contrast, Aristotle's law $\vec{F} = \mu \vec{v}$ is not. This is why we physicists follow Newton, but not Aristotle, abandoning him to the altar of the humanists. The key issue is whether the law of motion is first or second order in time.

An important special case occurs when $\vec{F} = -\vec{\nabla} V(\vec{x})$. Then $\vec{F}$ is unchanged, so that Newton's law $m\vec{a} = -\vec{\nabla} V(\vec{x})$ is T invariant.

In everyday life, friction and pain enter. Newton's law is modified phenomenologically to $m\vec{a} + \mu \vec{v} = -\vec{\nabla} V(x)$. The friction term $\mu \vec{v}$ manifestly breaks time reversal invariance, as a medieval peasant pushing a loaded cart along a muddy track knew only too well. Hence he believed Aristotle, and would have thought Newton was crazy.

Indeed, thus far in this book, we often invoke various symmetries without saying so. For instance, when I wrote down the diffusion equation $\vec{J} = -D\vec{\nabla} n$

in chapter I.5, I was implicitly saying that since $\vec{J}$ is a vector, then by rotational invariance, I have to construct a vector out of the density n, and $\vec{\nabla}$ is the only vector available. Note that if $\vec{J} = -D(\frac{\partial n}{\partial x}, \frac{\partial n}{\partial y}, a\frac{\partial n}{\partial z})$ with $a \neq 1$, then the z-direction is distinguished from the x- and y-directions, and so rotational invariance is clearly violated. We also see immediately that while P is respected, T is not, which is how it should be with diffusion. When combined with the equation of continuity, the resulting equation $\frac{\partial n}{\partial t} = D\nabla^2 n$ respects P but violates T.

A side remark. In Newtonian physics, time reversal invariance is easier to describe than space reflection. Interestingly, it is the other way around in more advanced physics. In particular, in the Schrödinger equation, $\frac{\partial \psi}{\partial t}$ appears with a factor of i, and thus time reversal must also be accompanied by complex conjugation, as was first noticed by Eugene Wigner.[6]

Possible versus probable

In contrast to our discussion of parity, we do not have a mirror for time. But we do have movies. Time reversal invariance is explained by taking a movie of a microscopic physical process and then playing the movie backward.[7] The issue is, using only the fundamental laws of physics, whether we can tell the difference.

We ask whether the process is possible, not whether it is probable. Big distinction!

Here the emphasis is on microscopic processes, not everyday processes.[8] Do the fundamental laws of physics contain an arrow[9] of time? There are of course many other arrows of time, most prominently the one associated with ever increasing entropy.[10]

The behavior of the electromagnetic field under T is readily determined, once again, by thinking about how they are generated, or equivalently, by looking at two of Maxwell's equations. Under time reversal, the current $\vec{J}$ is reversed, while charge remains unchanged. Thus,

$$\text{T}: \quad \vec{E}(\vec{x}, t) \to +\vec{E}(\vec{x}, -t), \quad \text{but} \quad \vec{B}(\vec{x}, t) \to -\vec{B}(\vec{x}, -t) \tag{5}$$

The equation $\vec{\nabla} \times \vec{E} = -\frac{1}{c}\frac{\partial \vec{B}}{\partial t}$, for example, is manifestly left unchanged.

By looking at (3) and (5), we see that P and T leave the energy density $\varepsilon \sim \vec{E}^2 + \vec{B}^2$ in an electromagnetic field unchanged but flip the direction of the Poynting vector (recall chapter II.1) $\vec{S} \sim c\vec{E} \times \vec{B}$, as you would expect. Indeed, if you didn't know (or remember) the expressions for ε and $\vec{S}$, you could obtain them by invoking rotational invariance, P, and T.

The equations of fluid dynamics

I want to further illustrate P, T, and rotational invariances with the equations of fluid dynamics, partly because the typical undergraduate is less familiar with these equations, and partly because I need them for subsequent chapters. As in electromagnetism, the dynamical variable in question (namely, the velocity of the fluid $\vec{v}(t, x, y, z)$) is a vector function of space and time.

Start with the Euler equation (this is derived in appendix ENS for the convenience of those readers not conversant with it, using the notation indicated there):

$$\frac{\partial \vec{v}}{\partial t} + (\vec{v} \cdot \vec{\nabla})\vec{v} = -\frac{1}{\rho}\vec{\nabla}P \tag{6}$$

When supplemented by the equation of continuity

$$\frac{\partial \rho}{\partial t} + \vec{\nabla} \cdot (\rho \vec{v}) = 0 \tag{7}$$

and by the equation of state $P(\rho)$, which characterizes the fluid, the Euler equation allows us to solve for $\vec{v}(t, x, y, z)$.

Evidently, (6) and (7) are invariant under rotation, with $\vec{v}$ transforming like a vector, and P and ρ like scalars.[11]

Let us now check P and T. Under P, we have, to repeat, $\vec{v} \to -\vec{v}$, $\vec{\nabla} \to -\vec{\nabla}$, and $t \to t$. Thus, all three terms in (6) flip sign, while both terms in (7) stay unchanged. Under T, we have $\vec{v} \to -\vec{v}$, $\vec{\nabla} \to \vec{\nabla}$, and $t \to -t$. Now the three terms in (6) stay unchanged, while both terms in (7) flip sign. Hence, fluid dynamics as described by (6) is indeed P and T invariant.

In the presence of external forces $\vec{F}$, we add $\vec{f} \equiv \vec{F}/\rho$ to the right side of (6). For example, for water waves on the earth's surface, we add $\vec{g}$, a vector pointing down and with magnitude $g = |\vec{g}|$. (While $\vec{g}$ looks like a vector, it actually does not transform under rotation, as already mentioned: it is a fixed quantity, due to our myopia, and generated by physics external to the situation at hand.) As remarked earlier, the presence of $\vec{g}$ breaks P.

It is important to understand, however, that $\vec{g}$ does not break T. Under T, $\vec{g} \to \vec{g}$, while, as just noted, all three terms in (6) also stay unchanged.

For later use, note that if the fluid is incompressible (which holds for water under ordinary conditions), that is, if ρ is constant, then (7) implies that

$$\vec{\nabla} \cdot \vec{v} = 0 \tag{8}$$

Exercises

(1) Show that $\vec{\nabla} = (\frac{\partial}{\partial x}, \frac{\partial}{\partial y}, \frac{\partial}{\partial z})$ transforms like a vector. Hint: You could always call the axis of rotation the z-axis; it is just a name.

(2) Show that the Schrödinger equation for a free particle, $i\frac{\partial \psi}{\partial t} = -\frac{\hbar^2}{2m}\vec{\nabla}^2 \psi$, is invariant under rotations.

(3) Show that Maxwell's equations are invariant under rotations.

(4) A physicist was lucky enough to have two girlfriends, named Lucy and Rita, whom he liked equally. (I am afraid that this story[12] is not going to be politically correct,

but you could change the genders and names to your liking.) The physicist lived on a train line precisely halfway between the homes of Lucy and Rita. (Thus, he found it convenient to think of everything happening in a (1+1)-dimensional spacetime and of Lucy and Rita as the left and right excitation, respectively, in the language of condensed matter physics.) Since he could never decide whom to visit, he simply left it up to chance. It so happened that on this line, trains went left and right at precisely the same frequency, once every hour. Our physicist thus went to the train station whenever the mood struck him and took the first train that came along. He figured that in the long run, he would end up visiting Lucy just as often as Rita. In fact, after some months, he found that he was visiting Rita nine times more often than Lucy! Unfortunately for our dumb physicist, both women got angry at him, Lucy for his not visiting her often enough, and Rita for his hanging around her too often. He ended up with no girlfriend. How did it happen that parity was spontaneously broken?

Notes

[1] See parts VII and VIII of *Group Nut*. Also see *Fearful*.

[2] For a more detailed discussion of symmetry, transformation, and invariance, see, for example, *Fearful* and *Group Nut*.

[3] See the story on p. 52 of *GNut* about how tensors transform.

[4] For example, M. Jackson's "man in the mirror" has his heart on the right side of his body, but nothing in the laws of physics says that he cannot function as he is.

[5] More precisely, their arguments change from (x, y, z, t) to (x', y', z', t').

[6] See, for example, *QFT Nut*, pp. 102–104.

[7] A. Zee, "Time Reversal," in *Mysteries of Life and the Universe*, edited by W. Shore, Harcourt, 1992, p. 176.

[8] Such as the breaking of an egg.

[9] One was discovered in the decay process of the so-called K mesons. See https://physicstoday.scitation.org/doi/abs/10.1063/PT.3.1774?journalCode=pto.

[10] A well known joke asks what would happen if you play a country and western song backward. Well, the guy gets his dog back, his pickup truck back, and finally his girl back, in that order.

[11] For example, $P'(t, x', y', z') = P(t, x, y, z)$, where the primed quantities pertain to the primed observer, and the unprimed quantities pertain to the unprimed observer, and where (x', y', z') is related to (x, y, z) by a rotation.

[12] Taken from my lectures on quantum Hall fluids, in *Field Theory, Topology and Condensed Matter Physics*, edited by H. B. Geyer, Springer-Verlag, 1994.

Galileo, viscosity, and time reversal invariance

Viscosity

From experience, we know that fluids can be "sticky," that is, viscous. Just as a Newtonian particle subject to friction will eventually come to rest, so will the flow of a viscous fluid. Viscosity is the analog of friction for fluids.

Make a movie of a viscous fluid coming to rest. Now play the movie backward. Of course we can tell the difference! Viscosity violates time reversal invariance. (Needless to say, the laws governing the interaction between the fluid molecules still preserve time reversal T. We have simply neglected other degrees of freedom, such as heat.)

To include viscosity, Navier and Stokes generalized Euler's equation (in the presence of an external force per unit mass $\vec{f}$)

$$\frac{\partial \vec{v}}{\partial t} + (\vec{v} \cdot \vec{\nabla})\vec{v} = -\frac{1}{\rho}\vec{\nabla}P + \vec{f} \tag{1}$$

given in the preceding chapter, to an equation now named after them

$$\frac{\partial \vec{v}}{\partial t} + (\vec{v} \cdot \vec{\nabla})\vec{v} = -\frac{1}{\rho}\vec{\nabla}P + \nu\nabla^2\vec{v} + \vec{f} \tag{2}$$

by adding the so-called viscosity term $\nu\nabla^2\vec{v}$. The coefficient of viscosity ν is a characteristic of the specific fluid being discussed.

Comparing the viscosity term with the Newtonian term $\frac{\partial \vec{v}}{\partial t}$, we obtain instantly the important result

$$[\nu] = \frac{L^2}{T} \tag{3}$$

Using dimensional analysis, you can readily obtain how the attenuation rate of infrasound depends on frequency, a currently fashionable area of research. (See exercise (4).)

In parallel with $ma = F$, some authors prefer to multiply (2) by ρ so as to obtain an equation for $\rho \frac{\partial \vec{v}}{\partial t}$, as mentioned in appendix ENS. The viscosity term then becomes $\rho \nu \nabla^2 \vec{v} \equiv \eta \nabla^2 \vec{v}$. These two trivially related coefficients, ν and $\eta \equiv \rho \nu$, are known as the kinematical and dynamical coefficient of viscosity respectively. Incidentally, ν and η for air and water,[1] the two most common fluids around, depend significantly on temperature.[2]

Time reversal and viscosity

Most elementary textbooks introduce the viscosity term by drawing a figure in which neighboring layers of the fluid move with different velocities in the same direction. But I think that it is most easily derived by violating time reversal invariance.

Under T, we noted in chapter VI.1 that all the terms in (2) besides the viscosity term $\nu \nabla^2 \vec{v}$ stay unchanged. In contrast, under T, $\nabla^2 \vec{v} \to \nabla^2(-\vec{v}) = -\nabla^2 \vec{v}$ clearly flips sign. Thus, the Navier-Stokes equation (2) is not time reversal invariant.

We see, literally in 5 seconds flat, that the viscosity term in (2) indeed describes viscosity.

In contrast, parity P is preserved by the Navier-Stokes equation. Under P, $\vec{x} \to -\vec{x}$, $\vec{\nabla} \to -\vec{\nabla}$, and $\vec{v} \to -\vec{v}$. Thus the viscosity term $\nabla^2 \vec{v} \to (-1)^3 \nabla^2 \vec{v}$ flips sign, just like the other terms in the Navier-Stokes equation, for example, $(\vec{v} \cdot \vec{\nabla})\vec{v} \to (-1)^3 (\vec{v} \cdot \vec{\nabla})\vec{v}$. We certainly do not expect viscosity to distinguish between left and right!

An immediate consequence is that viscosity requires the second power of $\vec{\nabla}$. In particular, a flow described by $v_x \propto y$, $v_y = 0$, $v_z = 0$ does not suffer viscous slowdown: $\nabla^2 \vec{v}$ vanishes in this case.[3] One way of seeing this is to note that $v_x \propto y$, $v_y \propto -x$, $v_z = 0$ corresponds to rigid rotation (as you could see by sketching the flow). In contrast, the viscosity term definitely affects the flow $v_x \propto y^2$, $v_y = 0$, $v_z = 0$.

A term that does not change sign under T is called T even, and a term that does is called T odd. Similarly, we refer to terms that are P even or P odd. Thus, the inertial term $\frac{\partial \vec{v}}{\partial t}$ is T even P odd, while the viscosity term $\nabla^2 \vec{v}$ is T odd P odd. That all terms in (2) are P odd implies that parity is conserved (that is, respected).

The reader might have realized that by the preceding discussion, the term $\vec{\nabla}(\vec{\nabla} \cdot \vec{v})$ would also be a candidate to describe viscosity. Indeed, this term is perfectly allowed but rarely included in elementary textbooks[4] on fluid dynamics. Why? Because, as was mentioned in chapter VI.1, for incompressible fluids (such as water), we have $\vec{\nabla} \cdot \vec{v} = 0$, and hence this proposed term vanishes. But by the same token, it must be included in certain situations, such as those involving shock waves and extreme compression. It contributes only if $\vec{\nabla} \cdot \vec{v}$ varies significantly from place to place.

Galilean invariance: a little cloud remaining still

Newtonian mechanics is certainly invariant under Galilean transformations; after all, Newton's work grew out of Galileo's work. I would venture to guess, however, that many theoretical physicists are more comfortable[5] with Lorentz invariance than with Galilean invariance, which is not much taught in the undergraduate curriculum. This is a pity.

Some of Galileo's arguments are quite beautiful. For example, he describes his observations below deck in a smoothly moving sailing ship. In his own words: "And if smoke is made by burning some incense, it will be seen going up in the form of a little cloud, remaining still and moving no more toward one side than the other."[6] Indeed, you verify Galilean relativity experimentally every time you pour yourself a drink while traveling by plane.

The Galilean transformation relating the time and space coordinates used by two observers in uniform relative motion is given by

$$t' = t \tag{4}$$

$$x' = x + ut \tag{5}$$

$$y' = y \quad \text{and} \quad z' = z \tag{6}$$

Derivation? Just common sense.

The notion of absolute Newtonian time is expressed by (4). Time goes by, and at the same rate for the two observers.

We see from (5) that the point $x = 0$, fixed for the unprimed observer, is moving forward according to $x' = ut'$ for the primed observer. And the point $x' = 0$, fixed for the primed observer, is moving backward according to $x = -ut$ for the unprimed observer. Thus, u denotes the relative velocity between the two observers. Meanwhile, the coordinates y and z are just going along for the ride.

We obtain immediately the Galilean law for the addition of velocities (which these days you can observe routinely while walking on the moving sidewalks in airports):

$$v' = \frac{dx'}{dt'} = \frac{d(x + ut)}{dt} = \frac{dx}{dt} + u = v + u \tag{7}$$

For simplicity, we assume that all motion is in the x-direction. Furthermore, acceleration is left unchanged:

$$a' = \frac{dv'}{dt'} = \frac{d(v + u)}{dt} = \frac{dv}{dt} = a \tag{8}$$

Hence, Newton's law of motion is Galilean invariant, while Aristotle's law is not.[7] Thus physics began.

Fluid dynamics is certainly Galilean invariant

Since fluid dynamics is derived from Newton's law of motion, it had better be Galilean invariant. You can readily check this. Given (4)–(6), it is simply a matter of working out how $\vec{v}(t, x, y, z)$, $\vec{\nabla} = (\frac{\partial}{\partial x}, \frac{\partial}{\partial y}, \frac{\partial}{\partial z})$, and $\frac{\partial}{\partial t}$ transform, and then plugging it all into (2). I will relegate the details to appendix Gal, but you can also try to work them out yourself. Very instructive exercise!

Interestingly, neither $\frac{\partial \vec{v}}{\partial t}$ nor $(\vec{v} \cdot \vec{\nabla})\vec{v}$ is Galilean invariant, but working together as a team, their noninvariant pieces under Galilean transformations cancel out.*

This also answers a typical student question when I introduce the viscosity term $\nabla^2 \vec{v}$. Why not simply add a term proportional to $\vec{v}$? The answer is that it is not Galilean invariant.

Einstein's insight

In hindsight, the Euler-Navier-Stokes equation (2) seems fairly obvious, as a simple application of Newton's equation $a = F/m$ (see appendix ENS). But in hindsight, almost everything seems obvious. Still, it took someone of Euler's caliber to take the first step. After (2) was written down, we can verify that it respects rotational and Galilean invariance, and parity P, but crucially not T, as we did earlier.

We can reverse the logic. Imagine that you do not know Euler's equation. Once you decide that a moving fluid can be described by the velocity field $\vec{v}(t, x, y, z)$, and given that you understand Newton, you try to write down an equation for $\frac{\partial \vec{v}}{\partial t}$. You could proceed as in appendix ENS. Or, you could require that P and T, rotational, and Galilean invariances are to be respected.

Indeed, the only term we can construct respecting P and T and rotational invariance is $(\vec{v} \cdot \vec{\nabla})\vec{v}$, as you can convince yourself. We can then write $\frac{\partial \vec{v}}{\partial t} + \lambda(\vec{v} \cdot \vec{\nabla})\vec{v}$ with some unknown coefficient λ. Imposing Galilean symmetry fixes λ to be 1, as you should verify. In other words, the so-called convective derivative $\frac{\partial \vec{v}}{\partial t} + (\vec{v} \cdot \vec{\nabla})\vec{v}$ in fluid dynamics follows from Galilean invariance. Furthermore, if we later drop T invariance, then the remaining three invariances determine for us how viscosity is to be included, as we have belabored above.

This reversal of logic represents a profound paradigm shift in the search for fundamental laws, as I mentioned in the prologue to part VI. In the late 19th century, hitherto unknown symmetry was deduced from the known laws, notably in electromagnetism. Starting with Einstein, but particularly during the second half of the 20th century, theoretical physicists often impose symmetry to determine (or at least help determine) the laws of physics. To me, this counts as one of Einstein's deepest insights.[8]

---

*This sort of cancellation is emblematic of more advanced physics, such as Yang-Mills theory and Einstein gravity.

Symmetry offers us mnemonics

A rather mundane use of the argument by symmetry is as a mnemonic, as already mentioned. Suppose you can't remember[9] how to write the viscosity term in the Navier-Stokes equation. Since $\frac{\partial \vec{v}}{\partial t}$ is T even P odd, you need to construct a term out of $\vec{v}$ and $\vec{\nabla}$ that is T odd P odd. You would be led by the nose to the viscosity term as the simplest possibility.

Similarly, if you don't remember what $\frac{\partial \vec{B}}{\partial t}$ equals in Maxwell's equations, you could readily figure it out using T, P, and rotational invariance.

From Galilean to Lorentz invariance

One of the greatest discoveries of physics, as the reader surely knows, was the realization that Maxwell's equations are not Galilean invariant. They were found to be Lorentz invariant.

Indeed, we have already exploited Lorentz invariance to promote our result for the nonrelativistic Larmor formula into the relativistic regime. An "easy" way of deriving Euler's equation (provided that you are fairly fluent with special relativity) is to write a Lorentz invariant equation for fluid motion and then take the nonrelativistic limit.[10] You can readily understand why. In Euler's equation, time t and space $\vec{x}$ are treated so asymmetrically that it literally hurts the eyes of those fluent with relativistic formalism. Incidentally, if you want to do early cosmology, you may need to know relativistic fluid dynamics.[11]

An interesting fact is that, from the group theoretic point of view, the algebra associated with the Lorentz transformation is considerably more appealing than the algebra associated with the Galilean transformation, as I have already remarked on in passing. To explain this remark[12] would take us far beyond the scope of this book.

Einstein recognized that Lorentz invariance is a property not only of electromagnetism, but of spacetime, and hence it must be imposed on all who live and play in spacetime, namely, all of physics. By imposing Lorentz invariance on mechanics, he found special relativity, and by imposing Lorentz invariance on gravity, he found general relativity. (See appendix Eg.)

The language for understanding symmetries

Group theory provides the language appropriate for discovering and understanding symmetries, as I just hinted at. To explore this statement would take an entire book.[13] We will certainly not be using any group theory in this book except in the most peripheral way. Here the discussion is limited to a couple of casual remarks.

With terrestrial gravity $\vec{g}$, the direction up and down is privileged compared to east, west, north, and south, as already mentioned. It appears that physics is invariant under only rotations around the vertical axis. This point had confused physicists for the longest time.

Historically, it was only after an arduous struggle that they eventually realized that the fundamental laws of physics were in fact invariant under the full rotation group $SO(3)$, defined as the set of all rotations in 3-dimensional Euclidean space. Using the language of group theory, we could say that Newton extended the symmetry group of physics from $SO(2)$ to $SO(3)$, and that later, Einstein extended $SO(3)$ to $SO(3, 1)$, the set of all "rotations" in (3+1)-dimensional Minkowskian spacetime.

Group theory has a somewhat limited impact on classical mechanics, but it really blossomed with the coming of the quantum era. Many thanks to the linear superposition principle in quantum physics!

Quantum mechanics is linear, while classical mechanics is not. This explains the somewhat paradoxical statement that, in a sense, classical mechanics is significantly more difficult* than quantum mechanics.[14]

States in quantum mechanics with the same energy are said to be degenerate. Usually, this degeneracy is due to the presence of a symmetry group, which transforms the degenerate states into linear combinations of one another. Thus, these states furnish what mathematicians call a "representation" of the group. Knowledge about representations tells us a great deal about quantum degeneracy.

This concludes our lightning survey of symmetries in physics.[15]

Exercises

(1) Work out how $\vec{v}(t, x, y, z) = (v_x, v_y, v_z)$ transforms under Galilean transformation, thus generalizing (7) and (8).

(2) Check that the viscosity term $\nabla^2 \vec{v}$ is Galilean invariant.

(3) A compressional wave in air is known as sound (of course). Let the equation of state for air be given by $P(\rho)$. Let $P = P_0 + \delta P$, $\rho = \rho_0 + \delta \rho$, and $\vec{v} = \vec{v}_0 + \delta \vec{v}$, with P_0 and ρ_0 constant in space and time, and $\vec{v}_0 = 0$. Treat δP, $\delta \rho$, and $\delta \vec{v}$ to first order, and determine the speed of sound in terms of $P(\rho)$. Neglect viscosity. Note that the result also follows from dimensional analysis.

(4) Determine the rate of attenuation of sound waves. Show that high-frequency waves are attenuated more. That makes total sense: The wave sloshes back and forth more often, making friction more effective. Low-frequency acoustic waves, known as infrasound (with frequencies far below the 20 hertz threshold of human hearing), are attenuated less and less. Infrasound with frequencies below 0.1 hertz can propagate around the earth. This fact makes infrasound monitoring an emerging area of research. Applications include early warnings of avalanches and volcanic activities.[16]

---

*Of course, classical mechanics is easier to grasp because it is closer to our everyday experiences. Furthermore, we can also say that nobody understands quantum mechanics, precisely because it is so remote from our everyday experiences and logic.

Notes

[1] Evidently, due to the huge difference in the densities of air and water, $\eta_{\text{water}} \gg \eta_{\text{air}}$, while $\nu_{\text{air}} \gg \nu_{\text{water}}$. I would say that η somehow accords more with our intuition and hence may be more suited for applications, while ν is more convenient for mathematical analysis.

[2] Denny, *Air and Water*, offers two useful tables, for ν on p. 63 and for η on p. 59.

[3] Hence I find some figures in some books explaining viscosity potentially misleading.

[4] It is, however, included in L. D. Landau and E. M. Lifshitz, *Fluid Mechanics*, Pergamon, 1959, pp. 48–49.

[5] I certainly am. One reason is that the Lorentz group is considerably more elegant than the Galilean group. See *Group Nut*, pp. 445–447.

[6] He also talks about fluttering butterflies. See *GNut*, p. 18.

[7] This is best seen by considering two particles interacting with a force that depends on the distance between them.

[8] See *Fearful*, chapter 6. I attribute this point of view to Einstein, but others may have espoused it earlier. Serious historians are welcome to correct me.

[9] The majority of physicists in this situation would reach for a book and flip around. But you, dear reader, are not part of this dull majority. Or are you?

[10] See, for example, *GNut*, pp. 233–234.

[11] See, for example, S. Weinberg, *Gravitation and Cosmology*.

[12] See, for example, *Group Nut*, chapter VII.2.

[13] See, for example, *Group Nut*.

[14] See, for example, *Group Nut*, part III.

[15] To go farther, see *Group Nut*.

[16] See the article by T. Feder about infrasound in *Physics Today*, August 2018, pp. 22–25.

Newton's two superb theorems

where is hell?

The distance between two masses

Newton, in proposing his law of universal gravity, had to show, crucially, that the force that pulls the apple to the ground is the same force that keeps the moon in orbit around the earth. Look at the moon floating serenely in the sky on a cloudless night, and you would agree that this is surely one of the most unbelievable propositions ever uttered by a human (which turns out to be true).

The gravitational force, as I reminded you back in chapter I.2, is derived from the gravitational potential

$$\phi = -\frac{GMm}{r} \tag{1}$$

with G Newton's gravitational constant, and r the distance between the two masses M and m. For the attraction between the earth and the moon, r is clearly the distance between them. The radii of the earth and of the moon are negligible compared to r. But for the attraction between the earth and the apple, it is not so clear what r should be. Is it the height of the apple tree? This hypothesis could be easily disproved just by holding an apple close to the ground.

The modern fly by night physicist, with centuries of hindsight behind him or her, could say that the only "relevant" length scale around is the radius of the earth. But not so fast! For Newton to compare the acceleration of the moon to the acceleration of the apple toward the earth, and thus to lay claim to the title "father of modern physics," he needs to know exactly which r to use in the comparison. (Note that, by comparing the two cases, Newton cleverly avoids having to know what[1] G is.)

Newton's two superb theorems

Newton spent almost 20 years proving what he called his two "superb theorems."

The apple is being pulled down by the patch of ground beneath your feet, stuff very close to the apple but comprising a small fraction of the entire earth. The rest of the earth, including the enormous amount of stuff on the other side of the world, is far away. Thus, to calculate the potential energy ϕ between the apple and the earth, we have to cut up the earth into a multitude of infinitesimal pieces and add up the contribution of each piece to ϕ.

How to do this posed a challenge to Newton, who had to invent integral calculus to tackle this problem (which these days could be given to students as homework). Newton arrived at his superb theorem, stating that the force F exerted by the earth on an object of mass m (an apple, a cannonball, or whatever) is as if the entire earth, with mass M, had been shrunk to a point located at the center of the earth. In other words, for the force $F = GMm/r^2$ on a terrestrial object, we should take for r the radius R of the earth.

Symmetry is essential

Putting together Newton's law of motion, law of gravitation, and first superb theorem, we obtain the acceleration due to gravity on the earth's surface:

$$g = \frac{GM}{R^2} \tag{2}$$

That the earth's gravity could be summarized in the universal acceleration g in everyday situations is highly nontrivial, as mentioned in chapter I.1.

Spherical symmetry is essential for the superb theorems, as you could imagine, and as you will see explicitly in appendix N. Newton had to idealize the earth as a perfectly round ball, and then mentally slice the solid ball into spherical shells. Clearly, since the contribution of each shell to potential energy adds up, we could focus on an individual shell. In other words, the problem of calculating the gravitational potential ϕ for a solid ball reduces to the problem of calculating the gravitational potential ϕ for a spherical shell.

The first superb theorem states that the gravitational force outside a spherical shell of mass M is the same as that for a point mass M located at the center of the shell. The second superb theorem states that the gravitational force inside a spherical shell vanishes.

That Newton took so long to complete his two superb theorems caused one of the most bitter fights in the history of physics. While he was off doing the math, so to speak, his rival, Robert Hooke (1635–1703), also came up with the law of gravity. Newton, disputing the claim, accused Hooke of not knowing the first superb theorem and thus that he could not possibly have calculated the force on the proverbial apple.

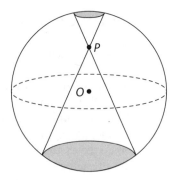

FIGURE 1. A plausibility argument for Newton's second superb theorem.

A famous saying of Newton's, something like "I could see farther than others because I was able to stand on the shoulders of giants," often quoted as an indication of his modesty, was apparently a nasty dig at Hooke, who was rather short.*

The failure of fly by night physics

Newton's two superb theorems provide a prime exhibit that fly by night physics does not suffice. An exact calculation is mandatory here. See appendix N. The closest I could come to a back of the envelope demonstration requires a knowledge of group[3] theory; in particular, the spherical symmetry implies that the rotation group $SO(3)$ plays an essential role. Ultimately, Newton's result could be traced to the grand orthogonality theorem. (Expand in spherical harmonics and argue that they all integrate to zero except the trivial monopole.)

However, it is easy to hand wave and render, say, the second superb theorem at least plausible.[4] See figure 1. The "cap" on the shell closer to the point P is smaller than the cap on the shell "on the other side of the world" farther from P, but it looks like its smaller size is compensated by the inverse square law.

Where is hell?

Before wrapping up this chapter, I must address an issue that may be burning you up. What motivated Newton to prove his second superb theorem?

Newton's second theorem addressed a central mystery of his time: Where is hell? While this is no longer a burning question of contemporary physics,

---

*This may well be apocryphal, but be that as it may, Sidney Coleman, my PhD advisor, a brilliant but exceedingly arrogant physicist, liked to quip "I could see farther than others because I was able to look over the shoulders of midgets."[2]

we can understand why it would puzzle physicists once upon a time. With a round earth, to imagine heaven up above our heads was no longer sensible; heaven would have to be a spherical shell wrapped around the world. It followed that hell must be in the center of a hollow earth. I think that most of my physicist colleagues would agree that this represents the simplest extension of an existing theory. A rudimentary understanding of volcanoes (plus a close reading of the Bible) provided strong observational evidence, confirming the theory for sure.

Furthermore, an erroneous calculation had convinced Newton that the earth was much less dense than the moon, which led his friend Edmond Halley (1656–1742), who by the way published Newton's *Principia* at his expense, to propose the hollow earth theory.[5] The idea may seem absurd to us, but not at that time. A location for hell had to be found. Every epoch in physics has its own top ten problems. It is conceivable that future generations will find our desperate attempt to quantize gravity absurd.

You now understand why Newton would even bother to attack this peculiar problem.

Incidentally, since there is no gravitational force in hell, the usual portrayal with leaping flames can't be right! Flames shoot up because gravity pulls the denser air surrounding the hot gas down.

Exercises

(1) Consider the ultimate subway: Suppose engineers were able to bore a tunnel through the earth connecting two cities, as shown in figure 2. The ultimate subway[6] would provide us with fast and (almost) free transport, assuming various technical problems could be solved, such as withstanding the searing heat near the center of the earth. Estimate the transit time.

FIGURE 2. The ultimate subway. Adapted from Zee, A., *Einstein's Universe: Gravity at Work and Play*. Oxford University Press, 2001.

(2) Taking the density of the earth to be uniform, calculate the transit time exactly, and show that it is universal (that is, independent of the location of the two cities). The dull function hypothesis triumphs! This remarkable fact is plausible. Consider traveling to a nearby destination. The train slides down a gentle slope into the ground. For comparison, consider traveling to the antipode of where you are: the train falls straight down and accelerates greatly, but it has to cover a much larger distance.

Notes

[1] Indeed, G was not measured until 1798, by Henry Cavendish (1731–1810) using equipment built and designed by his friend John Michell (1724–1793), now of black hole fame. Michell died before he could carry out the experiment.

[2] To my students, I say, "I could see farther than you guys because I have been looking at this stuff for many more years than you."

[3] See, for example, *Group Nut*.

[4] At least the undergrads in my classes seemed to accept this.

[5] N. Kollerstrom, "The Hollow World of Edmond Halley," *J. Hist. Astronomy* 23 (1992), p. 185. Witness the popularity of this idea in science fiction, notably Jules Verne's *A Journey to the Center of the Earth* (1864).

[6] Described in *Toy* on p. xxvi and explained without any math on pp. xxvii–xxx.

Part VII

Stars, black holes, the universe, and gravity waves

Stars

Stars idealized as nonrotating spheres of gas

One glorious achievement of 20th century physics is an understanding of how stars work. While various specialized textbooks[1] offer detailed explanations, here we explore how fly by night considerations can lead us to a qualitative, or even a semi-quantitative, picture of stellar structure.

For our purposes, stars are idealized as nonrotating spheres of gas.

We will start with a typical star, such as the sun, namely, a star on the main sequence,[2] neither a giant nor a dwarf.

Pressure counters the inward crush of gravity

To counter the inward crush of gravity, the gas in the interior of a star must press outward. For a star of mass M and radius R, we have by dimensional analysis, since pressure is a force per unit area,

$$P \sim \left(\frac{GM^2}{R^2}\right)/R^2 \sim \frac{GM^2}{R^4} \sim \frac{G(\rho R^3)^2}{R^4} \sim G\rho^2 R^2 \sim \frac{GM\rho}{R} \tag{1}$$

For convenience, we also express P in terms of the mass density $\rho \sim M/R^3$.

In reality, pressure $P(r)$, mass density $\rho(r)$, and so on all depend on r, the radial distance from the center of the star. In particular, we expect $P(r)$ to decrease steadily from some value at the stellar center to $P(R) = 0$ at the stellar surface. Indeed, this boundary condition determines R. We should also introduce the mass $M(r)$ enclosed inside the sphere of radius r centered on the star, with the evident boundary condition $M(R) = M$. By definition, $M(r)$

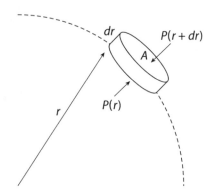

FIGURE 1. Pressure counters the inward crush of gravity.

can be expressed as an integral over $\rho(r)$, or equivalently, as the solution of the differential equation

$$\frac{dM(r)}{dr} = 4\pi r^2 \rho(r) \tag{2}$$

For our idealized star, it is actually easy to write down the exact equation for the pressure. Picture a spherical shell at radius r with thickness dr. See figure 1. Focus on a piece of the shell with area A, containing mass $\rho(r)Adr$. The outward force on this piece is then $(P(r) - P(r+dr))A = -\frac{dP}{dr}Adr$. Balanced against this is the inward pull of gravity. According to Newton's second superb theorem, only the mass $M(r)$ inside the spherical shell matters. Thus, $-\frac{dP}{dr}Adr = GM(r)(\rho(r)Adr)/r^2$, giving

$$\frac{dP(r)}{dr} = -\frac{GM(r)\rho(r)}{r^2} \tag{3}$$

We see that our fly by night equation (1) is consistent with (3) and (2) if we treat all functions of r as roughly constant and approximate derivatives like $\frac{dP}{dr}$ and $\frac{dM}{dr}$ by $\sim P/R$ and $\sim M/R$, respectively, so that $P/R \sim GM\rho/R^2$.

Luminosity, temperature gradient, and nuclear power

The photons produced in nuclear reactions deep inside the sun are eager to get out, perhaps ultimately to warm you some day. They are no different from the ink molecules described in chapter I.5, diffusing from a region of high density to a region of lower density.

And so, photons diffuse from the hotter inner region of the sun to the cooler outer region, scattering many times along the way, doing the random walk described in chapter I.5 and shown in figure 2. Denote the mean free path of the photons by $l(r)$. Consider a shell of thickness dr at radius r inside the sun. The number of steps N a photon has to take to traverse dr is determined by the

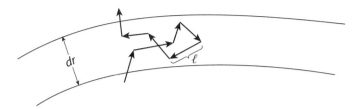

FIGURE 2. Random walking photons traverse a shell of layer dr.

random walk formula $\sqrt{N}l = dr$ given in chapter I.5, and hence $N = (dr/l(r))^2$. Each of these steps takes time $l(r)/c$, and hence the process takes time $\Delta t = (dr/l)^2 l/c = (dr)^2/lc$. (To make formulas look clearer, we often suppress the r dependence.)

The luminosity $L(r)$ at radius r is defined as the energy flowing per unit time through the shell at radius r. Denote the energy density at r by $\varepsilon(r)$. The inner part of the shell is hotter than the outer part, and hence has an excess energy density $d\varepsilon$. Just as the density diffusion current in chapter I.5 is proportional to the density gradient, the energy flow per unit area per unit time across the shell, namely $L/4\pi r^2$, should be proportional to the energy density gradient $\frac{d\varepsilon}{dr}$.

Let's first just use dimensional analysis: $[\frac{d\varepsilon}{dr}] = (E/L^3)/L = E/L^4$. The luminosity per unit area has dimension $[L/4\pi r^2] = (E/T)/L^2 = E/TL^2$. (Do not confuse the luminosity L with the generic length L used in dimensional analysis!) Thus, we need to multiply $\frac{d\varepsilon}{dr}$ by something with dimension $L^2/T = L(L/T)$ to get the luminosity per unit area. Since luminosity results from the diffusion of light, this quantity can only be the mean free path of the photon l multiplied by the speed of light c. Hence

$$\frac{L}{4\pi r^2} \sim -cl\frac{d\varepsilon}{dr} \qquad (4)$$

The minus sign is due to $\frac{d\varepsilon}{dr} < 0$.

Let's check this dimensional analysis result. Here is a fly by night derivation based on the picture of random walking photons described earlier. The excess energy on the two sides of the shell equals $\simeq (4\pi r^2 dr)d\varepsilon$, while the amount of time it takes for this energy to get across is $dt = (dr)^2/lc$. Thus, the luminosity equals this excess energy divided by the time to get across and hence is given by

$$L(r) \sim -\frac{(4\pi r^2 dr)d\varepsilon}{(dr)^2/lc} = -cl(4\pi r^2)\frac{d\varepsilon}{dr} \qquad (5)$$

which of course agrees with (4). (Actually, this expression for $L(r)$ is only off by a factor of $\frac{1}{3}$, which results from integrating some angular factor[3] $\cos^2\theta$. But that is outside the ken of the fly by night physicist.)

But we know the energy density of a hot gas of photons. As discussed way back in chapter III.5 on black body radiation (and later in chapter V.5), $\varepsilon(r)$ is given by the fourth power of the temperature $T(r)$. More precisely, $\varepsilon = aT^4$ (with $a = 4\sigma/c$ and the Stefan-Boltzmann constant $\sigma = \pi^2/60\hbar^3 c^2$).

Plugging $\frac{d\varepsilon}{dr} = 4aT^3\frac{dT}{dr}$ into (5), we obtain

$$\frac{dT}{dr} = \frac{1}{4aT^3}\frac{d\varepsilon}{dr} = -\frac{3L(r)}{16\pi r^2 cl(r)aT(r)^3} \tag{6}$$

This result looks complicated. But it isn't. Here is the back of the envelope version. Ignore the r dependence, differential equations, and all that. Just think of a ball of hot gas with radius R and temperature T. The time t it takes for a random walking photon to traverse R is determined by $R^2 \sim l(ct)$. Luminosity is $L \sim E/t$, with $E \sim aR^3T^4$. Hence,

$$L \sim E/t \sim aR^3T^4/(R^2/lc) \sim alRcT^4 = alR^2cT^3(T/R) \tag{7}$$

The last step merely expresses our feeling that the temperature gradient $\sim T/R$ is what matters. Ta da! We have obtained

$$T/R \sim L/(alR^2cT^3) \tag{8}$$

the poor person's version of (6).

Finally, the astrophysicist acknowledges that nuclear reactions power the star. Denote the energy produced per unit time by a unit mass of stellar stuff by $v(r)$. So the amount of energy dL produced per unit time by a shell of thickness dr is just $4\pi r^2 dr\rho v$. Thus

$$\frac{dL}{dr} = 4\pi r^2\rho(r)v(r) \tag{9}$$

To sum up, stellar structure is determined by (numerically) integrating these four coupled first order differential equations, (2), (3), (6), and (9), for $M(r)$, $P(r)$, $T(r)$, and $L(r)$, outward from $r = 0$. This is supplemented by knowledge from statistical physics, particle physics, and nuclear physics telling us how $P(r)$, $l(r)$, and $v(r)$ depend on $\rho(r)$, $T(r)$, and composition (namely, the relative abundance of electrons, protons, helium nuclei, etc.). For example, the mean free path[4] $l(r)$ is determined by the photon scattering cross section off electrons, protons, and various nuclei that might be present. Thus, knowing the four quantities M, P, T, and L at r, we know them at $r + \delta r$ with δr a suitably chosen step size.

We also need four boundary conditions, of which we already mentioned two: $P(R) = 0$, and $M(R) = M$, which define R and M. In other words, when $P(r)$, which steadily decreases as r increases, reaches 0, we have arrived at the surface of the star, and the value of $M(r)$ there gives the mass of the star. The other two boundary conditions to get us started are obviously $M(0) = 0$ and $L(0) = 0$.

Relationship between luminosity and mass of a generic star

In lieu of firing up the computer, let us now proceed to a back of the envelope understanding[5] of typical stars, namely, those on the main sequence. We already have two fly by night equations, (1) and (8), repeated here for your reading convenience:

$$P \sim \frac{GM\rho}{R} \tag{10}$$

and

$$L \sim alcRT^4 \tag{11}$$

Take the pressure as given by the ideal gas law $P \sim \rho T/m$ (derived back in chapter V.1). Then we have $T \sim Pm/\rho \sim GMm/R \propto M/R$ which is just the poor man's version of the virial theorem, stating the rough equality of kinetic energy and potential energy. Next, the mean free path is equal to $l = 1/n\sigma \propto 1/\rho\sigma$, with σ the relevant photon scattering cross section (for example, see chapter II.3) suitably averaged over the composition. If we take σ to be roughly constant for generic off-the-shelf stars on the main sequence, so that $l \propto 1/\rho \sim R^3/M$, then (11) yields

$$L \propto (R^3/M)R(M/R)^4 \sim M^3 \tag{12}$$

This simple dependence of luminosity on mass indeed holds[6] for main sequence stars for a range of M. By the way, in lieu of laboring over numbers, it may be numerically more accurate to simply refer to the standard we know and love, writing $L/L_\odot = (M/M_\odot)^3$.

Star birth

To do an actual calculation, astrophysicists would have to input some knowledge about $v(r)$ (defined in (9)), the energy produced per unit time per unit mass, from nuclear physics. Clearly, a detailed discussion would far transcend the scope and philosophy of this book. Instead, we will rely on a plausible assertion to arrive at a relation between the mass M and radius R of a typical main sequence star that does not depend on knowing $v(r)$.

First, a typical star like the sun burns at a nicely slow and steady rate, because the first step in energy production involves the process* $p + p \rightarrow d + e^+ + \nu_e$, which is governed by the weak interaction, as might be fairly well known even to readers of popular physics books.[7] Crucially, the Coulomb repulsion between two protons prevents them from getting close enough to interact weakly. Energy production can initiate only by quantum tunneling though the Coulomb barrier, as was first pointed out by George Gamow and as was discussed back in chapter III.1. The tunneling rate depends on the average kinetic energy of the protons, namely, the temperature of the star's interior.

Picture a gas cloud of mass M collapsing, until a critical temperature T^* is reached in the increasingly dense center of the cloud, thus igniting the weak process. The assertion is that T^* is determined largely by the details of nuclear and particle physics. Let us accept this fairly plausible statement.

Recall that the temperature T inside the cloud is proportional to M/R. As R decreases, T increases until it reaches T^*, at which point the cloud becomes a

---

*Two protons forming a deuteron and emitting a positron and an electron neutrino.

star! Setting M/R equal to T^*, we obtain

$$R \propto M \qquad (13)$$

The radii R of main sequence stars are roughly proportional to their masses M. The density varies like $\rho \sim M/R^3 \propto 1/M^2$. Low mass stars tend to be denser.

White dwarfs

Contrary to what one might naively think, quantum mechanics can manifest itself on macroscopic scales, indeed, on astrophysical scales. We present here a fly by night discussion of white dwarfs.

As a typical star, such as our sun, approaches the end of its life, having burned up its nuclear fuel, it could no longer generate the pressure $P \sim nT/m$ of a classical hot gas to hold itself up. The star slowly contracts. But then at some point, when the temperature drops beneath the Fermi energy E_F of the electron gas, the quantum Fermi degeneracy pressure kicks in.

Recall from chapters V.3 and V.4 that this equals $P \sim h^2 n^{\frac{5}{3}}/m_e$ (with m_e the electron mass). To save you the trouble of looking it up, I give a quick derivation here:

$$N/V \sim p_F^3/h^3 \implies p_F \sim h n^{\frac{1}{3}}$$

and

$$E/V \sim (p_F^3/h^3)(p_F^2/2m_e) \sim p_F^5/h^3 m_e \sim h^2 n^{\frac{5}{3}}/m_e$$

then

$$P = -\partial E/\partial V \sim E/V \sim h^2 n^{\frac{5}{3}}/m_e \qquad (14)$$

A star supported by the Fermi degeneracy pressure of electrons is known as a white dwarf. The number density of electrons equals the number density of protons, and so up to some numerical factor, $n \sim (M/R^3 m_p)$, with m_p the proton mass.* (Be sure to distinguish m_p from m_e.) Thus,

$$P \sim h^2 n^{\frac{5}{3}}/m_e \sim h^2 M^{\frac{5}{3}}/(R^5 m_p^{\frac{5}{3}} m_e) \qquad (15)$$

To get at the crux of the matter, and not to miss the forest for the trees, it is better to blank out the trees. So, let us suppress the various constants in (15) and simply write $P \sim \xi M^{\frac{5}{3}}/R^5$.

Henceforth in this chapter, the Greek letter ξ will stand for some combination of constants I won't bother to write, so as to avoid cluttering things up. (Here $\xi \sim h^2/Gm_p^{\frac{5}{3}} m_e$ evidently. Yes, we no longer need to display h^2 explicitly to remind us that the degeneracy pressure is quantum in origin.) Note that with this convention,[8] every time ξ appears, it might mean something different.

---

*Since $m_p \gg m_e$.

Physics now tells us to equate this pressure P to the inward crush of gravity, namely GM^2/R^4 as given in (1). Thus $\xi M^{\frac{5}{3}}/R^5 \sim M^2/R^4$. We learn that, surprise, the radius of a white dwarf

$$R \propto M^{-\frac{1}{3}} \tag{16}$$

decreases with its mass M! The $\frac{5}{3}$ in the Fermi degeneracy pressure gets beat by the 2 in Newtonian gravity. The more massive a white dwarf, the smaller it is. This is in sharp contrast to the behavior $R \sim M$ (see (13)) for main sequence stars.

Of course, we could also easily put back in the various constants and numerical factors. Doing that, we obtain[9]

$$R \simeq \frac{\hbar^2}{20 G m_p^{\frac{5}{3}} m_e} \left(\frac{Z}{A}\right)^{\frac{5}{3}} M^{-\frac{1}{3}} \sim 10^4 \text{ km} \left(\frac{Z}{A}\right)^{\frac{5}{3}} \left(\frac{M_\odot}{M}\right)^{\frac{1}{3}} \tag{17}$$

(For the meaning of Z and A, see exercise (1).) Thus, a solar mass white dwarf is about the size of the earth.

Chandrasekhar limit

Chandrasekhar noticed that this relationship, $R \propto M^{-\frac{1}{3}}$, has an important consequence: Massive white dwarfs have very high densities $\rho \sim M/R^3 \propto M^2$. Since $p_F \sim \hbar n^{\frac{1}{3}}$, the higher the density, the higher the Fermi pressure p_F will be, and thus at some high mass M, the electrons on the surface of the Fermi sphere have momentum $p_F \gtrsim mc$ and become relativistic.

The energy $\varepsilon(p)$ of a relativistic particle with momentum p and mass m is given by $\varepsilon(p) = \sqrt{(pc)^2 + (mc)^2}$. Again, to avoid clutter so as to see the essence of the matter more clearly, let us use units with $c = 1$. Then $\varepsilon = \sqrt{p^2 + m^2} \simeq p + \frac{m^2}{2p} + \cdots$ in the ultrarelativistic limit $p \gg m$.

So, for highly massive white dwarfs, we should replace the expression $E = \frac{V}{\hbar^3} \int_{p_F} d^3 p \, \frac{p^2}{2m}$ from chapter V.4 for the energy E of a Fermi gas by (with m the electron mass; we omit the subscript e for later convenience)

$$E = \frac{V}{\hbar^3} \int_{p_F} d^3 p \, \varepsilon(p) \simeq \frac{V}{\hbar^3} \int_{p_F} d^3 p \, \left(p + \frac{m^2}{2p} + \cdots\right) \tag{18}$$

First, let us look at the extreme relativistic limit, in which we keep only the leading term in the integral (or equivalently, set the electron mass m to 0). Then, by dimensional analysis, $P \sim E/V \sim p_F^4/\hbar^3 \sim \hbar n^{\frac{4}{3}}$. The 5 in (15) "magically" turns into a 4, and so

$$P \propto M^{\frac{4}{3}}/R^4 \tag{19}$$

Now, when we equate P to the inward crush of gravity GM^2/R^4, we get $\xi M^{\frac{4}{3}}/R^4 \sim M^2/R^4$, the powers of R match, and so the radius R drops out altogether! Something seems wrong.

Before reading on, how would you resolve this?

In $\varepsilon \simeq p + \frac{m^2}{2p} + \cdots$, we have to keep the next-to-leading term. That term, $\frac{m^2}{2p}$, is two powers of p down from the leading term p, and thus (18) now gives us, instead of (19),

$$\frac{E}{V} \propto p_F^4 + \xi p_F^2 \propto \frac{M^{\frac{4}{3}}}{R^4} + \xi' \frac{M^{\frac{2}{3}}}{R^2} + \cdots \tag{20}$$

(with ξ and ξ' yet more constant factors that we are too lazy to work out).

A little red light just went on

Caution! Ever since chapter I.1, from the very beginning of this book, we have intentionally driven calculus teachers everywhere nutty by canceling numerator and denominator instead of differentiating (thus $\frac{d\theta}{dt} \sim \frac{d\theta}{dt} \sim \frac{\theta}{t}$ in the pendulum problem). Well, this is okay as long as θ varies simply like a power of t; we are just off by some numerical factor. Indeed, we have already used this several times in this chapter and have gotten away with it, for example, writing $P = -\partial E / \partial V \sim E / V$.

But even, and especially, flying by night, we have to pay attention to the little red light on the instrument panel. Here, in (20), E is the sum of two terms. Even so, writing $-\partial E / \partial V \sim E / V$ might be okay if we are off "merely" by a relative numerical factor between the two resulting terms. But no, not here. We see from (20) that, with $V \sim R^3$, the two terms in E go like $V^{\frac{1}{3}}$ and $V^{-\frac{1}{3}}$, respectively, and we are going to miss a relative minus sign in the expression for the pressure P. And that is not okay to miss, as it turns out!

So, better be honest and differentiate rather than divide, thus obtaining the pressure

$$P \propto \frac{M^{\frac{4}{3}}}{R^4} - \xi'' \frac{M^{\frac{2}{3}}}{R^2} + \cdots \tag{21}$$

with a crucial relative minus sign. (It is understood that in my "fuzzy" notation, all the ξs are positive.) Equating P to the inward crush of gravity GM^2/R^4, we finally obtain

$$\frac{\left(\eta M^{\frac{4}{3}} - M^2 \right)}{R^4} = \eta' \frac{M^{\frac{2}{3}}}{R^2} \tag{22}$$

(I ran out of primes, so I switched from ξ to η, ha ha.)

This determines the dwarf's radius R as a function of its mass M. But, following Chandrasekhar, we see, even more interestingly, that since the right side of (22) is manifestly positive, the quantity $\left(\eta M^{\frac{4}{3}} - M^2 \right)$ must be positive: There is an upper bound, the Chandrasekhar limit, to how massive a white dwarf can be!

Again, putting in all the constants, we obtain the maximum mass of a white dwarf:

$$M_{\text{Ch}} \simeq 0.1 \left(\frac{Z}{A} \right)^2 \left(\frac{hc}{Gm_p^2} \right)^{\frac{3}{2}} m_p \tag{23}$$

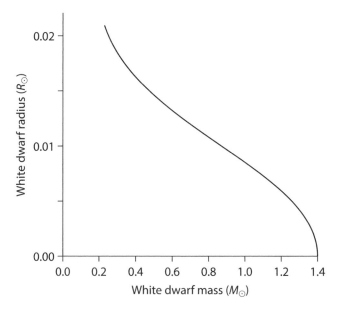

FIGURE 3. The radius of a white dwarf as a function of its mass plotted in units of $R_\odot$ and $M_\odot$. For a sense of scale, note the radius of the earth $\sim 0.01 R_\odot$. The issue is whether the relation $R \sim M^{-\frac{1}{3}}$ in (16) would continue forever as M increases or hold only for $M \lesssim M_\odot$. Chandrasekhar showed that a white dwarf cannot be more massive than $1.4 M_\odot$.

A more precise calculation yields the celebrated result $M_{\mathrm{Ch}} \simeq 1.4 M_\odot$.

Thus, for a white dwarf, a plot of the radius R as a function of mass M is as shown in figure 3.

Neutron stars

During the late life stages of stars with masses $M \gtrsim 8 M_\odot$, nuclear reactions involving ever larger nuclei are ignited as gravitational contraction leads to ever higher densities. Eventually, the density at the core is so high that electrons are pressed hard against the protons, triggering the weak interaction process $e^- + p \to n + \nu_e$. A burst of neutrinos ν_e is emitted, and the outer parts of the star are blown off in a supernova explosion, leaving behind a ball of neutrons. A neutron star is born.

The pressure is now provided by the Fermi degeneracy pressure of the neutron gas. The job of holding up the star passes from the electrons to the neutrons. We can simply modify the pressure given in (15), replacing the electron mass m_e by the neutron mass m_n, which for our purposes equals the proton mass, and setting $Z/A = 1$. Equate this to the crush of gravity GM^2/R^4, as given in (1). Since $m_n/m_e \sim 2{,}000$, we expect a neutron star to be about a thousand times smaller than a white dwarf of comparable mass. Numerically,

$$R \sim 14 \text{ km} \left(\frac{1.4 M_\odot}{M} \right)^{\frac{1}{3}}.$$

Exercises

(1) Show how the dependence of R on Z and A in (17) comes about. Regard the white dwarf as made of various nuclei, each containing Z protons and $A - Z$ neutrons.

(2) Estimate the mean free path for a photon in the sun.

(3) Estimate how long it takes a photon produced near the center of the sun to get out.

Notes

[1] For example, D. Maoz, *Astrophysics in a Nutshell*, or D. Clayton, *Principles of Stellar Evolution*. I follow Maoz's clear and concise treatment here.

[2] Recall that a main sequence star runs on proton fusion in its core.

[3] Viz., $\int_0^{\pi/2} d\theta \sin\theta \cos^2\theta = \int_0^1 d\cos\theta \cos^2\theta = \int_0^1 du\, u^2 = \frac{1}{3}$.

[4] Stellar astrophysicists customarily talk about opacity $\kappa(r)$, defined by $\kappa(r) = 1/(\rho(r)l(r))$ instead of the mean free path $l(r)$. Of course, $l(r)$ (and hence $\kappa(r)$) depends on local conditions at r, such as the composition. We will not go there.

[5] I follow Maoz, *Astrophysics in a Nutshell*, p. 46.

[6] Needless to say, all sorts of professional astrophysicists might jump up and down at this point, exclaiming ifs and buts. For very massive stars, the luminosity approaches what is known as the Eddington luminosity $\propto M$.

[7] I first heard about this by reading those written by George Gamow.

[8] I once wrote a paper with F. Wilczek stating that Greek letters such as ξ and η will denote numbers of order 1. We were too lazy to work them out. Somewhat to our surprise, the editor okayed it.

[9] Maoz, *Astrophysics in a Nutshell*, p. 75.

Collapse into black holes

How Einstein upset the struggle between mass density and pressure

Stars depend on the balance between the crush of gravity and resistance of pressure, as discussed in chapter VII.1. The higher the mass density ρ pushing inward, the higher will be the pressure P pushing outward needed to reach an equilibrium.

In Newtonian physics, this is no big deal, just an ongoing struggle between mass density ρ and pressure P; we have a balanced football game between offense ρ and defense P.

Pressure versus gravity

But as soon as Einstein ushers us into the relativistic world, the ball game changes entirely.

To understand why, go back to elementary physics, where the concept of pressure is introduced as a force per unit area: thus $[P] = [F]/L^2 = [F]L/L^3 = E/L^3$. The last step follows since force times distance is work done, which is energy. In other words, pressure also has the dimension of an energy density and can be thought of as such, as I have mentioned repeatedly.

The equation $P \sim \frac{GM^2}{R^4}$ in chapter VII.1 can be rewritten as

$$PR^3 \sim \frac{GM^2}{R} \sim G\rho^2 R^5 \tag{1}$$

and regarded as an approximate equality between pressure energy PR^3 and gravitational energy GM^2/R.

Relativistic stars and black holes

But Einstein taught us that $m = E/c^2$, and thus in relativistic physics, pressure also acts like a mass density. Dimensionally, $[P] = [\rho]c^2$.

Now we get into a vicious cycle in the struggle between mass density ρ and pressure P. As ρ increases, P increases to keep up, but this increase in P contributes a mass density P/c^2. Thus, the effective mass density is given by something like

$$\rho \to \rho_{\text{effective}} \sim \rho + \frac{P}{c^2} \qquad (2)$$

There is even more mass density to counteract than we had naively thought, so P has to increase, but this leads to even more mass density. The result is that under some circumstances, pressure is utterly defeated. The star collapses and inexorably, a black hole forms.

In our football analogy, when the game goes relativistic, some of the players on defense (that is, the pressure side of the equation) suddenly start going over from time to time to help the offense. Sure enough, the defense cannot hold and gets crushed.

When the going gets tough, the tough go over to the other side!

The sophisticated and the naive

Let's listen in on how two students might reason out this relativistic effect (2) of pressure also contributing to mass density.

The sophisticated and mathematical (typically an unthinking grad student) would say, "Ah, I recall from my course on general relativity that ρ and P are both components* of the same stress energy tensor[1] $T^{\mu\nu}$. Stress has to do with pressure, right? Under a Lorentz transformation, the different components get all mixed up; that's why it is called a tensor."

The naive and physical (allegedly a smart junior) considers an observer seeing a box of gas moving with velocity $v \ll c$. "The box has a kinetic energy density $\frac{1}{2}\rho v^2$, but that's not all: The box is also Lorentz contracted. Recall that the contraction is of order v^2/c^2 because of those ubiquitous $\sqrt{1 - \frac{v^2}{c^2}}$ factors in special relativity. Evidently, somebody has to do work squeezing the box, something like a pressure P times the amount of contraction v^2/c^2, and so we should add the work density Pv^2/c^2 to the kinetic energy density $\sim \rho v^2$. Thus, the effective ρ is actually $\sim (\rho + \frac{P}{c^2})$. Aha, we have (2)!"

In relativistic physics, P appears on both sides of the balance, the outward push as well as the inward pull.

---

*Indeed, ρ is the time-time component and P/c^2 makes up the three diagonal space-space components.

Pressure ends up on the losing side

While the physics behind (1) and (2) is pretty clear, it might still be worthwhile to scribble something on the back of an envelope. Imagine starting with $c = \infty$ and so no special relativity. Then slowly decrease c. Thus, plug (2) into (1), and ignore terms going like $\frac{1}{c^4}$. We obtain the criterion for collapse to be

$$\left(\frac{1}{GR^2} - \frac{2\rho}{c^2} \right) P \lesssim \rho^2 \tag{3}$$

When the quantity $(\frac{1}{GR^2} - \frac{2\rho}{c^2})$ is positive, we can always crank up the pressure P to avoid collapse. But when it is negative, that is, when $\rho \gtrsim \frac{c^2}{GR^2}$, no matter how large we crank the pressure up, we can't avoid collapse. Eliminating ρ in favor of M, we recover (of course) the Michell-Laplace criterion $GM/c^2 \gtrsim R$ from chapter I.2. Physics is consistent, as always.

Note

^{1} Also known as the energy momentum tensor. See appendix Eg.

The expanding universe

The "Great Debate"

While many of our fellow citizens are becoming jaded by cosmic news, I find it amazing how much we have learned about the universe in a little over a hundred years. A crucial reason for our remarkable progress is the observationally verified cosmological principle, stating that on scales much larger than galaxies, the universe is homogeneous and isotropic. There is neither a special location nor a special direction. This principle descended intellectually from Copernicus's bold assertion that the earth was not at the center of the universe.

For the longest time, the precise opposite of the cosmological principle would appear to hold. Our home galaxy, the Milky Way, was thought to comprise the entire universe. Modern cosmology started with Vesto Slipher's and Milton Humason's measurements of the redshifts of the galaxies. A landmark event was the "Great Debate" in 1920 between Harlow Shapley and Heber Curtis on the nature of spiral nebulas. Shapley opined that they were small local features within the Milky Way, while Curtis held that they were separate galaxies far outside our own. Had the former been right, the universe could hardly be homogeneous and isotropic.

Einstein gravity and the universe

Einstein gravity empowered physicists to discuss the universe. For those readers unfamiliar with Einstein gravity, I give an outrageously brief account in appendix Eg merely to give you a whiff of this profound theory. All we need for our purposes here are three takeaways from that appendix:

1. The geometry of spacetime is characterized by a metric $g_{\mu\nu}$ defined by $ds^2 = g_{\mu\nu}(x)dx^\mu dx^\nu$.
2. The field equation determining the curvature of spacetime has the form

$$R^{\mu\nu} - \frac{1}{2}g^{\mu\nu}R = (\cdots\partial\cdots\partial\cdots)^{\mu\nu} = GT^{\mu\nu} \qquad (1)$$

where the dots are constructed out of $g_{\mu\nu}$ and its matrix inverse.
3. The stuff that fills the universe can be described by an energy momentum tensor $T^{\mu\nu}$.

One goal of cosmology is to find out what the universe is filled with, and thus how the universe expands. The logic is reversible: Observing the universe enables us to deduce what it is filled with. Thus was dark energy discovered.

In summary, you specify $T^{\mu\nu}$, then solve (1) for the metric $g_{\mu\nu}$ describing the universe.

Please don't worry if you are not on a first name basis with (1). All you need to know for the rest of this chapter is that the dynamical variable is the metric of spacetime and that the equation governing it involves two powers of ∂.

Conservation of derivatives

I would like to elaborate on that last remark about the two powers of ∂. Newton's equation for determining the gravitational potential $\vec{\nabla}^2\phi \sim G\rho$ in terms of the mass density ρ involves two spatial derivatives. As explained in appendix Eg, Einstein's field equation (1) for determining the spacetime metric (which generalizes ϕ) in terms of the energy momentum tensor $T^{\mu\nu}$ (which generalizes ρ) must also involve two derivatives, albeit temporal as well as spatial, as befitting a relativistic theory. We know that for weak gravitational fields, Einstein's equation must reduce to Newton's equation. Hence, the two powers of ∂ in (1).

Derivatives cannot simply appear out of nowhere: they are conserved. (Some derivatives could become negligible compared to others. For example, the temporal derivatives of a weak gravitational field are much smaller than the spatial derivatives. Hence Newton's equation for determining the gravitational potential as given above does not involve the temporal derivative $\frac{\partial}{\partial t}$.)

Some undergraduates in my course were surprised by this conservation law. The reason was that they were taught to actually execute the derivation without delay. For example, upon seeing $\frac{d}{dx}\sin x$, they immediately replaced it by $\cos x$. The derivative manifestly evaporated.

But here we are not merely calculating: we are discussing the internal logic of physical theories as manifest in the corresponding equations of motion. Later, in chapter IX.4, we will again invoke the conservation of derivatives.

Closed, flat, or open

That the universe is homogeneous and isotropic allows only three possibilities: space is (a) closed like the surface of a sphere, (b) flat, or (c) open like the surface of a hyperboloid. It should be emphasized that we are referring to the curvature of space here, not the curvature of spacetime.

Until fairly recently, cosmology textbooks had to deal with all three cases. Then observations showed that space is very close to flat, and textbooks became simpler. We can start with the metric (with $c = 1$)

$$ds^2 = -dt^2 + a^2(t)(dx^2 + dy^2 + dz^2) \tag{2}$$

The expression $(dx^2 + dy^2 + dz^2)$ in (2) indicates that space is Pythagorean and flat.

Hubble expansion and the scale factor: space flat, spacetime not

The dimensionless function of time $a(t)$ is called the scale factor and determines whether the universe is expanding or contracting.

Consider two points in space with the same y, z coordinates, but with their x coordinates differing by some length L. At time t, the physical distance between the two points equals

$$\Delta s = \int_1^2 ds = \int_1^2 a(t)dx = a(t)\int_1^2 dx = a(t)L \tag{3}$$

The fractional rate at which this distance is changing, known as the Hubble "constant," is given by

$$H(t) \equiv \frac{1}{\Delta s}\frac{d\Delta s}{dt} = \frac{1}{a}\frac{da}{dt} \equiv \frac{\dot{a}}{a} \tag{4}$$

The distance L has dropped out.[1]

At any given t, the 3-dimensional space we live in is flat, as was just emphasized. However, spacetime is definitely not flat, as indicated by (2): ds^2 for spacetime is not proportional to $-dt^2 + dx^2 + dy^2 + dz^2$. Indeed, the variation of $a(t)$ with respect to t measures the deviation from a flat spacetime. So, space is flat, but spacetime not.

Energy density and pressure

To do cosmology, we plug into the metric, as indicated by (2), $g_{tt} = -1$, $g_{xx} = g_{yy} = g_{zz} = a^2(t)$, with all other components of $g_{\mu\nu}$ equal to 0, into the formula for determining the curvature of spacetime, so as to obtain the left side of (1).

Next, the right side. Recall the brief discussion in appendix Eg of the energy momentum tensor $T^{\mu\nu}$, which appears on the right side of Einstein's field equation. The energy momentum tensor consists of the energy density[2] ρ and pressure P. Thus, if you tell us what you are filling the universe with (in other words, if you specify the ρ and the P of the stuff), we can calculate by using (1) how the scale factor $a(t)$ is changing with time.

At first sight, Einstein's field equation (1) consists of numerous equations as the indices μ, ν range over t, x, y, z or $0, 1, 2, 3$. However, since no special direction is favored in an isotropic universe, by the same argument given in the last section of appendix Eg, you would not be surprised to learn that there are only two equations, corresponding to the time-time component and the space-space component.[3]

You are worried. Two equations are still one too many! There is only one unknown function $a(t)$ to solve for.

But physics is consistent, of course, thanks to energy momentum conservation: Energy density ρ and pressure P are related by

$$d(\rho a^3) = -P d(a^3) \tag{5}$$

which you might recognize as the first law of thermodynamics $dE = -PdV$ at constant entropy, volume being proportional to a^3. So Einstein's field equation (1) contains only one[4] differential equation for one unknown function $a(t)$.

Cosmic expansion without drudgery

At this point, hard working fly by day physicists would plug the metric (2) for spacetime into Riemann's formula for the curvature tensor to extract the Ricci tensor $R^{\mu\nu}$ and the scalar curvature R. Then, for whatever stuff is specified to fill the universe at a given epoch, they would solve for $a(t)$ using (1). Exactly that is done in any number of textbooks.[5]

But we fly by night physicists want to avoid the drudgery of calculating $R^{\mu\nu}$ and R. With some clever arguments, we can actually get pretty far in understanding the expansion of the universe.

First, we argue long and hard in appendix Eg that the left side of (1), as indicated, contains two, and only two, spacetime derivatives. The Ricci tensor $R^{\mu\nu}$ and the scalar curvature R consist of some nonlinear messes constructed out of $g_{\mu\nu}$, but never mind the general formulas. Here they can only have the form $\sim \cdots \partial \cdots \partial \cdots$, with the dots constructed out of $a(t)$. But since $a(t)$ depends on time t only, and not on space, each ∂ can only end up as $\frac{d}{dt}$. Thus, only linear combinations of $\ddot{a} \equiv \frac{d^2 a}{dt^2}$ and $(\dot{a})^2 = (\frac{da}{dt})^2$ (with coefficients that could depend on a) can appear. Incidentally, since $a(t)$ manifestly does not depend on x, y, z, the spatial derivatives in Einstein's equation have disappeared.

Second, suppose we let $a(t) \to \lambda a(t)$. We can absorb λ by letting* $\vec{x} \to \vec{x}/\lambda$, leaving (2) unchanged. Thus, physics must not change upon $a(t) \to \lambda a(t)$. This

---

*We are surely free to change our coordinate choice without changing the physics.

means that the quantities to be set equal to the physical ρ and P can only be $\frac{\ddot{a}}{a}$ and $(\frac{\dot{a}}{a})^2$.

Remarkably, Einstein's equations (1) for the universe end up having the form of some linear combinations of $\frac{\ddot{a}}{a}$ and $(\frac{\dot{a}}{a})^2$ (with numerical coefficients) equal to some linear combinations of ρ and P.

For whatever stuff is specified to fill the universe at a given epoch, solve for $a(t)$. Even more simply, with two equations, we can take linear combinations, so that one equation has only $(\frac{\dot{a}}{a})^2$ on the left side. Only an ordinary first order differential equation to solve!

Third, let us think about what $(\frac{\dot{a}}{a})^2$ could be proportional to. There are three possibilities: (1) ρ only, (2) P only, or (3) a linear combination of ρ and P.

Trying to guess what the equation for $a(t)$ should look like

At this stage, the equation for $(\frac{\dot{a}}{a})^2$ is generic, since we have not specified what ρ and P are. Now consider a universe filled with nonrelativistic matter, that is, matter that moves slowly compared to c (which, as you recall, has been set to 1). Pressure, as Boyle and other greats mentioned in part V explained to us, is due to the constituent of matter zipping around. By definition, in nonrelativistic matter, the kinetic energy of the constituents is much less than their rest energy, so that P is negligible compared to ρ.

But with $P = 0$ and $\rho \neq 0$, gravity still operates to cause the universe to contract or expand (the two processes are the "same" in that they are related by time reversal, as discussed in chapter VI.1). We don't want $\dot{a} = 0$ in this case, and so we can rule out $(\frac{\dot{a}}{a})^2$ being proportional to P only.

And thus the fly by night physicist, chanting an incantation to the god of simplicity, ventures to guess that $(\frac{\dot{a}}{a})^2 \propto \rho$.

That the square of $\dot{a}$ appears in $(\frac{\dot{a}}{a})^2 \propto \rho$ implies invariance under time reversal $t \rightarrow -t$. Incidentally, this explains one feature that puzzles some students: The more stuff the cosmos has, the faster it expands. Resolution: Think of time reversing the expansion and picture contraction.

We also kind of suspect that, since P implies a force and since $ma = F$, the acceleration $\frac{\ddot{a}}{a}$ should be driven by both P and ρ.

Indeed, the fly by day physicists,[6] panting and sweating from their heavy lifting, tell us that the two equations are

$$\left(\frac{\dot{a}}{a}\right)^2 = \frac{8\pi G}{3}\rho \tag{6}$$

and

$$\frac{\ddot{a}}{a} = -\frac{4\pi G}{3}(\rho + 3P) \tag{7}$$

As explained above, there is only one equation, which we choose to be (6). Differentiating (6) with respect to t and using energy momentum conservation (5), we obtain (7).

We don't really need the precise form of these two equations; I present them here merely to show you what they look like. In fact, we will now proceed by writing (6) as

$$\dot{a}^2 \propto \rho a^2 \tag{8}$$

Remarkably, the simple equation (8) governs cosmic expansion. Of course, this is largely due to the perfect cosmological principle. Homogeneity and isotropy impose powerful constraints on the universe.

Three different fillings to choose from: matter, radiation, and dark energy

The universe was filled with different kinds of stuff at different stages in its life. To gain a first understanding, cosmologists customarily assume, for the sake of simplicity, that one kind of stuff dominates. We will see later that this assumption is often justified.

As a first example, fill the universe with nonrelativistic matter, such as a bunch of electrons, nucleons, nuclei, atoms, whatever, moving slowly compared to c. The nature of the matter doesn't even matter, as long as it is not moving fast. The pressure P is then negligible, as explained above. Since the matter is just sitting around, the energy density ρ is entirely composed of the mass density. Think of the matter enclosed in a box. As the box expands by a factor a, the energy density ρ decreases like $1/a^3$, merely because the volume of the box has increased like a^3. This also follows immediately by setting $P = 0$ in (5).

Thus, $\rho = K/a^3$ with a constant K that we don't care about for our purposes here, but it depends on how many different kinds of atoms (just to be specific) are around and their masses. Plugging this equation into (8), we obtain

$$\dot{a}^2 \propto 1/a \tag{9}$$

I trust that anybody reading this book is able to solve simple differential equations such as this.

Instead, in the fly by night spirit of this book, let us employ "calculus for clever dummies," as we already did way back in the very first chapter of this book. We do what every calculus teacher would warn us against doing, namely, canceling the d in the "numerator" and "denominator" of $\frac{da}{dt}$, that is, setting

$$\frac{da}{dt} \rightarrow \frac{\not{d}a}{\not{d}t} \rightarrow \frac{a}{t} \tag{10}$$

Simplifying (9) by this "poor man's calculus" to $a^3 \propto t^2$, we obtain[7]

$$a \propto t^{\frac{2}{3}} \qquad \text{matter} \tag{11}$$

Astonishing, no? The universe expands.
You aren't astonished? Einstein was.

We can even deduce how fast the universe is expanding. Amazing, yes?

As another example, fill the universe with relativistic matter, consisting of particles zipping around at the speed of light c. The prototypical example is a photon gas, discussed in chapters III.5 and IV.2. Indeed, in the very early universe, when it was very hot, with the temperature T far exceeding the masses of the elementary particles, everybody (quarks, leptons, and whatnots) is moving around with speed c like photons.

In chapter IV.2, using entropy conservation and dimensional analysis, we deduced that the energy density of a box of photons goes like the inverse $\frac{4}{3}$ power of the volume of the box. The volume of the proverbial cosmic box varies like a^3. Thus, $\rho = K/a^4$ with a constant K that depends on how many different kinds of particles were zipping around (with fermions and bosons counted differently, as you might have guessed from chapters V.3–V.5).

Plugging this into (8), we have $\dot{a}^2 \propto 1/a^2$. Again, using "calculus for clever dummies," we obtain immediately

$$a \propto t^{\frac{1}{2}} \qquad \text{radiation} \qquad (12)$$

Our third example is about the mysterious dark energy. While nobody knows exactly what dark energy is, observational evidence suggests that the long-sought-for cosmological constant offers the most likely explanation.[8] For our purposes here, we will simply assume that this is the case and that $\rho = \Lambda$ is a constant. Then (8) immediately yields

$$a \propto e^{Ht} \qquad \text{dark energy} \qquad (13)$$

with H a constant determined by Λ.

Note that with ρ constant, (5) implies that $P = -\rho = -\Lambda$, the infamous negative pressure that causes the expansion of the universe to accelerate.

You might have realized that once we figure out how $\rho(a)$ depends on a, then we can solve (8) by simply recognizing that it corresponds exactly (up to some irrelevant constants) to the elementary problem of a point particle in classical mechanics moving in a 1-dimensional potential $V(a) = -\rho(a)a^2$ with total energy equal to 0.

Universe dominated by different stuff in different eras

Since radiation density goes like $\rho_r \propto \frac{1}{a^4}$, while matter density goes like $\rho_m \propto \frac{1}{a^3}$, as the universe expands and a increases, the radiation dominated era eventually gives way to a matter dominated era. See figure 1.

But going back to the early universe, as $a \to 0$, we watch radiation increasingly dominating over matter. Our present matter dominated universe used to be radiation dominated.

Since the temperature of radiation $T \sim 1/a$ (see chapter IV.2), as we go back to the early universe, T keeps on increasing. High energy photons go on a rampage. In the early universe, atoms and molecules were dissociated into nucleons

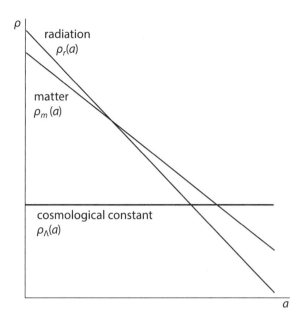

FIGURE 1. A schematic log-log plot of ρ versus the universe's scale factor a for radiation, matter, and the cosmological constant. As the universe evolves, matter eventually dominates over radiation. As the universe evolves further, the cosmological constant, which had been insignificant all along, eventually dominates over matter. The cosmic coincidence puzzle is "Why now, when we are around?" From Zee, A. *Einstein Gravity in a Nutshell*. Princeton University Press, 2013.

and electrons, and even earlier, nucleons in turn were dissociated into quarks and gluons.[9] Eventually, the temperature formally reached infinity, and our equations become singular. We have reached the mysterious Big Bang.

We defined the cosmological constant to be, duh, a constant. Thus, in the early universe, as $a \to 0$, with $\rho_r \propto \frac{1}{a^4}$ and $\rho_m \propto \frac{1}{a^3}$, the contribution $\rho_\Lambda = \Lambda$ of dark energy (assuming that it is the cosmological constant, of course) is totally negligible compared to ρ_r and ρ_m. In contrast, as the universe expands, dark energy eventually dominates matter. See figure 1.

Cosmic coincidence

As functions of the scale factor a, the two lines in figure 1 representing $\rho_m(a)$ and $\rho_\Lambda(a)$ intersect in what amounts to an instant in the history of the universe. Why now, when we humans are around? This puzzling question is known as the cosmic coincidence problem.[10]

At present, ρ consists of approximationly 68% dark energy, 27% dark matter, and 5% luminous matter. As far as cosmic expansion is concerned, dark matter and luminous matter can be lumped together. Thus, we happen to live in an era bearing witness to a cosmic struggle between dark energy and dark matter.

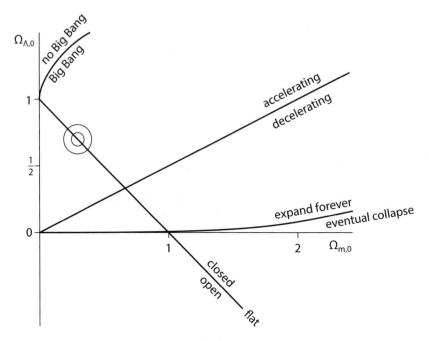

FIGURE 2. The fate of the universe. This cosmic diagram describes the overall history of the universe according to how much matter and how much dark energy, respectively, the universe contains at present. The variables $\Omega_{m,0}$ and $\Omega_{\Lambda,0}$ on the two axes are dimensionless measures of ρ_m and ρ_Λ, respectively, evaluated in the present epoch. The fate of the universe depends on $\Omega_{m,0}$ and $\Omega_{\Lambda,0}$. For example, a universe with $(\Omega_{m,0},\ \Omega_{\Lambda,0}) = (2, \frac{1}{2})$ is closed and will expand forever but at a decelerating rate. Our universe is located inside the circles, the sizes of which indicate uncertainty and our confidence in the observational data. We see that it is flat and accelerating. By the way, this diagram is fun and fairly easy to construct. From Zee, A. *Einstein Gravity in a Nutshell.* Princeton University Press, 2013.

To understand the expansion now, we have to put two different kinds[11] of ρ in (8): $\rho_m(a)$ and $\rho_\Lambda(a)$. Solving, we obtain a diagram depicting the fate of the universe. See figure 2.

We could go quite a bit further into cosmology with this sort of fly by night approach,[12] but it would take us far beyond the scope of this book.

A Newtonian mnemonic

Remarkably, a pseudo-derivation of the central equation (6) of Einsteinian cosmology can be concocted using Newtonian mechanics. In a Newtonian universe filled with a constant mass density (never mind that such a universe does not really make sense), consider a large sphere of radius $R(t)$ and an infinitesimal unit mass on the surface of the sphere. The unit mass has kinetic energy $\frac{1}{2}\dot{R}^2$ and, by Newton's superb theorems, potential energy $-G(4\pi R^3/3)\rho/R$.

By energy conservation, its total energy $\frac{1}{2}\dot{R}^2 - G(4\pi R^3/3)\rho/R$ should be conserved. Calling this constant $-\frac{1}{2}k$, we obtain (6) for $k=0$ and even understand where the $8\pi/3$ comes from!

For $k=-1$, the total energy is positive, indicating that the Newtonian sphere could expand indefinitely, roughly corresponding to an open universe. For $k=+1$ and negative total energy, the sphere would ultimately have to yield to gravity and contract.

I do not take this Newtonian pseudo-derivation[13] seriously but value it as a highly useful mnemonic that can also serve to motivate pedagogically the subtle physics contained in Einstein gravity.

Exercises

(1) The discussion in the text indicates that we could model both matter and radiation with the generic equation of state $P=w\rho$, where evidently $w=0$ for matter and $w=\frac{1}{3}$ for radiation. Solve for the behavior of $a(t)$ for arbitrary w.

(2) Show that your result for exercise (1) also covers the cosmological constant.

(3) Relate (6) to our discussion of the collapse of a dust cloud in chapter I.2.

(4) Solve (9) correctly, and verify that we obtained the power of t correctly in every case.

Notes

[1] Incidentally, the conventional normalization is to set $a(t_0) = 1$, with t_0 the time at present. The redshift z commonly used by astronomers is then related to $a(t)$ by $a(t) = \frac{1}{1+z}$. A large redshift $z > 0$ corresponds to a time when the universe was smaller by a factor $a(t) < 1$.

[2] Unfortunately, physicists change notation depending on the context. In appendix Eg, energy density is denoted by ε to distinguish it from the mass density.

[3] Recall in particular that $T^{\mu\nu}$ has only two nonvanishing components: T^{00} and $T^{11} = T^{22} = T^{33}$.

[4] Mathematically, this is due to the Bianchi identity.

[5] See, for example, *GNut*, pp. 357–359.

[6] See, for example, *GNut*, pp. 493–494.

[7] Some readers might recognize this as just dimensional analysis. Whah, some other readers might say, you told us that a is dimensionless! Simply regard (9) as a mathematical equation $\dot{a}^2 = 1/a$. Denote the dimension of a by A. Then $[\dot{a}^2] = (A/T)^2 = [1/a] = 1/A$. So $A = T^{\frac{2}{3}}$.

[8] See *GNut*, chapters VI.2 and X.7.

[9] See *GNut*, chapter VIII.3.

[10] See *GNut*, p. 751.

[11] The resulting equations are still quite manageable. See *GNut*, chapter VIII.2.

[12] See "A short course in cosmology," pp. 465–492 in A. Zee, *Unity of Forces in the Universe*.

[13] See the discussion on p. 475 of S. Weinberg, *Gravitation and Cosmology*.

Power radiated in gravity waves ▬▬▬

Two massive objects orbiting around each other

The detection of gravity waves[1] captivated the world in late 2016. Remarkably, it is not that hard[2] to figure out roughly how much power was radiated away in gravity waves by two black holes[3] of masses m_1 and m_2 orbiting each other, before they merged or reached relativistic speeds, whichever came first.

Three length scales enter: the sizes r_{bh} of the two black holes (taken to be comparable), their separation r from each other, and the wavelength λ of the waves emitted. We will stay in the pointlike[4] regime $r_{\text{bh}} \ll r$, which holds until the black holes "touch" and then merge, at which time nonlinear effects raise their ugly heads, and numerical relativity becomes necessary. From our understanding of electromagnetic wave emission, $\lambda \sim c/\omega \sim cT \sim c(r/v)$ (with T the orbital period and v the orbital velocity). Thus, we are in the long wavelength regime $\lambda \gg r$ as long as the black holes move with nonrelativistic speeds.

To fix our thoughts, let me break with the rest of this book and first state the precise fly by day result for the energy $\mathcal{E}$ per unit time radiated away by the two black holes in a circular orbit around each other:

$$\frac{d\mathcal{E}}{dt} = \frac{32}{5} \frac{G^4}{c^5 r^5} (m_1 m_2)^2 (m_1 + m_2) \tag{1}$$

We fly by night physicists surely would not be able to get the dependence on the mass ratio m_1/m_2, at least not on the first pass, and certainly not the $\frac{32}{5}$. So we will assume that $m_1 \sim m_2 = m$.

For this problem, our old standby, good old dimensional analysis, cannot help us right off the bat. Writing $\frac{d\mathcal{E}}{dt} \sim \frac{G^\alpha m^\beta}{c^\gamma r^\delta}$ we have four unknowns,

α, β, γ, and δ, to determine but with only three equations matching powers of M, L, T. To use dimensional analysis, you would have to guess one of these four unknowns. If you could do that, all power to you! (As we will see, with the clarity of hindsight, you might could have done it, as the saying goes.) Alternatively, you could try to argue what the G dependence must be, but that is not so easy: In particular, in (1), ω or the orbital period has already been eliminated using Kepler's law, which involves G.

Einstein's field equation in the weak field limit

Unless you are fluent with Einstein gravity, you should now go read the lightning review of general relativity given in appendix Eg. See also the executive summary in chapter VII.3. Strikingly, to fly by night, we need precious little of Einstein's splendid theory.

Einstein's fundamental result says that the curvature of spacetime is determined by the energy momentum tensor $T_{\mu\nu}$. We learn in appendix Eg that the field equation determining the spacetime metric $g_{\mu\nu}$ has the form*

$$E_{..} = R_{..} - \frac{1}{2}g_{..}R \sim GT_{..} \tag{2}$$

(Here $E_{..}$, known as the Einstein tensor, denotes a particular combination of the Ricci curvature tensor and the scalar curvature that appears in Einstein's equation, and the dots represent some indices we are suppressing.) All of this is "explained" (ha ha) in appendix Eg.

All we need to know here is that $R_{..}$ involves two powers of spacetime derivatives $\partial_\lambda \equiv \frac{\partial}{\partial x^\lambda}$ acting on $g_{\mu\nu}$. (All right, I show you some indices, but they will be gone soon enough.) Denote the deviation of $g_{\mu\nu}$ from the flat Minkowski metric $\eta_{\mu\nu}$ by $h_{\mu\nu}$, namely, $g_{\mu\nu} = \eta_{\mu\nu} + h_{\mu\nu}$. Since $\eta_{\mu\nu}$ consists "merely" of a bunch of 1s, 0s, and $-$1s, the ∂s can act only on $h_{\mu\nu}$. In the weak field limit, that is, for $h_{\mu\nu}$ small, the curvature tensor approaches

$$R_{..} \to \partial^2 h_{..} + O(h^2) \tag{3}$$

Incidentally, when people wax poetic about small ripples[5] of spacetime in the popular press, they are referring to $h_{..}$, a measure of the amplitude of the ripples.

Since Einstein gravity is a relativistic theory par excellence, we have set $c = 1$; thus, for example, ∂^2 is short for[6] $\nabla^2 - \frac{\partial^2}{\partial t^2}$. Later, we will restore c using dimensional analysis. (Note that quantum mechanics does not enter in this problem, and so Planck's constant and units are not relevant here.)

Einstein's field equation thus reduces to[†]

$$\partial^2 h_{..} \sim GT_{..} \tag{4}$$

---

*A more detailed, but still schematic, version is given in chapter VII.3 but it is not needed.

†On the left side of this equation, we should actually write $\bar{h}_{..}$ instead of $h_{..}$ where $\bar{h}_{\mu\nu} \equiv h_{\mu\nu} - \frac{1}{2}\eta_{\mu\nu}h$, with $h \equiv \eta^{\mu\nu}h_{\mu\nu}$. Also, the harmonic gauge has been chosen. I am suppressing all technical details here.

in the weak field limit. It is worth repeating that this generalizes the Newton-Poisson equation $\nabla^2 \phi \sim G\rho$ with the well known solution $\phi \sim G \int d^3x' \rho(\vec{x}')/ |\vec{x} - \vec{x}'|$.

Thus, as we learned in chapter II.2 and appendix Gr, or reasoning simply by analogy, we can solve (4) with no work to obtain

$$h_{ij} \sim G \int d^3x' \frac{T_{ij}(\vec{x}')}{|\vec{x} - \vec{x}'|} \sim \frac{G}{R} \int d^3x' T_{ij}(\vec{x}') \tag{5}$$

Here $R \equiv |\vec{x}|$ denotes the distance from us to the orbiting black holes (not to be confused with the curvature $R_{..}$, of course). Certainly, $R >>>> r$, the distance between the two black holes (namely, the typical value of $|\vec{x}'|$ in (5)). Pulling $R = |\vec{x}|$ out of the integral is an excellent approximation.

We are assuming implicitly that the source $T_{..}$ in (4) is oscillating at a characteristic frequency ω, which, as was already noted, is the inverse of the orbital period of the two black holes. I have also suppressed the oscillatory factor $e^{-i\omega t} e^{ikR}$ that was so crucial in chapter II.2. In the units we are using, with $c = 1$, $k = \omega/c = \omega$. The Lorentzian ∂^2 in (4) has been effectively replaced by the Newtonian ∇^2 in order to obtain (5), which you should compare with its electromagnetic analog, (5) in chapter II.2. The appearance of the integral over T_{ij} parallels the appearance of the integral over J_i in electromagnetism. Believe me, things are pretty much the same as in our discussion of the dipole radiation of electromagnetic waves.

Einstein gravity carries one more index than electromagnetism

To leading approximation, electromagnetic waves are emitted by dipoles, not by monopoles (aka charges) as a direct consequence of charge conservation $\partial_\mu J^\mu = \partial_0 J^0 + \partial_i J^i = 0$. For your reading convenience, I repeat from chapter II.2 the mathematical steps involved (being sloppy with the distinction between upper and lower indices here!): $\int d^3x \, J_i = \int d^3x \, J_j \delta^i_j = \int d^3x \, J_j \partial_j x^i = -\int d^3x \, (\partial_j J_j) x^i = \int d^3x \, (\partial_0 J_0) x^i = \partial_0 \int d^3x \, x^i J_0 = i\omega \int d^3x \, x^i J_0$, namely, the dipole moment of the source times the oscillation frequency. (The third equal sign follows upon integration by parts.)

Well, Einstein gravity "merely" involves one more index than electromagnetism. Thus, to massage $\int d^3x \, T_{ij}$ in (5) into shape, we simply have to play the same trick we just played on $\int d^3x \, J_i$ twice, but invoking energy momentum conservation $\partial_\mu T^{\mu\nu} = \partial_0 T^{0\nu} + \partial_i T^{i\nu} = 0$ instead of charge conservation. Schematically,[7]

$$\int d^3x \, T_{ij} \sim \partial_0^2 \int d^3x \, x^i x^j T_{00} \sim \omega^2 D^{ij} \tag{6}$$

with the second moment of the energy or mass distribution of the source defined by

$$D^{ij} \equiv \int d^3x \, x^i x^j T_{00}(x) \tag{7}$$

Compare this with the expression for the dipole moment $\int d^3x \, x^i J_0$ given above. We are implicitly assuming, as was mentioned earlier, that J_0 and T_{00} represent monochromatic sources with frequency ω (namely, the frequency of the electromagnetic wave and gravity wave detected).

The need for a quadrupole moment

Plugging (6) into (5), we obtain the amplitude of the wave $h_{..}$, but we are still a long way from being done. We have yet to calculate the energy going into our detector on earth. This will be done in the next section, but let us anticipate the final result. What happens is that the second moment D^{ij} of the mass distribution of the source appearing in (6) and defined in (7) gets replaced by the quadrupole moment Q^{ij} of the mass distribution of the source defined by

$$Q^{ij} \equiv \int d^3x \left(x^i x^j - \frac{1}{3}\delta^{ij}\vec{x}^{\,2} \right) T_{00}(x) = D^{ij} - \frac{1}{3}\delta^{ij}D \tag{8}$$

(Here $D = \delta^{ij}D^{ij}$ is the trace of the second moment.)

The very neat physical reason for why this happens can be traced all the way back to the first of Newton's two superb theorems discussed in chapter VI.3. In Newtonian gravity, the gravitational potential outside a spherical mass distribution of total mass M at a distance R away from the center of the distribution is given by GM/R. An analogous theorem in Einstein gravity states that the spacetime metric outside a spherical mass distribution of total mass M at a distance R away from the center of the distribution is fixed, determined completely by GM/R. Thus, a spherical mass distribution cannot radiate.

We see from (7) that $D^{ij} = \frac{1}{3}\delta^{ij}D$ for a spherical mass distribution, since the angular average of $x^i x^j$ is equal to $\frac{1}{3}\delta^{ij}\vec{x}^{\,2}$. But $D \propto \int d^3x \, r^2 T_{00}(x)$ is manifestly positive and does not vanish (and hence D^{ij} also does not vanish). In contrast, the quadrupole moment Q^{ij} is constructed to vanish for a spherical mass distribution.

It makes physical sense, then, that in this leading order, the energy radiated in gravity waves does not vanish only if the quadrupole moment Q^{ij} does not vanish.[8]

Historically, Einstein stated erroneously in his 1916 paper that a spherically symmetric mass distribution could radiate, but came to the correct conclusion that a quadrupole moment Q^{ij} was required in a 1918 paper, prompted[9] by a correspondence with the Finnish physicist Gunnar Nordström.

Energy carried by a gravity wave: determining it in three "different" ways

Next, as promised, we relate the energy radiated per unit time through a unit area to the amplitude h of the gravity wave. We fly by night physicists are going to do it three different ways.

It is easiest to reason by analogy with electromagnetism. By dimensional analysis, we obtained way back in chapter II.1 that the energy radiated per unit time through a unit area by an electromagnetic wave is given by the Poynting vector $\vec{S} \sim c\vec{E} \times \vec{B}$. Since E and B are given by a spacetime derivative acting on the vector potential A_μ (that is, $E \sim B \sim \partial A$), we have $S \sim (\partial A)^2$ with $c = 1$. As you may recall (or, if you don't, look it up in the table of dimensions given in this book!), A has dimension of $M^{\frac{1}{2}}L^{\frac{1}{2}}/T$, that is, $(M/L)^{\frac{1}{2}}$ in the units we are using. As a check, $[S] = (M/L)/L^2 = M/L^3$ indeed.

In Einstein gravity, h plays the role corresponding to the role played by A in electromagnetism, and we expect that the energy radiated per unit time through a unit area to be given by something like $(\partial h)^2$. But oopsy daisy, h is dimensionless.[10] So, to get something with the same dimension as $S \sim (\partial A)^2$, we must divide $(\partial h)^2$ by something with dimension $((M/L)^{-\frac{1}{2}})^2 = L/M$ and characteristic of gravity. Care to guess what that might be?

Use good old dimensional analysis again. Well, from Newton's law, the potential energy $E \sim GM^2/r$. With $c = 1$, mass M and energy E have the same dimension.* Hence $[E] = [G][E]^2/L$, and so $[G] = L/E = L/M$, implying that $[(\partial h)^2/G] = (1/L^2)(E/L) = E/L^3$, precisely an energy per unit time per unit area with[11] $c = 1$.

For those readers comfortable with the action, I give a slightly more sophisticated argument in an endnote.[12] That's the second way.

For the third way, we invoke Einstein's field equation (2). By the time the gravity wave reaches earth, it has long been propagating in empty spacetime, so that (2) reduces to $E_{\mu\nu} = 0$. Now expand in powers of $h_{..}$, and shove the quadratic terms to the right side:

$$E_{\mu\nu}^{(1)} = -E_{\mu\nu}^{(2)} + O(h^3) \tag{9}$$

Let us write this as $E_{\mu\nu}^{(1)} \simeq Gt_{\mu\nu}$ where by high school algebra $t_{\mu\nu} \equiv -E_{\mu\nu}^{(2)}/G$. Comparing this with (2), we see that, to this leading order, we can regard $t_{\mu\nu}$ as some kind[13] of effective energy momentum tensor of the gravity wave. By construction, $E_{\mu\nu}^{(2)} \sim (\cdots \partial \cdots h \cdots \partial \cdots h)$, namely, a mess of terms with indices all over the place, but as explained in appendix Eg and in chapter VII.3, all we care about is that it contains two ∂s and two hs. Thus, schematically, the energy momentum tensor of the gravity wave is given by something like $\frac{1}{G}(\partial h)^2$, in agreement with our fly by night result obtained earlier.

In summary, the energy per unit time carried by an electromagnetic wave and by a gravity wave passing through a unit area is given by (with $c = 1$)

$$S \sim (\partial A)^2 \text{ electromagnetic wave;} \qquad S \sim \frac{1}{G}(\partial h)^2 \text{ gravity wave} \tag{10}$$

---

*Again, I am using the same letter E for energy and for the dimension of energy, as I have been doing throughout this book.

Power radiated by a quadrupole

Incidentally, while our gravity wave detectors are enormous on human scale, they are vanishingly small compared to the vast distance to the orbiting black holes. Thus, the gravity wave arriving on earth is comfortably described as a plane wave.

The fly by day calculation of the power passing through a unit area is a bit lengthy, but conceptually parallels the analogous calculation for electromagnetic waves. We plug the precise expression for $h_{\mu\nu}$ into the precise expression for $\vec{S}$, sum over the two polarizations of the wave (just as for electromagnetic waves), integrate over directions, and so on. After all this is done,[14] we witness the quadrupole moment emerging, as dictated by the physics.

Let us leave that show of strength to the fly by day people. We fly by night physicists press on, armed with this physical knowledge that gravity waves are emitted by quadrupole moments Q. Suppressing all indices, we obtain, from (5), (6), and (8),

$$h \sim G\omega^2 Q/R \qquad (11)$$

The ripples h reaching us are proportional to Newton's constant G and the quadrupole moment Q of the source, and they are inversely proportional to the distance R to the source. The two extra powers of ω are due to the tensorial character of Einstein gravity. All makes sense, no?

The crucial content of (11), as was explained in chapter II.2, is the implicit oscillatory factor $e^{i(\vec{k}\cdot\vec{x}-\omega t)}$. To get the power radiated per unit area, all we have to do is plug h into (10):

$$(\partial h)^2/G \sim \omega^2 h^2/G$$
$$\sim \omega^2(G\omega^2 Q/R)^2/G$$
$$\sim G\omega^6 Q^2/R^2 \qquad (12)$$

In the first step, we converted ∂ to ω. In the second, we plugged in our result (11) for h.

Multiplying this by the surface area[15] $4\pi R^2$ of the sphere of radius R, we obtain the desired power radiated $\frac{d\mathcal{E}}{dt} \sim G\omega^6 Q^2$. The quadratic dependence on Q is expected, but the 6th power of ω might be a bit of a surprise.

Restoring the speed of light

At this point, it is convenient to restore c by appealing to dimensional analysis. First, look at the result for $\frac{d\mathcal{E}}{dt}$ cited in the preceding paragraph. Work out the dimensions:

$$[G\omega^6 Q^2] = \left(\frac{L^3}{MT^2}\right)\left(\frac{1}{T}\right)^6 (ML^2)^2 = \frac{ML^7}{T^8} \qquad (13)$$

Interesting that such high powers appear. Next, note that $[\frac{d\mathcal{E}}{dt}] = M(L/T)^2/T = ML^2/T^3$.

We see that a factor of c^5 is needed to match* dimensions:

$$\frac{d\mathcal{E}}{dt} \sim \frac{1}{c^5} G \omega^6 Q^2 \tag{14}$$

As a check, the powers of c had better appear in the denominator, not the numerator, so that as $c \to \infty$, the rate of energy emission vanishes, as it should.

It is instructive to compare this with the rate of energy emission by a dipole in electromagnetism derived in chapter II.2: $\frac{d\mathcal{E}}{dt} \sim \omega^4 p^2 / c^3$. Recall that the electromagnetic dipole moment p equals ed with d the size of the dipole. For comparison, the gravitational quadrupole moment of two massive objects of mass m separated by distance d circling each other is of order $Q \sim md^2$. (A triviality: the electromagnetic coupling e is contained in p, while G is customarily not absorbed into Q.) The important physics is contained in the differing frequency dependence, ω^6 versus ω^4, due to the quadrupole versus the dipole.

Calculating the quadrupole moment: Newtonian orbital mechanics

It remains only to calculate the quadrupole moment Q of the orbiting black holes and the orbital frequency ω, an entirely Newtonian exercise, and plug into (14). It is important to realize that although gravity waves do not exist in Newtonian gravity, in the regime we are considering, before the final embrace, Newton still rules the orbit!

The quadrupole moment Q is in fact supremely easy to estimate by eyeballing (8): $Q \sim mr^2$, with r the separation between the black holes and m their masses $\sim \int d^3x \, T(x)$. The frequency ω is determined by a Newtonian exercise we went through eons ago in chapter I.2: $Gm^2/r \sim mv^2 \sim m\omega^2 r^2$, which implies $\omega^2 \sim Gm/r^3$. Indeed, we recognize this relation between the frequency ω of the gravity wave and the separation r between the black holes as none other than Kepler's law.

Power radiated in gravity waves by two black holes orbiting around each other

Plugging all this in, we finally reach our goal:

$$\frac{d\mathcal{E}}{dt} \sim \frac{G^4 m^5}{c^5 r^5} \tag{15}$$

This fly by night result is consistent with (1), of course.

---

*That powers L and T both match provides a bit of a check.

It might be worthwhile to summarize the steps silently (that is, without any prose or yak, but with $c = 1$) and remind ourselves of the physics:

$$h_{..} \sim G \left(\int d^3x \, T_{..} \right) / R \sim G \, \partial_0^2 \int d^3x \left(x^i x^j - \frac{1}{3} \delta^{ij} \vec{x}^2 \right) T_{00} / R \sim G \omega^2 Q / R$$

$$(\partial h)^2 / G \sim G \omega^6 Q^2 / R^2$$

$$Q \sim m r^2$$

$$\omega^2 \sim Gm / r^3$$

$$\frac{d\mathcal{E}}{dt} \sim R^2 \left((\partial h)^2 / G \right) \sim G \omega^6 Q^2 \sim G (Gm/r^3)^3 (mr^2)^2 \sim G^4 m^5 / r^5 \qquad (16)$$

Various powers in the expression for the power radiated

I think that it would have been fairly hard to guess, when we first started, the power of m for example, so as to be able to exploit dimensional analysis immediately. As I said, with the clarity of hindsight, we could have figured out, or at least understood, the power 5. Of the 5, 2 come from the source (for both Newton and Einstein, gravity is sourced by mass, and we had to square to get the power* radiated), and 3 come from converting ω^6 to m^3 using Newtonian mechanics. Of the ω^6, 4 come from the $\partial^2 / \partial t^2$ required by quadrupole emission (remember that h has to be squared) and 2 come from the two derivatives in the gravitational analog of the Poynting vector in (12).

Of course, since various quantities in physics are interrelated, the expression for the power radiated can be written in numerous different ways. For example, if you prefer, you could write

$$\frac{d\mathcal{E}}{dt} \sim \frac{1}{G} \left(\frac{Gm}{cr} \right)^5 \sim \left(\frac{v}{c} \right)^5 \left(\frac{v^5}{G} \right) \qquad (17)$$

At first glance, this looks a bit peculiar, with v raised to the 10th power. A quick check, however, shows that $\frac{v^5}{G}$ does have the dimension of a power: $\frac{v^5}{G} = \frac{MT^2}{L^3} (\frac{L}{T})^5 = (M(\frac{L}{T})^2) \frac{1}{T}$.

My own favorite grouping would be to write

$$\frac{d\mathcal{E}}{dt} \sim \left(\frac{Gm}{c^2 r} \right)^4 \left(\frac{mc^2}{r/c} \right) \qquad (18)$$

Recall from chapter I.2 the remark that Kepler's law follows immediately from the dimension $[Gm] = L^3 / T^2$, so that for a black hole, Gm/c^2 is a length characterizing its size. Thus, the factor $\epsilon \equiv (Gm/c^2 r)$ measures the size of the two orbiting black holes as compared to the radius of the orbit, a ratio assumed to

---

*I trust that you are not confused by the two uses of the English word "power."

be tiny to begin with. Thus, the two orbiting black holes radiate away a fraction ϵ^4 of their rest masses in the time it takes for light to traverse the orbit.

You can now (ta dah!) calculate the energy emitted by all kinds of sources, for example, a vibrating and rotating neutron star.[16]

Exercises

(1) Write down the angular momentum J radiated away per unit time in gravity wave by our monochromatic quadrupole source with frequency ω.

(2) Estimate the time it takes for the two black holes to collide with each other.

Notes

[1] For a discussion of whether the term "gravity wave" or "gravitational wave" is preferable, see my *On Gravity*, p. 5. I am aware that 19th century physicists had already coopted the term "gravity wave" for the water waves to be discussed in chapter VIII.1.

[2] Easy to say now! In fact, the calculation of the energy carried by a gravity wave generated bitter controversies lasting for decades. I urge you to read the fascinating account by D. Kennefick, *Traveling at the Speed of Thought*, Princeton University Press, 2007. In particular, see the chronology given on p. 142.

[3] With the fly by night approach used in this chapter, only the masses of the two objects enter, but we say "black hole" just to be definite. The discussion also applies to neutron stars.

[4] Standard textbooks on Einstein gravity treat only point particles moving along geodesics. Incidentally, the leading correction due to the finite size of the black holes can be obtained using an effective field theory approach. See chapter N.1 in *QFT Nut* (second edition) and the references cited therein.

[5] Some readers might find pp. 31–33 of my semi-popular book *On Gravity* helpful.

[6] This differential operator is also denoted by $\Box$ in chapter II.2 and appendix Gr.

[7] We have $\int d^3x \, T_{ij} \sim \int d^3x \, T_{kj} \partial_k x^i \sim \int d^3x \, x^i \partial_k T_{kj} \sim \partial_0 \int d^3x \, x^i T_{0j}$. Next, repeat these steps for the index j. For more details, see, for example, *GNut*, pp. 569 and 576.

[8] To leading order: We are certainly not worrying about the octupole and higher moments here in a fly by night book.

[9] See Kennefick, *Traveling at the Speed of Thought*.

[10] Because by definition, the spacetime metric $g_{\mu\nu} = \eta_{\mu\nu} + h_{\mu\nu}$ and both $g_{\mu\nu}$ and $\eta_{\mu\nu}$ are dimensionless.

[11] Again, a reminder that $\hbar$ is not in the game, as if you needed to be reminded.

[12] I will discuss the action briefly in part IX. If you have never heard of the action, one place to start is *GNut*, chapter II.3. Recall that the action of a point particle $S = \int dt \, \frac{1}{2}mv^2$, being an integral over energy, has the dimension of ET. For a field theory such as electromagnetism or gravity, $S = \int dt \, d^3x \mathcal{L}$, and thus the action density, otherwise known as the Lagrangian density $\mathcal{L}$, has dimensions $ET/L^3T = E/L^3$, namely energy per unit volume. In units with $c = 1$, this is the same as energy per unit area per unit time. For Einstein gravity, the action equals $S = \int d^4x R/G$ as explained in appendix Eg, and hence R/G is an energy per unit area per unit time, which reduces to $\sim (\partial h)^2/G$ in the weak field limit.

[13] The technical term is energy momentum pseudotensor.

[14] In particular, M. Maggiore, *Gravitational Waves*, vol. 1, performs this calculation in commendable detail.

[15] I included the factor of 4π just to show you that, ha ha, I can write down an exact formula when I have to.

[16] My very first research project, suggested to me by J. A. Wheeler when I was an undergrad, was to calculate gravity wave emission from a vibrating and rotating neutron star. In hindsight, Wheeler was motivating me to read the appropriate chapters in Landau and Lifshitz's book on classical fields. My task was simply to understand the

formula given in the book, plug in numbers, and prepare a joint paper for publication. But then M. Goldberger and S. Treiman said to me, "We came to rescue you from Wheeler's clutches. You have to start learning quantum field theory!" Thus the paper was never published but, due to Wheeler's influence at the time, it was widely cited. A sample: J. A. Wheeler, *Annual Review of Astronomy and Astrophysics*, 1966; D. W. Meltzer and K. S. Thorne, *Astrophysical Journal*, 1966; J. B. Hartle, *Astrophysical Journal*, 1967; C. Hansen and S. Tsuruta, *Canadian Journal of Physics*, 1967; K. S. Thorne and A. Campolattaro, *Astrophysical Journal*, 1967; S. Detweiler, *Astrophysical Journal*, 1975; Ramen K. Parui, *Astrophysics and Space Science*, 1993.

Interlude

Math medley 3

A glimpse of random matrix theory

In several areas of physics, notably random matrix theory,[1] we have to integrate over matrices. For definiteness, consider the following integral

$$I = \int d\varphi \; e^{-N \, \text{tr} \, V(\varphi)} \tag{1}$$

over all possible N by N hermitian matrices φ. Here $V(\varphi)$ denotes a polynomial in φ, for example, $V(\varphi) = \frac{1}{2}m^2\varphi^2 + g\varphi^4$, and tr is the trace, as usual.

An N by N hermitian matrix φ has N real diagonal elements. It also has $N(N-1)/2$ complex elements above the diagonal, which by hermiticity are the complex conjugate of the $N(N-1)/2$ complex elements below the diagonal, thus amounting to $2 \times N(N-1)/2 = N(N-1)$ real variables altogether. Thus, in total, φ has $N + N(N-1) = N^2$ real variables.

The integral over φ in (1) is defined to be the multiple integral over these N^2 real variables, exactly what you would have thought. No mystery at all to what integration over matrices means.

A nasty Jacobian?

First, as you should recall* from either a course on linear algebra or a course on quantum mechanics, a hermitian matrix can always be diagonalized by a unitary matrix:

---

*Recall also that a hermitian matrix is such that $\varphi^\dagger = \varphi$, and that a unitary matrix U satisfies $U^\dagger U = I$ with I the identity matrix.

$$\varphi = U^\dagger \Lambda U \qquad (2)$$

with U an N by N unitary matrix and Λ an N by N diagonal matrix with diagonal elements equal to λ_i, $i = 1, \ldots, N$. Now you see that the trace in the integrand in (1) is going to be a big help, since

$$\operatorname{tr} V(\varphi) = \operatorname{tr} V(U^\dagger \Lambda U) = \operatorname{tr} U^\dagger V(\Lambda) U = \operatorname{tr} U U^\dagger V(\Lambda)$$
$$= \operatorname{tr} V(\Lambda) = \sum_k V(\lambda_k) \qquad (3)$$

(For the second equality, think of V as a polynomial, as indicated in the example above.)

Thanks to that crucial trace, the integrand in (1) is rigged in such a way as to not depend on U. So, be happy and change integration variables from φ to U and Λ:

$$Z = \int dU \int \left(\prod_i d\lambda_i \right) \mathcal{J} \, e^{-N \sum_k V(\lambda_k)} \qquad (4)$$

with $\mathcal{J}$ the Jacobian. Since the integrand does not depend on U, we can throw away the integral over U. It just gives some overall constant.[2]

So far so good, but how do we evaluate the Jacobian $\mathcal{J}$? And here I come to the main point of this math medley and why I would even include this stuff in a book like this.

To calculate $\mathcal{J}$ would seem like quite a mathematical challenge. But remarkably, by a combination of some body English and dimensional analysis, we will be able to determine $\mathcal{J}$ without breaking a sweat.

Applying dimensional analysis to matrix integrals

Dimensional analysis? You might be puzzled. Isn't this just a matter of doing an integral? There is no physical dimension anywhere in sight. But having read math medley 2, you are not all that surprised.

First, the change of variables $(\varphi \to \lambda_i, U)$ is reminiscent of the change of variables from Cartesian to spherical coordinates $(x, y, z \to r, \theta, \phi)$.

Indeed, U is analogous to the angular coordinates θ, ϕ, and if, in an integral over 3-dimensional space, the integrand does not depend on θ or ϕ, the integral over them merely produces an overall factor of 4π, just some constant, like the integral over U in (4).

You have known since childhood that this change of variables $d^3x \to \sin\theta \, d\theta \, d\varphi \, dr \, r^2$ produces a Jacobian $J = r^2 \sin\theta$. You might have also noticed that J vanishes at $\theta = 0$ and π, namely, at the north and south poles, respectively. But did you ask why?

Well, the change of coordinates $(x, y, z \to r, \theta, \phi)$ is strictly speaking ill defined* at the north and south poles, and so something must go berserk there, as manifested in J vanishing.

Guessing the Jacobian

Same reasoning applies here! The change of integration variables from φ to Λ and U in (2) is ill defined when any two of the λ_is are equal. (In quantum mechanics, this is called a "degeneracy.") When that happens, the corresponding 2 by 2 submatrix in φ at some stage is not only diagonal but also proportional to the identity. The unitary matrix U does not "know quite what to do." Thus, when any two of the λ_is are equal, the Jacobian $\mathcal{J}$ in (4) must vanish.

So, $\mathcal{J}$ must be proportional to $(\lambda_m - \lambda_n)$ for any $m \neq n$. Since the λ_is are created equal, interchange symmetry[†] (or in common parlance, democracy) dictates that

$$\mathcal{J} = \left(\prod_{m>n} (\lambda_m - \lambda_n) \right)^{\beta} \tag{5}$$

with β some positive real number.

But what is β? This is where dimensional analysis gets to shine! Assign some dimension to the matrix φ, let's say length L to be definite. From (2), we see that λ has dimension L, since U, being unitary, is dimensionless. With N^2 matrix elements, $d\varphi$ has dimension L^{N^2}. However, in (4), $(\prod_i d\lambda_i)\mathcal{J}$ has dimension $L^N L^{\beta N(N-1)/2}$. Therefore, $N^2 = N + \beta N(N-1)/2$, which determines β to be 2.

Dyson gas

We have thus found that, up to an uninteresting overall constant,

$$Z = \int \left(\prod_i d\lambda_i \right) \left(\prod_{m>n} (\lambda_m - \lambda_n) \right)^2 e^{-N\sum_k V(\lambda_k)} \tag{6}$$

This looks like a formidable integral and cannot be done explicitly for an arbitrary V. What would you do next?

Freeman Dyson had the tremendous insight of inviting the Jacobian into the exponential, so as to write (6) as

$$Z = \int \left(\prod_i d\lambda_i \right) e^{-N\sum_k V(\lambda_k) + \sum_{m>n} \log(\lambda_m - \lambda_n)^2} \tag{7}$$

---

*What is the longitude of the north pole? Did you ask your geography teacher that? (While reading this book, you have already driven your calculus teacher crazy. I ought to write a book about how to drive your other teachers crazy.)

[†]Recall math medley 1.

Does this remind you of anything?

Dyson pointed out that in this form $Z = \int (\prod_i d\lambda_i) e^{-NE(\lambda_1, \dots, \lambda_N)}$ is just the partition function of a classical 1-dimensional gas consisting of N atoms. Think of λ_i, a real number, as the position of the ith atom in 1-dimensional space. The energy of a configuration,

$$E(\lambda_1, \dots, \lambda_N) = \sum_k V(\lambda_k) - \frac{1}{N} \sum_{m>n} \log(\lambda_m - \lambda_n)^2 \qquad (8)$$

consists of two terms with clear physical interpretations: The gas is confined in a potential well $V(x)$, and the atoms repel[3] one another with the 2-body potential $-\frac{1}{N} \log(x-y)^2$. Note that the situation is easier than usual: The atoms are not even moving around. There is no kinetic energy term. Regard the atoms as infinitely massive.

Large N approximation

Thus far, everything we have done goes through for finite N. Clearly, except for the harmonic potential $V(\lambda) = \frac{1}{2}m^2\lambda^2$, there is no way we can evaluate (6) analytically, for $N = 7$, say.

But in the large N limit, we can evaluate Z using statistical mechanics. Fortunately, in most applications, physicists are interested in the limit $N \to \infty$. Note that we have scaled things so that the two terms in E are of the same order in N, since each sum counts for a power of N. Note also that if we regard Z as the partition function, N corresponds to the inverse temperature, so that the problem for $N \to \infty$ reduces to calculating, for example, the density of the Dyson gas at zero temperature. Remember that this corresponds to the density of eigenvalues of the matrix φ.

I won't be able to go into the methods for evaluating Z here, but remarkably, the Dyson gas analogy already enables us to draw useful conclusions. For instance, for the "double well" potential $V(\varphi) = -\frac{1}{2}m^2\varphi^2 + g\varphi^4$ (note minus sign), we can conclude that the eigenvalues of the random matrices generated by the corresponding distribution will cluster around $\pm\sqrt{m^2/4g}$, the position of the two wells in $V(\lambda)$.

Random matrix theory

If you have followed frontline theoretical physics at all, you know that over the past few decades, the large N approximation has played a starring role in many developments such as the so-called AdS/CFT gauge-gravity duality.[4]

Eugene Wigner (1902–1995), Nobel Prize 1963, introduced random matrix theory in 1954 while thinking about nuclear physics. In stark contrast to other physicists who were busy proposing various complicated Hamiltonians to describe large atomic nuclei and solving for their eigenstates, Wigner took a fly by night tack and said that these nuclei had such a large number (the N in our problem) of energy levels that it might be better to simply replace these complicated Hamiltonians by random N by N hermitian matrices. While random matrix theory cannot possibly predict the energy levels of a specific atomic nucleus, of course, it has proved to be quite successful in describing statistical properties, such as the distribution of the spacing between energy levels in complex nuclei.

Random matrix theory has blossomed over the decades into an extremely fruitful subject, with implications for many fields, including mathematics, economics, finance,[5] condensed matter physics, string theory, and even biophysics.[6] You may even be able to think of the large matrices that might be relevant for each subject. For example, in finance, one might be interested in the correlation matrix of the prices of N different stocks.[7]

Compared to the other work a day nuclear physicists in the 1950s, Wigner proposed such a strikingly original approach that I often think of him as the commander of the squadron of fly by night physicists.[8]

Notes

[1] See, for example, chapter VII.4 of *QFT Nut*.

[2] In the language of quantum field theory, U corresponds to the unphysical gauge degrees of freedom—the eigenvalues $\{\lambda_i\}$ are the relevant degrees of freedom. See *QFT Nut*, chapter VII.1.

[3] Note that this corresponds to the repulsion between energy levels in quantum mechanics if we think of φ as the Hamiltonian of some system in quantum mechanics.

[4] See, for example, J. McGreevy, *Advances in High Energy Physics* 723105 (2010); *GNut*, chapter IX.11.

[5] See, for example, the work of J.-P. Bouchard. He financed me to live in Paris for a year.

[6] H. Orland and A. Zee, *Nuclear Physics* B620 (2002), pp. 456–476.

[7] Joël Bun, Jean-Philippe Bouchaud, and Marc Potters, *Physics Reports*, 666 (2017), pp. 1–109.

[8] An astonishing recent discovery (in 2019) is the connection between quantum gravity and random matrices. See P. Saad, S. Shenker, and D. Stanford, "JT Gravity as a Matrix Integral," arXiv:1903.11115.

Part VIII

From surfing to tsunamis, from dripping faucets
to mammalian lungs

Water waves

Water waves are awesome

I am guessing that in your daily life, you have seen water waves (on the surface of a pond, for example) a lot more often than falling apples or colliding billiard balls. You also appreciate that water waves display a bewildering variety of behavior, from soothing ripples to pounding surfs.

Waves and dispersion

Let us start by reviewing some basic concepts about waves in general, not restricted to water waves. A wave is characterized by its period T and wavelength λ, namely, its variations in time and space, respectively. As in chapter I.1, introduce circular frequency[1] $\omega \equiv 2\pi/T$. Also, recall the wave number vector $\vec{k}$, defined to point in the direction that the wave is propagating, with magnitude $k \equiv 2\pi/\lambda = |\vec{k}|$. Then the wave can be written variously as $\cos(\omega t - \vec{k} \cdot \vec{x})$, $\sin(\omega t - \vec{k} \cdot \vec{x})$, or in complex notation,* $e^{i(\omega t - \vec{k} \cdot \vec{x})}$. Often, it is convenient to simply write k for $\vec{k}$. The context should make things clear. We will also refer to ω simply as frequency. Much of this has already been discussed in connection with electromagnetic waves in part II.

You understand that the vector $\vec{k}$ is a more basic concept than the elementary wavelength λ, which does not transform naturally under rotations. However, our brains evolved to perceive period and wavelength more directly than frequency and wave number. (Recall the remark about x versus $1/x$ in

---

*As usual, it is understood that either the real or imaginary part is taken.

chapter III.5.) Thus, we will often end up interchanging these two reciprocal sets of variables.

Dispersion relation: group versus phase speed

How the frequency $\omega = \omega(\vec{k})$ depends on $\vec{k}$ is known as a dispersion relation.

Perhaps you are already familiar with the concepts of phase velocity and group velocity. I review these in appendix Grp for those readers who are not. In summary, for a wave with dispersion $\omega(\vec{k})$, phase speed or phase velocity[2] equals

$$v_p \equiv \frac{\omega}{k} \tag{1}$$

and group speed or group velocity equals

$$v_g \equiv \frac{d\omega}{dk} \tag{2}$$

The phase speed merely states that in one period T, a wave (say, $\cos(\omega t - kx)$), with a definite value of k, has traveled from crest to crest through a distance λ and thus has a speed $v_p = \lambda/T = \omega\lambda/2\pi = \omega/k$. The group velocity, in contrast, determines how a wave packet (formed by superposing waves with different values of k centered around some characteristic k_*) moves. The derivative $\frac{d\omega}{dk}$ is to be evaluated at k_*.

Note that v_p and v_g are the only[3] two reasonable expressions with the dimension* of a speed: $[\omega/k] = (1/T)/(1/L) = L/T$.

For $\omega \propto k^\alpha$,

$$v_g = \alpha v_p \tag{3}$$

The group moves faster than the phase for $\alpha > 1$ and the other way around for $\alpha < 1$.

If $\alpha = 1$, the wave is said to disperse linearly. Electromagnetic waves and sound waves disperse linearly, and thus for them, $v_g = v_p$. For an electromagnetic wave, $\omega = ck = c|\vec{k}|$. As everybody knows, the wave propagates with a universal speed c independent of k.

For electromagnetic waves, $\vec{k}$ is 3-dimensional, but for waves on the surface of a body of water, $\vec{k}$ would naturally be a 2-dimensional vector. Earth's gravity privileges the direction known as up and down, which we will call, as usual, the z-axis. The wave vector $\vec{k}$ then lies in the x-y plane. By rotational invariance, we can always choose the x-axis to be along $\vec{k}$, in which case $\vec{k} \cdot \vec{x} = kx$, giving the wavelength as $\lambda = 2\pi/k$.

For much of this chapter, we will consider plane waves with a definite value of k, suppressing the vectorial character of $\vec{k}$.

---

*Once again, the generic time used in dimensional analysis is not to be confused with the period of a wave with a definite k.

FIGURE 1. Gravity pulling down on a crest.

Ocean waves

Consider water waves on the ocean. Picture a crest (see figure 1). The force of gravity pulling down on a small volume of water under the crest is proportional to the density of water ρ, while the inertial mass of this quantity of water is also proportional to ρ. Thus, once again, by the celebrated equality of gravitational mass and inertial mass, ρ cancels out. In the dispersion relation, only g, the acceleration due to gravity on the earth's surface, can appear.

The fly by day* physicist would now write down the Euler equation[4] discussed in chapter VI.2, impose the appropriate boundary condition, invoke some linear approximation, and solve.[5]

In contrast, the fly by night physicist appeals to dimensional analysis: $[\omega] = 1/T$, $[k] = 1/L$, and $[g] = L/T^2$. The only possibility is

$$\omega^2 \sim gk \tag{4}$$

We have derived the dispersion of water wave in one line.

Surprise! The dispersion relation is not linear, in contrast to the more familiar electromagnetic wave and sound wave.

Into shallow water

Now notice that the group velocity

$$v_g = \frac{d\omega}{dk} \sim \sqrt{g/k} \sim \sqrt{g\lambda} \tag{5}$$

diverges as $k \to 0$ or $\lambda \to \infty$, thus exceeding any speed limit, including the speed of light! The same is true of the phase velocity $v_p = \frac{\omega}{k} = \frac{1}{2}v_g$.

Clearly, something breaks down in this long wavelength limit. Think for a moment before reading on.

Wavelength long compared to what? (As I have remarked earlier in this book, physics students should always ask this "compared to what" question.) For λ large, another length scale enters, namely, the depth or height h of the water. See figure 2. The preceding discussion was implicitly in the deep water regime (I said ocean, didn't I?), in which $h \gg \lambda$ and thus drops out of the game.

---

*As I emphasized in the preface, I also love this kind of fly by day approach, enjoying the delicious triumph of determining $\omega(k)$, with all the numerical factors correct.

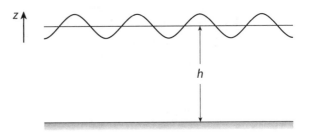

FIGURE 2. For shallow water, the depth h enters the analysis.

But if h becomes relevant, then dimensional analysis tells us that $\omega^2 \sim gkF(hk)$ with some unknown function $F(\xi)$.

All we know thus far is that $F(\xi) \to 1$ for large $\xi \equiv hk = 2\pi h/\lambda$, that is, as $h \to \infty$ with k fixed. Now we want to know about the opposite limit: how $F(\xi)$ behaves as $\xi = hk \sim h/\lambda \to 0$, that is, in the shallow water regime. Dimensional analysis is no help, since F is already a function of the dimensionless ratio h/λ.

Appeal to the god of simplicity

Let us instead appeal to the god of simplicity, supplemented by a flying guess.

A moment ago, we had already dismissed the possibility $F(\xi) \to$ some constant as $\xi \to 0$ on physical grounds. Lightning fast water waves? Not!

So, what is the next simplest[6] guess? As $\xi \to 0$, suppose $F(\xi) \to \xi$. (An undergrad might[7] ask, "How about $F(\xi) \to \xi^2$?" The riposte would be, "Physically, as $\xi \to 0$, we could argue that $F(\xi)$ vanishes, but give me a good reason why its derivative should also vanish.")

Thus, we claim that in the shallow water regime,

$$\omega^2 \sim (gk)(hk) \sim gh\vec{k}^2 \tag{6}$$

In this regime, water waves disperse linearly, just like electromagnetic waves and sound waves, with a speed (phase and group) given by $v_g = v_p \sim \sqrt{gh}$.

Crossover between the deep and shallow water regimes

The crossover between the deep and shallow water regimes can be determined by equating (4) and (6): $gk \sim ghk^2$, that is, $\lambda \sim h$. As expected, the relevant physics changes when the wavelength becomes comparable to the depth. See figure 3.

Individual elements of water in the wave are moving around like crazy, rushing this way and that. As we go down into the water, at what depth is the

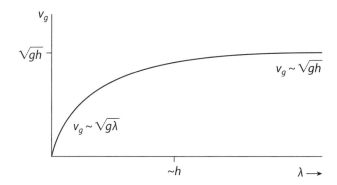

FIGURE 3. Group speed of water waves changes according to whether the wavelength λ is much larger or smaller than the depth h.

presence of a wave on the surface no longer felt? Intuition tells us that the length scale must be set by λ. Experience diving, or watching underwater nature films on television, suggests that the wave motion is exponentially attenuated: A couple of wavelengths beneath the surface, all is calm. What is happening at the surface hardly disturbs the creatures deep down below. In shallow water, however, the bottom sets some sort of boundary condition. So, h appears in (6) but not in (4).

To summarize:

$$\text{deep water:} \quad \omega^2 \sim gk, \qquad \text{shallow water:} \quad \omega^2 \sim ghk^2 \tag{7}$$

The art of interpolating: experience and sense

Let us continue our fly by night approach. Care to guess what $F(\xi)$ might be?

All we know is that $F(\xi) \to \xi$ as $\xi \to 0$, and $F(\xi) \to 1$ as $\xi \to \infty$. Mathematically, there are of course an infinite number of possible functions with these two limits. But invoking the dull function hypothesis, two educated guesses come to mind: $F(\xi) = \xi/(1+\xi)$ and $F(\xi) = \tanh \xi \equiv (e^\xi - e^{-\xi})/(e^\xi + e^{-\xi})$.

I am not saying that we could really decide, without doing some real work, which of these two possibilities is correct. But I am saying that experience and sense could help us quite a bit. Let's say you are offered the two functions in the preceding paragraph. Which one would you choose?

Well, while some undergraduates might choose the more elementary function, from my experience, the more sophisticated students would say that with all these waving exponentials $e^{i(k_x x + k_y y)}$ flying around, some decaying and growing exponentials might sneak in also. My earlier remark about those underwater nature films suggesting that wave motion is exponentially attenuated just a few wavelengths below the surface supports this possibility. So, the hyperbolic tangent might not be so outlandish. Let's see how this comes about.

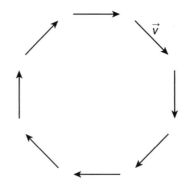

FIGURE 4. Vorticity and curl.

Laplace everywhere

I now interrupt the narrative to do a quick review of incompressible irrotational fluid flow.

First, as discussed in chapter VI.1, the incompressibility of water means that the flow velocity $\vec{v}$ satisfies $\vec{\nabla} \cdot \vec{v} = 0$. (Note that here we are inside the water, so to speak, and thus $\vec{v}$ is a 3-dimensional vector depending on $\vec{x} = (x, y, z)$, in contrast to the vector $\vec{k}$ considered thus far. As before, z denotes the vertical coordinate perpendicular to the surface of the water. Do not confuse $\vec{v}$ with either the group or phase velocity of the wave!)

Second, the flow is irrotational. In other words, vortices do not play a role in the problem. If you recall the notion of curl, you would see that vorticity $\vec{\Omega}$ is defined as the curl of the velocity field: $\vec{\Omega} \equiv \vec{\nabla} \times \vec{v}$. If you don't, simply draw* a picture of some fluid flowing around a point (see figure 4), and mentally compute $(\vec{\nabla} \times \vec{v})_z = \frac{\partial v_x}{\partial y} - \frac{\partial v_y}{\partial x}$. Saying that the flow is irrotational means $\vec{\Omega}$ vanishes.

Some readers might find helpful a well known analogy between incompressible irrotational flow and electromagnetism: Let $\vec{v} \to \vec{B}$, so that incompressibility $\vec{\nabla} \cdot \vec{v} = 0$ corresponds to absence of magnetic monopoles $\vec{\nabla} \cdot \vec{B} = 0$, and the absence of vorticity corresponds to absence of current and electric field $\vec{\nabla} \times \vec{B} = 0$.

The absence of vorticity, $\vec{\nabla} \times \vec{v} = 0$, implies that a potential ϕ exists such that $\vec{v} = \vec{\nabla}\phi$. But then incompressibility $\vec{\nabla} \cdot \vec{v} = 0$ requires ϕ to satisfy

$$\nabla^2 \phi = \frac{\partial^2 \phi}{\partial x^2} + \frac{\partial^2 \phi}{\partial y^2} + \frac{\partial^2 \phi}{\partial z^2} = 0 \tag{8}$$

Note that this governs what happens beneath the surface of the water.

---

*Another pet peeve: the typical physics undergraduate should draw more. You can master a lot of physics with diagrams! Think Feynman diagram, Penrose diagram, etc. A colleague remarks that he has the opposite pet peeve. Without equations, diagrams tell us little.

The bottom line is that Laplace's equation pops up, as it often does all over the place.

You know the reason, now that you have read chapters VI.1 and VI.2. In physics, symmetries pretty much dictate what equations are allowed to pop up. Rotational invariance[8] demands the Laplacian $\nabla^2 = \vec{\nabla} \cdot \vec{\nabla}$. So (8) is just about the only possible equation.[9]

What Laplace tells us: three negative numbers cannot sum to zero

So Laplace tells us in (8) that the three quantities $\frac{\partial^2 \phi}{\partial x^2}$, $\frac{\partial^2 \phi}{\partial y^2}$, and $\frac{\partial^2 \phi}{\partial z^2}$ add up to 0, which means that they can't all be positive or all be negative.

Given that $\vec{v}$ and hence ϕ is oscillatory in the x, y directions (that is, $\frac{\partial^2 \phi}{\partial x^2}$ and $\frac{\partial^2 \phi}{\partial y^2}$ are proportional to some negative constant times ϕ), we are forced to set $\frac{\partial^2 \phi}{\partial z^2}$ to some positive constant times ϕ. In other words, the z dependence of ϕ (and hence of $\vec{v}$) must be exponential, like some linear combination of, schematically, e^{+kz} and e^{-kz}. Indeed, I already snuck in the phrase "exponentially attenuated" as we go deep. So, we better have some exponentials flying around.

They are in fact flying around. For $\phi \propto e^{i(k_x x + k_y y)}$, Laplace's equation tells us

$$\frac{\partial^2 \phi}{\partial z^2} = -\left(\frac{\partial^2 \phi}{\partial x^2} + \frac{\partial^2 \phi}{\partial y^2} \right) = +(k_x^2 + k_y^2)\phi = +k^2 \phi \qquad (9)$$

The art of imagining ourselves calculating

For definiteness, let us choose coordinates such that z increases as we go deep down, with $z = 0$ at the surface. For ocean waves, with h effectively infinite, the e^{+kz} solution is excluded. In general, for finite h, both e^{+kz} and e^{-kz} are admitted, and we determine the correct linear combination by setting the z component of $\vec{v}$ to zero at the bottom, that is, at $z = h$.

So, I would not be surprised if a linear combination of e^{+kz} and e^{-kz} (that is, some hyperbolic function of z) pops up in a serious fly by day calculation. Somewhere along the way, when we fit to the boundary condition at $z = h$, this hyperbolic function turns into $\tanh hk$, with k required to appear for dimensional reasons. Of the two choices offered, a more sophisticated undergrad would go with $F(\xi) = \tanh \xi$, as I said, even though it "looks more complicated" than $F(\xi) = \xi/(1 + \xi)$. That turns out to be exactly right.

You hear exponential attenuation, you think hyperbolic!

We thus obtain

$$\omega^2 \sim gk \tanh hk \qquad (10)$$

I hope you enjoy watching the art of imagining doing a calculation without actually doing it.

It so happens that (10) is "exact." (In other words, we can replace the symbol $\sim$ by an equal sign.) Of course, we fly by night physicists wouldn't know this, unless some kindly fly by day types tell us. But what we do know is that, while there may well be factors of 2 here and there, we aren't missing any factors of 2π. This relates to the wisdom, already remarked on in chapter I.1, of using ω and k instead of period T and wavelength λ. I have seen books in which (10) is written in terms of T and λ with factors of 2π all over the place. The point is that when you plug a wave like $\sin(\omega t - \vec{k} \cdot \vec{x})$ into the fluid equation (given in appendix ENS), there aren't any factors of 2π lurking behind the bushes. Perhaps fly by night physics qualifies as an art form, or at least requires some sense.

Incidentally, with the dispersion in (10), the resulting expression for the group velocity $v_g = \frac{d\omega}{dk}$ involves two terms, so that the expression for the phase velocity $v_p = \frac{\omega}{k}$ is simpler. But in the two regimes where the power laws (7) hold approximately, the relation (3) between v_g and v_p still holds.

Loss of symmetry

This "calculation without drudgery" also explains our initial surprise at seeing the first power of k in the dispersion relation (4), contrary to our experience with electromagnetism and quantum mechanics. The reason is that we don't have our beloved and familiar 3-dimensional rotational invariance. The k actually comes from an equation of the form $\frac{\partial^2 \phi}{\partial t^2} + g\frac{\partial \phi}{\partial z} = 0$, which is perfectly invariant under rotations of the x- and y-axes into each other, but not under 3-dimensional rotations.

Much has been made of simplicity and beauty in our search for the fundamental laws of physics.[10] This chapter allows me to make two observations. First, the simplicity that fundamental physicists were astonished to find in the second half of the 20th century is in the laws of physics themselves, not in how they are manifested under restricted and local circumstances (for instance, a condensed state of nucleons and electrons sloshing around under gravity on a large planet, large compared to the sentient beings studying the waves). Second, we admire and appreciate the awesome variety of phenomena generated by the comparatively few laws of physics.

Exercises

(1) Calculate the phase and group velocities of deep water waves. Which is larger?

(2) Calculate the phase and group velocities of shallow water waves. Which is larger?

Notes

[1] I have already mentioned the advantage of ω over f, the general public's frequency.

[2] A colleague remarks that some poor freshmen spend hours learning the distinction between speed and velocity, and here I am confounding them some more.

[3] At this point, some would ask: what about $\frac{d^2\omega}{dk^2}$? This expression, being equal to $\frac{d}{dk}\frac{d\omega}{dk}$, has the wrong dimension.

[4] We don't even need the Navier-Stokes equation, since viscosity hardly plays a role.

[5] Actually not that difficult to do. See, for example, Trefil, *Introduction to the Physics of Fluids and Solids*, p. 68, and Landau and Lifshitz, *Fluid Mechanics*, pp. 37–39.

[6] The possibility of $F(\xi) \to \infty$ hardly bears consideration.

[7] Indeed, in class, I can almost always count on some wise guy asking me this.

[8] The definition $\vec{v} = \vec{\nabla}\phi$ tells us that ϕ is even under parity and odd under time reversal.

[9] You might wonder about the time dependence of ϕ: The answer is that we have already taken it out, having implicitly written $e^{i\omega t}\phi$.

[10] Indeed, whole books have been written on this topic. See, for example, *Fearful*.

A physicist at the seashore

Waves at the seashore

A physicist goes to the beach and wonders why waves always come in parallel to the shore. As we can see from figure 1, it is due to refraction. Indeed, you are able to watch the waves come in by depending on the refraction of light as it goes through the lens in your eyeball that your sainted mother made for you. As the wave goes into shallow water, v_g decreases (as you may recall from chapter VIII.1 or by looking at figure 3 later in this chapter). The wave slows down and is thus obligated to bend, just as light going into your lens does. The laws of physics hold always.

Travel time for ocean waves to reach the beach

Waves at the beach are typically generated by distant storms at sea rather than locally. Suppose that at the beach, you observe waves coming in every 6 seconds. Furthermore, you heard that there was a storm at sea 500 km away. How long have those waves been traveling?

Since some fly by day types told us in chapter VIII.1 that we can replace the $\sim$ sign in the dispersion relation we obtained by the $=$ sign, we might as well do that for numerical estimates in this chapter. Let me remind you that we had

$$\omega^2 = gk \tanh hk \tag{1}$$

and more usefully,

$$\text{deep water}: \quad \omega^2 = gk, \qquad \text{shallow water}: \quad \omega^2 = ghk^2 \tag{2}$$

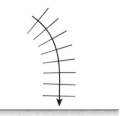

FIGURE 1. Refraction of water waves at the seashore.

Since these waves have been out at sea most of the time until they approach shore, we use the deep formula $\omega^2 = gk$, which converts to $\lambda = gT^2/2\pi \simeq$ $(10 \text{ m/sec}^2)(6 \text{ sec})^2/6 \simeq 60$ m $\ll$ depth of ocean. The group velocity is then $v_g = \frac{1}{2}\sqrt{g/k} = \frac{1}{2}\sqrt{g\lambda/2\pi} \simeq \frac{1}{2}\sqrt{10 \times 60/6} \text{ m}^2/\text{sec}^2 \simeq 5$ m/sec. Thus, these waves have been traveling for $\tau \simeq 500 \times 10^3/5 \text{ sec} \sim 1$ day.

Beating between waves

You might have noticed that waves at the beach often come in sets, with a large wave followed by smaller waves and then larger waves, and then the cycle repeats. Looking at the figure in appendix Grp, you would see that this is more or less due to a beating phenomenon between two waves close to each other in frequency and wave number. The number of crests between one large wave and the next large wave is shown there to be given by $\sim (\frac{1}{2\Delta k}/\frac{1}{k}) = k/2\Delta k$.

From observing the number of waves between two large ones, you could estimate the area over which the storm occurs. The reasoning goes as follows. Call the distance from shore to the storm L, and the size (that is, the length scale, not the strength) of the storm ΔL. See figure 2. A wave from the "back" has to move faster than a wave from the "front" in order to get to shore at the same time, thus $\frac{L+\Delta L}{v_g+\Delta v_g} \sim \frac{L}{v_g}$, that is, $\frac{\Delta L}{L} \sim \frac{\Delta v_g}{v_g}$, from which we can solve for ΔL. Rough equality of quasi-logarithmic derivatives holds. For ocean waves, $v_g = \frac{1}{2}\sqrt{g/k}$, and thus $\Delta v_g/v_g \sim -\Delta k/2k$. Ignoring irrelevant signs, we find that $k/2\Delta k \sim L/4\Delta L$. Let's say that we observe 10 smaller waves between two large ones. Then $L/\Delta L \sim 40$. If the storm occurred 400 km away, then its width is approximately 10 km.

Waves slowing down

As a wave comes to shore, from deep water to shallow water, its dispersion changes, and correspondingly, so does the group velocity v_g of the wave. See figure 3.

This figure looks deceptively like figure 3 in chapter VIII.1, but there v_g was plotted as a function of wavelength λ; here it is plotted as a function of depth h.

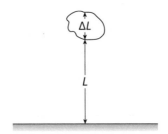

FIGURE 2. A storm with diameter $\sim \Delta L$ at a distance L from shore; this figure is evidently not to scale.

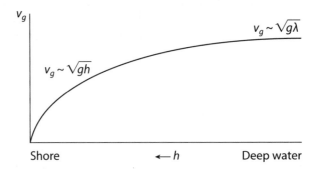

FIGURE 3. The group velocity v_g of a water wave decreases as it comes to shore.

The figure explains what we see qualitatively. Incoming water waves slow down and pile up. The amplitude of a wave thus increases until it breaks.

To be more quantitative, denote by ε the energy per unit area contained in a wave. (By area, we mean the area parallel to the surface of the earth.) Let ζ, positive or negative, denote the vertical displacement of the wave, namely, its amplitude. By elementary physics, the potential energy of a mass m raised to a height of ζ is equal to $mg\zeta$, and so the potential energy of the mass of water raised by the wave equals $\rho(A\zeta)g\zeta$, with A the area under the wave. Up to some factor of 2, we thus obtain $\varepsilon \sim \rho g \zeta^2$. The energy flux F (namely, the energy delivered by the wave per unit length of wave front per unit time), with dimension of E/LT, is by dimensional analysis given by ε times the group velocity, hence $F \sim \varepsilon v_g$.

By energy conservation, F stays the same as the wave comes to shore, implying that $\zeta^2 \propto \varepsilon \propto 1/v_g \propto 1/h^{\frac{1}{2}}$. Thus, we obtain

$$\zeta \propto \frac{1}{h^{\frac{1}{4}}} \tag{3}$$

The vertical displacement of the wave increases as h decreases, as expected.

The implicit approximation here is that, while (2) and (1) were derived with h treated as constant, we can still apply them where h, though changing, is changing slowly on the scale of λ. Evidently, we would expect more dramatic effects when h changes abruptly. In real life, the near-beach submarine terrain, in spite of eons of wave action, is hardly expected to be smooth everywhere. Besides a steady offshore wind, surfers favor locales with a sudden change in h.

In the linear regime

The reader might be wondering how we have gotten this far without even mentioning the equation governing fluid flow, as given and explained in appendix ENS. The answer is that we have been working in the linear regime, treating the amplitude ζ of the wave as small compared to the wavelength λ. In this regime, all physical quantities are oscillating gently according to $e^{i(\omega t - \vec{k} \cdot \vec{x})}$.

Denote the velocity field of the fluid by $\vec{v}(t, \vec{x})$. Note that $\vec{v}$, which specifies the velocity v of an infinitesimal element at the point $\vec{x}$ at time t, is not to be confused* with the group velocity v_g of the wave, nor with the phase velocity v_p, which describe the collective motion of the water. The fluid flow equation, which is just Newton's $\vec{a} = \vec{F}/m$ in disguise, equates the acceleration of the infinitesimal fluid element $\frac{\partial \vec{v}}{\partial t} + (\vec{v} \cdot \vec{\nabla})\vec{v}$ to the driving force per unit mass. (Readers needing a review should turn to appendix ENS.)

As you probably know, the notorious difficulty of fluid dynamics is due to the term $(\vec{v} \cdot \vec{\nabla})\vec{v}$. While the term $\frac{\partial \vec{v}}{\partial t}$ is meekly linear in $\vec{v}$, this term is defiantly not.

When water waves go nonlinear

In the linear regime, various quantities oscillate nicely. For example, the vertical displacement varies like $\zeta(t, \vec{x}) = \zeta_0 e^{i(\omega t - \vec{k} \cdot \vec{x})}$. What about the fluid velocity $\vec{v}$? Call the direction of the wave the x-axis and the vertical direction the z-axis (as we have been doing). For simplicity, suppress the dependence on y in the perpendicular direction. Then $\vec{v}(t, x, z) = (v_x(t, x, z), 0, v_z(t, x, z)) = \vec{v}_0(z)e^{i(\omega t - kx)}$, with $\vec{v}_0$ decreasing exponentially in z as we go beneath the surface.

Ordinarily, the velocity v is considerably less than v_g and v_p, in accordance with common observations out on the sea or a lake: a wave can go by rapidly with some small debris in the water bobbing around slowly. The undergrads in my class can relate more easily to a stadium wave: a stadium wave zooms around the sports stadium much faster than the humans in it could move.

So, when are we forced out of the linear regime? Clearly, things go nonlinear when $(\vec{v} \cdot \vec{\nabla})\vec{v}$ becomes comparable to or larger than $\frac{\partial \vec{v}}{\partial t}$. This condition is analyzed in appendix ENS. Not surprisingly, we learn there that the linear approximation fails when the fluid velocity v becomes comparable to or larger than v_g and v_p. We also learn that this condition can also be written as the amplitude of the wave ζ becoming comparable to or larger than the wavelength λ, which also accords with our intuition.

---

*As you have already been warned in chapter VIII.1.

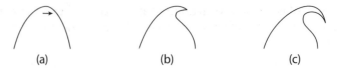

FIGURE 4. As the water in the crest moves faster than the wave can move, at some point the wave curls over and breaks.

After this long excursion, we are back at the beach looking at incoming waves. As the wave approaches shore, ζ increases, and the wave starts to go nonlinear, with $v \gtrsim v_g, v_p$. At the crest, water is trying to move faster than the wave can move, as shown in part (a) of figure 4. Compelled to move ahead, the crest curls over, and the wave breaks (parts (b) and (c) of figure 4). Under some circumstances, the tip of the wave can curl over so much that it forms a quasi-circular tunnel that surfers delight in.

To close this section, I want to show you that we can easily estimate v by using dimensional analysis plus a little bit of sense. We expect v to depend on the vertical displacement ζ, the frequency ω, and the acceleration g due to gravity. So write $v \sim \zeta^a \omega^b g^c$. Right off the bat, it seems that dimensional analysis would not suffice, since only L and T are involved but not M. We get only two equations $a + c = 1$ and $b + 2c = 1$ for our three unknowns. But our earlier remarks about debris bobbing around slowly indicates that for small ζ, we expect $v \propto \zeta$, so that $a = 1$. Thus, we obtain[1]

$$v \sim \omega \zeta \qquad (4)$$

The condition $v \gtrsim v_g$ for the wave to go rogue therefore translates to $\zeta \gtrsim \lambda$, as we already know.

With your experience playing around with sines and cosines, you would not be at all surprised if I tell you that, for $v_x = C e^{kz} \sin(\omega t - kx)$ with C some constant, then $v_z = C e^{kz} \cos(\omega t - kx)$. Indeed, this form is consistent with the incompressibility $\vec{\nabla} \cdot \vec{v} = 0$ of water. Thus, beneath the surface, individual elements of water just trace out circles with radius decreasing exponentially with depth. Well, come on, everything in classical physics makes sense.

Tsunami

In the area where I live, signs at the seashore warn people to move to higher ground in case of tsunami, a Japanese word meaning "harbor wave."[2] I suppose that there are similar signs all around the Pacific rim. Tsunamis are caused by earthquakes (for example, the Cretan tsunami[3] of 365, which was well documented in antiquity). But for definiteness, let's say an earthquake occurs in Japan, generating a tsunami that traverses the Pacific and hits the west coast[4] of North America.

Normally, we think of the ocean as exceedingly deep, but counterintuitively, for studying tsunamis, we should use the shallow water approximation. The average depth of the Pacific Ocean is "only" a bit over 4 km. On the other hand, the wavelength of a tsunami is set by some characteristic length scale of the earthquake fault, which may be of order $\sim 10^2$ km or more. Thus, we should use $v_g \sim \sqrt{gh}$ from the shallow water regime in (2).

You are already plugging in some numbers? But wait! For us to see that this is a highly respectable speed, we don't even have to do that. Equating the kinetic energy of a falling object to its potential energy per unit mass $\frac{1}{2}v^2 = gh$, we see that the tsunami speed is just the terminal speed of an object falling from a height of several kilometers in the absence of air resistance. That's pretty fast!

Dutifully plugging in numbers, I get $v_g \sim \sqrt{gh} \sim \sqrt{10 \times 4000}$ m/sec = 200 m/sec ~ 720 km/hour. This is comparable to the speed of a commercial jet liner.[5] So the time to cross the Pacific is of order 10 hours.

Exercise

(1) Ships at sea reportedly do not even notice a tsunami going by. Explain.

Notes

[1] For a detailed analysis, which is not that difficult, by the way, since we are talking about linear physics, see for example Landau and Lifshitz, *Fluid Mechanics*, pp. 37–39.

[2] The Chinese word means "sea roaring."

[3] https://en.wikipedia.org/wiki/365_Crete_earthquake.

[4] Consider Crescent City's claim to fame: https://en.wikipedia.org/wiki/Crescent_City, California.

[5] Hence there is no need, for anybody with experience with long distance flights, to look up the width of the Pacific Ocean (as is done in some books I look at) to have a sense of how long it would take a tsunami to traverse the Pacific.

Surface tension and ripples

New physics kicks in at large k

High energy physicists often ask for larger and larger accelerators, arguing that new physics would kick in at high energies and momenta, as already mentioned in chapter IV.4. According to de Broglie, or the uncertainty principle, a beam with momentum p would allow us to probe physics at a distance scale $\sim \hbar/p$.

Water waves offer a parable of this worldview, except that exorbitant funds are not needed. We only have to take a closer look. Do we expect the dispersion $\omega^2 \sim gk$ to continue to hold at high k (and hence high ω)? Or will we find new physics?

You bet. New physics, known as ripples, kick in. At large k (that is, small wavelength λ), the dispersion $\omega^2 \sim gk$ fails. For waves with wavelengths smaller than a certain characteristic λ_c to be determined below, surface tension becomes important. Ripples (namely, waves at high k) are known as capillary waves.

Surface tension

Surface tension, denoted by γ, is characterized by an energy per unit area:

$$[\gamma] = E/L^2 \tag{1}$$

(More precisely, surface tension is a property of the two fluids sharing the interface. If left unspecified, we presume that one of the fluids is air.) For example, at room temperature, for water and air, $\gamma \simeq 72$ erg/cm^2, for mercury and air, $\gamma \simeq 550$ erg/cm^2.

Fluids, including water, seek to minimize their surface area. Think of a water droplet. Or a soap bubble.

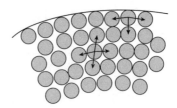

FIGURE 1. The molecules at the center of the social scene have more friends than those on the periphery.

Amusingly, surface tension originates in the molecular character of water. Since molecules attract one another, each molecule seeks to have as many neighbors as possible. Nobody wants to be on the surface. I tell my class that the typical undergrad, similarly, does not want to be on the periphery of a social group (figure 1).

Surface area therefore costs energy. In principle, physicists studying water waves could have discovered new physics in the form of molecules by going to ever higher wave number k, just like particle physicists going to higher $\hbar k$.

How ripples disperse

Picture the crest of a wave. The curved surface extends over a larger area than would be the case were the surface flat. For small enough ripples, surface tension dominates over gravity, which we will now ignore. Referring back to the discussion in chapter VIII.1, we see that ρ, the density of water, will play a role if gravity is no longer the driving force.

Our task is to relate ω and k, neither of which involves mass. Thus, the dimension of mass contained in γ and ρ must cancel out, which implies that the ratio γ/ρ is what appears in the dispersion relation. Now observe that $[\gamma/\rho] = (E/L^2)/(M/L^3) = (E/M)L = (L/T)^2 L = L^3/T^2$. We conclude that, in the surface ripple regime,

$$\omega^2 \sim \frac{\gamma}{\rho} k^3 \tag{2}$$

Interestingly, ω^2 now goes like k^3, not k.

The appearance of a k^3 may be puzzling (at least it was to me at first sight). Once again, as explained in chapter VIII.1, this is due to the earth's gravity breaking 3-dimensional rotational invariance into 2-dimensional rotational invariance. The z-direction, otherwise known as up and down, is privileged. A bit of thought might convince us that it must come from a term like $\frac{\partial}{\partial z}(\frac{\partial^2}{\partial x^2} + \frac{\partial^2}{\partial y^2})\phi$ (with ϕ some velocity potential). And indeed it does.[1]

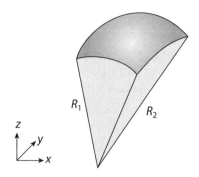

FIGURE 2. Laplace's law.

Cross over from waves into ripples

The fly by night physicist guesses reasonably that, once out of the shallow water regime, waves behave like[2]

$$\omega^2 \sim gk + \frac{\gamma}{\rho}k^3 = gk\left(1 + \frac{\gamma}{\rho g}k^2\right) \tag{3}$$

The cross over from waves to ripples occurs at a critical wavelength

$$\lambda_c \sim \sqrt{\frac{\gamma}{\rho g}} \tag{4}$$

Numerically, for water, $\lambda_c \sim \sqrt{\frac{72 \text{ ergs/cm}^2}{1 \text{ g/cm}^3 \cdot 10^3 \text{ cm/s}^2}} \sim 0.3$ cm

Laplace's law

For definiteness, focus on the crest of a capillary wave. Due to surface tension, the pressure P_w on the water side exceeds the pressure P_a on the air side. To find the pressure difference $\Delta P = P_w - P_a$, the fly by night physicist once again appeals to dimensional analysis. See figure 2.

The difference ΔP is resisted by surface tension and the curvature of the wave crest. So write $\Delta P \propto \gamma$. Since $[\Delta P] = E/L^3$ and $[\gamma] = E/L^2$, we need to divide γ by a length. Well, we're looking at you, radius of curvature. Surface being a 2-dimensional entity, there are two radii of curvature R_1 and R_2, one along the x direction and the other along the y direction so to speak, as indicated in the figure. We thus obtain Laplace's law

$$\Delta P = P_w - P_a = \gamma\left(\frac{1}{R_1} + \frac{1}{R_2}\right) \tag{5}$$

That marquis has got to be the smartest marquis ever! Notice that we can even allow ourselves the luxury of using an equal sign here: we can take Laplace's law as the definition of γ.

How did Laplace know to combine the two radii of curvature as in (5)? For instance, a student might wonder about $\frac{1}{\sqrt{R_1 R_2}}$ instead of $(\frac{1}{R_1} + \frac{1}{R_2})$, which is certainly allowed by dimensional analysis alone. We can rule out that option by taking the limit $R_1 \ll R_2$, in which case the surface becomes flat in one direction. The problem should reduce to an essentially 1-dimensional problem with $\Delta P \to \gamma/R_1$. (Recall that the usefulness of taking limits was discussed back in part I.)

A somewhat "deeper" answer is provided by rotational invariance, as we will now see. Let's denote the vertical displacement of the water by $\zeta(x,y)$ as in chapter VIII.2.

What are the two radii of curvature at (x, y)? Let's start out simply. Instead of a surface, think of a curve described by $\zeta(x)$. From elementary calculus, we know that $\frac{d\zeta}{dx}$ measures the slope, and $\frac{d^2\zeta}{dx^2}$ measures how the curve curves. Now jump from curve to surface. By democracy (or more seriously, by rotational symmetry), we know that the sum $\frac{\partial^2 \zeta}{\partial x^2} + \frac{\partial^2 \zeta}{\partial y^2}$ (at least for ζ small) must describe curvature, namely, the sum of the inverses of the two radii of curvature. Right dimension? Check: $[\frac{\partial^2 \zeta}{\partial x^2}] = L/L^2 = 1/L$.

Thus, the pressure difference in Laplace's law is given by none other than the Laplacian:

$$\Delta P = \gamma \nabla^2 \zeta \tag{6}$$

That marquis would be a shoo-in for the Nobel Prize if it had existed back then!

Water striding insects

For a wave with the dispersion $\omega \propto k^\alpha$ with $\alpha \leq 1$, the phase velocity is greater than or equal to the group velocity $v_g = \alpha v_p$, but for capillary waves, $\alpha = \frac{3}{2}$, and so $v_g = \frac{3}{2} v_p > v_p$. It has been reported[3] that water striders, insects who take advantage of surface tension to walk on water, use this fact to communicate with one another. They can even tell if the wave is produced by a male or female. (Wait, we could do the same with sound waves. So perhaps no need to be impressed.)

The magic of water waves

As we have seen in this and the preceding chapters, the physics of water waves is in many ways richer than that of electromagnetic waves. In particular, Lorentz invariance fixes the dispersion of electromagnetic waves to be $\omega = ck$ for all k, and that's that. In contrast, the dispersion of water waves exhibits three different regimes (see figure 3). There are waves in the middle of the sea, as happy as can be. But when the wavelength exceeds the depth, the wave starts scraping the bottom, and so, sensing the boundary condition, it is obliged to change its behavior. In the other extreme, when the wavelength gets small,

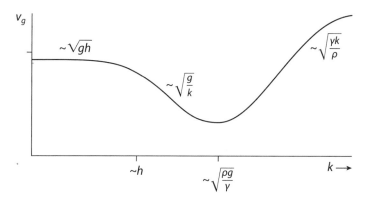

FIGURE 3. Schematic plot of the group velocity of water waves, exhibiting three regimes.

gravity cedes its control to surface tension, and the wave becomes a ripple, vaguely reminding us of water's molecular origins.

Let us now look more closely at the transition. In the absence of surface tension, v_g (which, within a factor of order 1, is the same as v_p) would decrease to 0 like $\sim 1/\sqrt{k}$ with increasing k. But in reality, molecular attraction pulls v_g back up, so that there is actually a minimum velocity

$$v_{\min} \sim \left(\frac{g\gamma}{\rho}\right)^{\frac{1}{4}} \tag{7}$$

Since this results from the contest between gravity and surface tension, we could have readily obtained $v_{\min}$ by dimensional analysis. Let's check: $[v_{\min}] = \left(\frac{L}{T^2}\frac{E}{L^2}\frac{L^3}{M}\right)^{\frac{1}{4}} = \frac{L}{T}$. Indeed.

All this is within the linear regime, of course. In a howling wind with water waves going nonlinear, the very concept of dispersion relation gets blown away.

Exercise

(1) Estimate $v_{\min}$.

Notes

[1] See, for example, Landau and Lifshitz, *Fluid Mechanics*, p. 238.
[2] This relation happens to be exact. See Landau and Lifshitz, *Fluid Mechanics*, p. 238.
[3] M. Denny, *Air and Water*, p. 287.

From dripping faucets to mammalian lungs and water striders

Water drops

We learned about surface tension and capillary waves in chapter VIII.3. Here we mention some examples of them at work in everyday life.

Observe a slowly dripping faucet. We see a water drop slowly filling up, pregnant with possibilities. Surface tension wages a mighty, but ultimately futile, struggle against gravity. Finally, gravity wins. A drop of water, almost spherical but not quite, falls.

Can you estimate the radius r of the droplet? The force holding the water against a fall is of order γr. Against this tugs the ever-present gravity, with a force $(\rho r^3)g$. Equating these, the fly by night physicist obtains

$$r_{\text{droplet}} \sim \sqrt{\frac{\gamma}{\rho g}} \tag{1}$$

Perhaps you recognize that this is just the critical wavelength λ_c at which ripples start to dominate water waves, as discussed in chapter VIII.3. Indeed, if we were to quibble over factors of 2 and such, we might venture to guess that critical radius r_{droplet} is about a quarter of the wavelength λ_c.

This is not a book about experimental physics, but I can't resist asking you how to measure the radius of a droplet coming out of a slowly dripping faucet.[1]

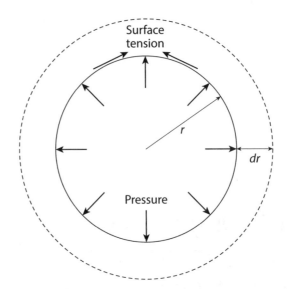

FIGURE 1. An alveolus being expanded from radius r to $r + dr$. The tube bringing air into the alveolus is not shown. Adapted from Denny, M. *Air and Water: The Biology and Physics of Life's Media*. Princeton University Press, 1993.

Entertaining children in restaurants

One trick that little kids like to play in restaurants involves sucking a favorite beverage up a straw[2] and then capping the top end of the straw with a thumb. Hold it over some victim and suddenly let go of the thumb; the fluid falls out in a fast downward stream. Always a fun thing* to do and see!

Estimate the maximum radius of the straw for which this is possible. Clearly, there is a maximum. Do not invert a glass of water in a restaurant! Surface tension would be no match against gravity in that case.

Mammalian lungs

You might recall from an elementary course on biology that mammalian lungs, including ours, have a treelike structure, branching repeatedly and ending in small spherical cavities called alveoli. With each inhalation, the alveoli are filled with fresh air, and oxygen is delivered by diffusion into the capillaries that surround the alveoli.[3]

Work is done in expanding an alveolus from radius r to $r + dr$ against the surface tension of the fluid (some kind of pulmonary surfactant secreted by various cells) covering the alveolus. An idealization of the process is depicted in figure 1.

---

*And an opportunity to tell them about molecules.

Since we know by heart the volume and surface area of a sphere, we might as well include the factors of 4π and whatnot. Thus, the work done is given by $PdV = Pd(\frac{4\pi}{3}r^3) = 4\pi Pr^2 dr$, while the increase in surface energy equals $\gamma d(4\pi r^2) = 4\pi(2\gamma r dr)$. Equating, we obtain

$$P = \frac{2\gamma}{r} \tag{2}$$

Here we can afford to write equal signs!

We see an evolutionary trade-off here. For a given total volume of alveoli, $r^3 \propto 1/N$, with N the number of alveoli. The smaller the alveoli are, the more surface area ($\sim Nr^2 \propto 1/r$) is available for oxygen exchange. But then more work is involved in breathing. As alluded to above, evolution has also come up with surfactants that reduce the surface tension and lessen the effort to breathe.

When we wrote down Laplace's law $\Delta P = \gamma(\frac{1}{R_1} + \frac{1}{R_2})$ in chapter VIII.3, we indulged in the luxury of writing an equal sign. Since $R_1 = R_2 = r$ for a sphere of radius r, we see, by comparing with (2), that we actually got the coefficient right.

Notes

[1] If you say that you would grab a ruler and a magnifying glass, then you are not destined to be an experimentalist. The correct answer is to count the number N of drops falling into a glass and measure the volume of water in the glass in the large N approximation.

[2] As I write this, plastic straws are being outlawed in some developed countries in the fight against ocean pollution. So this scene might soon fade into history. Paper straws?

[3] See, for example, https://en.wikipedia.org/wiki/Pulmonary_alveolus#Diseases. I quote: "A typical pair of human lungs contains about 700 million alveoli, producing 70 m^2 of surface area. Each alveolus is wrapped in a fine mesh of capillaries covering about 70% of its area."

Drag, viscosity, and Reynolds number

Stokes drag

What is the force F needed to keep a sphere of radius a moving with velocity v through an incompressible fluid of density ρ and (kinematical) viscosity ν?

By appealing to Galileo invariance, we can also state the problem as the pressure needed to keep a fluid flowing steadily past a sphere sitting at rest. This is actually the preferred way of looking at the problem. The boundary conditions are that the fluid velocity approaches the prescribed velocity at spatial infinity and vanishes on the surface of the sphere (see figure 1).

This problem was famously solved by Stokes[1] in 1851.

Instead of the exact answer, the fly by night physicist would now try to obtain F using dimensional analysis. Note that the mass (or equivalently, the density of the sphere) does not enter. We have to push the fluid past the sphere; whether the sphere is hollow or filled with lead does not matter. In the second way of stating the problem, the sphere simply provides a boundary condition.

At first sight, dimensional analysis would be inadequate, since we have four variables, with dimension $[a] = L$, $[v] = L/T$, $[\rho] = M/L^3$, and $[\nu] = L^2/T$, and so we are one equation short. Indeed, note that the combination $Re \equiv va/\nu$ is dimensionless, known as the Reynolds number. Thus, any answer we find could be multiplied by an unknown function of Re.

Reynolds number

The notion of Reynolds number[2] is not restricted to Stokes's problem. More generally, recall the Navier-Stokes equation derived in appendix ENS. Consider the ratio of the inertial term $(\vec{v} \cdot \vec{\nabla})\vec{v}$ to the viscous term $\nu \nabla^2 \vec{v}$:

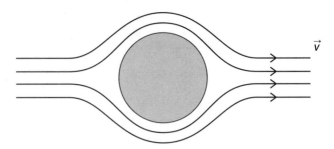

FIGURE 1. Fluid flow around a sphere.

$$\frac{\text{inertial}}{\text{viscous}} \sim \frac{v^2/l}{vv/l^2} = \frac{vl}{v} \tag{1}$$

Here l denotes some length and v some velocity characteristic of the flow.

The Reynolds number

$$Re \equiv \frac{vl}{v} \tag{2}$$

thus measures the relative importance of inertia to viscosity.

For $Re \ll 1$ (that is, for low velocity or high viscosity), the flow is laminar (that is, regular and amenable to analytic treatment). Stokes's solution is strictly for low Reynolds number.

In contrast, for $Re \gg 1$ (that is, for high velocity or low viscosity), the flow tends to become turbulent and nasty.[3] As surely you have heard, mastering turbulence is one of the outstanding unsolved problems of physics (and engineering). One way of remembering this is to picture the flow of highly viscous fluid like maple syrup, which is rarely if ever turbulent.

By definition, the Reynolds number is not a precise quantity. At this point, invariably, questions about how one defines a characteristic length are asked. The answer is that usually it is self-evident; in Stokes's problem, it is the radius of the sphere, and it does not matter whether you take the radius or the diameter. All we care about is whether $Re \ll 1$ or $Re \gg 1$. From my experience, some undergrads have trouble with fuzzy concepts[4] such as the Reynolds number. To gain some feel for Re, it is probably best to go through a few numerical examples, as in the exercises.

Determining the drag by dimensional analysis

Given this background, we will now fly by night to get to Stokes's formula for the drag on a sphere even though we are confronted by four variables.

First, with $[F] = ML/T^2$, we immediately deduce that $F \propto \rho$, since ρ is the only one of the four containing M.

Now the crucial observation is that $\vec{F}$ is actually a vector, and since $\vec{v}$ is the only other vector around, we must have $\vec{F} \propto \vec{v}$. (In other words, we are using rotational invariance and the isotropy of the fluid.) By assumption, we are dealing with low velocity, and thus F depends on v linearly.

With $[\rho v] = (M/L^3)(L/T) = M/L^2T$, in order to get two inverse powers of T to form the force F, we are compelled to multiply ρv by v. Finally, matching the power of L, we obtain

$$F \sim \rho v a v \tag{3}$$

Stokes determined the overall coefficient to be $6\pi \simeq 20$, and so in this example, dimensional analysis cannot be trusted numerically. However, dimensional analysis does produce the interesting result that the drag force is proportional to a, rather than to the "frontal" area $\sim a^2$. Physically, the fluid can flow around the sphere, which certainly is not like a flat disk of radius a facing the fluid head on.

Terminal velocity

In elementary physics, a falling object enjoys a constant acceleration g and so would soon attain an astounding velocity,[5] but of course in real life, there is air resistance, which may or may not be important, depending on the size and mass of the object.

Let ρ_b denote the density of the falling body, and ρ_f the density of the fluid (here air). The falling body reaches a terminal velocity v_t when Stokes drag, which grows with velocity, equals the downward pull of gravity: $\rho_f v a v_t \sim mg \sim \rho_b a^3 g$, thus giving the terminal velocity:

$$v_t \sim \left(\frac{\rho_b}{\rho_f}\right)\left(\frac{ga^2}{v}\right) \tag{4}$$

If the fluid is water, then buoyancy must be taken into account in calculating the terminal velocity, but for air, buoyancy is normally negligible.

Be cautious plugging numbers into (4)! See exercise (2). You have to first check that the Reynolds number is small enough for Stokes drag to be applicable. With turbulent flow for Reynolds number of order 1 and larger, you would have to consult empirical formulas and plots.[6]

Since gravity scales like a^3, while drag goes like a, terminal velocity decreases with size like a^2. So teeny bits of stuff, such as dust and bacteria, drift down extremely slowly. Fine dust in air pollution can remain suspended for a long time.

The density of most biological organisms is essentially that of water, and so for falling through air, in (4), $\rho_b/\rho_f = \rho_{\text{water}}/\rho_{\text{air}} \simeq \frac{1\,\text{g/cm}^3}{1\,\text{kg/m}^3} = 10^3$. Since $g/v_{\text{air}} \simeq \frac{10^3\,\text{cm/sec}^2}{1.5\,\text{cm}^2/\text{sec}} \simeq 6 \times 10^2/\text{cm sec}$, we obtain $v_t \simeq (60\,\text{m/sec})(a/\text{mm})^2$.

Breaking up the problem

The fly by night physicist learned an important lesson from this calculation of the terminal velocity, a lesson equally valuable in many life situations. Break the problem into separate problems, if at all possible!

If we were told to calculate the terminal velocity of a falling object, we might have been flustered by the presence of a dimensionless ratio (ρ_b/ρ_f) and three quantities a, g, and ν involving L and T out of which we needed to construct a velocity. It might appear that we would have to fly by day.

Life at low Reynolds number

A much celebrated paper[7] in (relatively) contemporary physics is Ed Purcell's "Life at Low Reynolds Number," in which the reader is introduced to the strange world inhabited by microscopic organisms. Due to their small sizes, they are faced with the difficulty of locomotion at low Reynolds numbers. Purcell asked the reader to imagine swimming in a pool filled with maple syrup while being forbidden to move any part of your body faster than say 1 cm per minute.

Exercises

(1) What is the Reynolds number for (a) a human walking through air, (b) a bird flying, (c) a mosquito flying, and (d) sap flowing through the xylem of a tree?

(2) Estimate the terminal velocity of falling (a) bacteria and (b) raindrops.

(3) How is the expression for terminal velocity given in (4) modified if buoyancy is important? Apply this to plankton.

(4) Since you know the volume of a sphere and since I told you about the 6π in Stokes's result, you could determine the overall coefficient in the expression for terminal velocity given in (4).

Notes

[1] The textbook solution takes three pages (see pp. 64–66) in *Fluid Mechanics*, by Landau and Lifshitz, so it is not exactly an easy problem for undergraduates.

[2] According to Wikipedia, "The concept was introduced by Sir George Stokes (1819–1903) in 1851, but the Reynolds number was named by Arnold Sommerfeld in 1908 after Osborne Reynolds (1842–1912), who popularized its use in 1883." Well, Stokes had enough stuff named after him, but still, he looked quite angry. See https://en.wikipedia.org/wiki/Reynolds_number.

[3] These days, you can easily access many images of the transition from laminar to turbulent flow on the web.

[4] Strange in a way, since, like me at their age, they surely have taken courses outside of physics and mathematics.

[5] Recall our discussion of tsunamis in chapter VIII.2.

[6] For example, Denny, *Air and Water*, pp. 115–116.

[7] E. M. Purcell, *American Journal of Physics* 45 (1977).

Part IX

From private neutrinos to charm

Prologue to Part IX

I debated with myself for a long time about whether to include the material in part IX. The stuff about general relativity in chapter VII.3 is already a bit of a stretch. I chatted with undergrads, including myself as I was when I was an undergrad. I remember my younger self impatient to skip over thermodynamics and such, and, by simply ignoring various courses (such as a dreaded course on optics) required for graduation, jumping ahead to the "good stuff," neutrinos and quarks and all that. But in the end, it was chats with real undergrads,[1] not with my undergrad self, that convinced me to include a part IX, at the risk of losing some readers. I suggest that readers already a bit shaky with the rest of this book should skip this last part[2] and perhaps come back to it eventually.

Education, at least in physics, mandates that we tell the young and gullible only that which is true, precise, and proven. But on the frontiers of research, the intrepid is often confronted with the false or at least the possibly false, the nebulous, and the implausible. Particle physics[3] in its golden period, from the 1950s until the 1970s, during which the strong and weak interactions were elucidated and unified with electromagnetism, offers many vivid examples of the fly by night approach. The reason was that the theory was totally unknown until the early 1970s, and therefore much of the best work consists of inspired guesses. There were no equations to be solved, even if you wanted to solve one. Nobel Prizes were earned on leaps of faith. Under such circumstances, fly by night physics is often the only guide.

I offer you here three out of a wealth of examples, which I describe as (1) the weak interaction crying out (chapter IX.2), (2) each with a private neutrino (chapter IX.3), and (3) the smoothness of charm (chapter IX.4).

But first I have to sketch some essential tools of the trade. As Feynman said when he tried to explain strange particles to alleged freshmen in his famous lectures: "I have to cut some corners." I also inevitably have to cut a bunch of corners. Fortunately, by the very nature of this book, cutting corners is the name of our game. In places I am compelled to simply state a result. You just have to take my word for it, but I will always refer you to a derivation in some textbook, or at least give some handwaving argument to make the stated result sound plausible. On the other side of the coin, I necessarily have to fill you in on the historical background a bit. I have to be brief and can only encourage you to read more about this arduous but fascinating struggle.

I admit that the going might be tough for some undergrads. Just keep in mind that my goal is to give you a flavor, an impressionistic understanding, merely to whet your appetite so as to entice you to read more detailed treatments.

Notes

[1] Including Ashley Ong, my undergraduate assistant. See the preface.

[2] One primitive on Amazon thought that the last part of *Group Nut* was too advanced for him. Obvious response: Why not read a more elementary book? Alternatively, skip that part and come back to it later. Or hack that part off and give it to a friend.

[3] For an introduction suitable for the general public, I recommend Y. Nambu, *Quarks: Frontiers in Elementary Particle Physics*, World Scientific, 1985. For a more detailed survey, see, for example, *Fearful*.

A lightning introduction to particle physics and quantum field theory

Yes. I was seeing something in space and time. There were quantities associated with points in space and time, and I would see electrons going along, scattered at this point, then it goes over here, scatters at this point, so I'd make little pictures of it going. That's what those things were. Emits a photon, the photon goes over here—. ... And I did think consciously: "Wouldn't it be funny if this really turns out to be useful, and the *Physical Review* would be all full of these funny-looking pictures? It would look very amusing." [R. P. Feynman, talking about how he invented his diagrams]

Particle physics units

Before I can talk about particle physics, even in the loosest fly by night manner, I have to tell you about units. Since particle physics is both relativistic and quantum, practitioners customarily choose units such that $c = 1$ and $\hbar = 1$. Setting $c = 1$ means that distance and time have the same dimension: $[L] = [T]$. Time can be used to measure distance,[1] and vice versa. Next, set $\hbar = 1$. Since $[\hbar] = (ML/T)L = ML = 1$, we see that in these units, length has the dimension of an inverse mass.

To summarize,

$$[L] = [T] = 1/[M] \tag{1}$$

I trust that you can see the tremendous utility of these units, which I might call "particle physics units." We can choose to express everything in terms of a mass M or in terms of a length L. High energy physicists have traditionally used mass, or equivalently, energy. All physical quantities have dimension of M to some power.[2]

Since gravity plays no role in traditional particle physics, we do not need to go full Planckian.

Electromagnetism in fundamental units

I exclaimed back in chapter II.1 that the basic quantities in electromagnetism have some rather peculiar fractional dimensions: $[e] = M^{\frac{1}{2}} L^{\frac{3}{2}}/T$ and $[E] = [B] = M^{\frac{1}{2}}/L^{\frac{1}{2}} T$. But in particle physics units, these weird dimensions of electromagnetism disappear. We now have

$$[e] = M^{\frac{1}{2}} L^{\frac{3}{2}}/T = 1 \tag{2}$$

and

$$[E] = [B] = M^{\frac{1}{2}}/L^{\frac{1}{2}} T = 1/L^2 = M^2 \tag{3}$$

so that e is dimensionless (and numerically $\simeq 0.3$) and the electromagnetic field has dimension of an inverse length squared, or equivalently, a mass squared.

Perhaps not surprisingly, most particle theorists* are not even aware, or have forgotten, that e and the electromagnetic field have strange dimensions. I had to be reminded by Dyson. Arguably, e has to be dimensionless in sensible units if it were to claim to be fundamental.[3]

Recall from chapter II.1 the energy density $\varepsilon \sim \vec{E}^2 + \vec{B}^2$ and the energy flux $\vec{S} \sim c\vec{E} \times \vec{B}$ of an electromagnetic field in empty space. In fundamental units, these expressions, which are quadratic in the electromagnetic field, all have dimension of M^4, which equals either M/L^3 or $M/(L^2 T)$, thus corresponding to either energy per unit volume or energy per unit area per unit time.

Also recall that the electrostatic potential $\Phi = A_0$ can be packaged with the vector potential $\vec{A}$ to form the electromagnetic potential $A_\mu = (A_0, \vec{A})$, written as a 4-vector. Since $\vec{B} = \vec{\nabla} \times \vec{A}$, we have $[\vec{A}] = L[B] = LM^2 = M$, and so the vector potential $\vec{A}$ has dimension of mass.

Since the potential A_μ is a 4-vector, the field strength $F_{\mu\nu} = \partial_\mu A_\nu - \partial_\nu A_\mu$ is an antisymmetric 4-tensor whose components are the familiar $\vec{E}$ and $\vec{B}$. The field strength has dimension $[F] = M^2 = [E] = [B]$, in accordance with (3).

---

*Experimentalists are of course different: They deal with "real" electromagnetism all the time.

Physics is where the action is: Lagrangian and Lagrangian density

I have mentioned in passing* the Lagrangian, the Lagrangian density, and the action, concepts basic to modern theoretical physics. I also mentioned that these concepts will not be needed until part IX. Well, that time is now.

My purpose here is to merely give you a flavor of these concepts, and thus I limit myself to a few bare bone remarks. I certainly do not expect anybody to learn[4] these concepts here.

Back in chapter II.1, I stated (without much discussion) that the Lagrangian of a Newtonian point moving in 1-dimensional space with its position denoted by q is given by $L = \frac{1}{2}m\dot{q}^2 - V(q)$, namely, the difference of its kinetic energy and potential energy. Evidently, L has the dimension of energy. The action S is defined to be the integral of the Lagrangian over time: $S = \int dt L$. If someone tells us what $q(t)$ is, that is, a complete record of where the particle is at any given time t (namely, the history of the particle), we can plug $q(t)$ into the integral $S = \int dt L$ and obtain a real number, known as the action of that particular history. Given some other $q(t)$, we obtain some other value of S. In other words, S is a functional of the function $q(t)$.

The profound insight of Euler and of Lagrange is that the particle "chooses" the history or the path $q(t)$ that extremizes[5] the action S. This action principle amounts to a startling reformulation of Newtonian mechanics.

In classical physics, the action principle represents an elegant, but optional, formalism. One could spend one's entire physics career happily using some version of $F = ma$, without ever engaging the action. But with the advent of quantum mechanics and of quantum field theory, the action now plays a central role.

Action is dimensionless

To proceed further, I have to tell you about the dimension of the action in particle physics units. Turns out the action doesn't have any, as befits something so fundamental to physics.

But first let us find the dimensions of the action in "everyday" units. From the definition $S = \int dt L$, we see that the action has dimension (in any units) of $[S] = ET$, which is in fact the same dimension as that of $\hbar$.

In the Dirac-Feynman formulation[6] of quantum physics, the most logical and fundamental of the three formulations,[7] the probability amplitude associated with a particular path or history is given by $e^{iS/\hbar}$. Elegantly simple, no?

Planck's constant $\hbar$ is what we should measure the action S by. We now understand what $\hbar$ means: It is the quantum of action.

---

*Notably in chapter II.1 on electromagnetism, in an endnote in chapter VII.4 on gravity wave emission, and in appendix Eg on Einstein gravity.

In particle physics units, with $\hbar$ set equal to 1, the action S is dimensionless:

$$[S] = 1 \tag{4}$$

Always remember that physics is where the action is, and that the action is dimensionless.

Lagrangian density

In a field theory such as electromagnetism, physics is local in spacetime, and so the Lagrangian L is given by an integral of a Lagrangian density $L = \int d^3x\,\mathcal{L}$, as was mentioned in chapter II.1. The action is then the integral of $\mathcal{L}$ over spacetime:

$$S = \int dt\, L = \int dt \int d^3x\, \mathcal{L} = \int d^4x\, \mathcal{L} \tag{5}$$

Since $[S] = 1$, $[\mathcal{L}] = M^4$ in fundamental units.

That the action S, in these units, is a number associated with a particular history implies that it must be Lorentz invariant. Note that in a field theory, specifying a history means specifying the spatial configuration of the field as a function of time. Hence $\mathcal{L} \sim (\vec{E}^2 - \vec{B}^2)$ is also Lorentz invariant, which immediately implies that, in contrast, the energy density $\varepsilon \sim (\vec{E}^2 + \vec{B}^2)$ is not.[8] But we already know this. Suppose a box contains a certain amount of energy. Then to an observer moving by, the box Lorentz contracts,[9] and the energy increases (like the time component of a 4-vector). Thus, the energy density ε is not invariant.

It follows that the Lorentz invariance of $(\vec{E}^2 - \vec{B}^2)$ should also be manifest; indeed, in relativistic notation, $\mathcal{L} = -\frac{1}{4}F_{\mu\nu}F^{\mu\nu}$ (I might as well include the correct numerical coefficient here). Schematically, $\mathcal{L} \sim (\partial A)^2$.

For our purposes, we only need to know that $\mathcal{L}$ is quadratic in ∂ and quadratic in A.

Quantum field theory: rejecting a half-assed approach

You might already have encountered, in a course on advanced quantum mechanics, the emission and absorption of electromagnetic radiation by an atom transitioning between two states. You might also recall that the interaction Hamiltonian (or equivalently, the interaction Lagrangian) is given by $\int d^3x\, e\rho(x)\Phi(x) = \int d^3x\, e\psi^\dagger(x)\psi(x)A_0(x)$, with charge density $e\rho(x)$ given in terms of the Schrödinger wave function ψ of the electron in the atom (and its conjugate $\psi^\dagger$) and the electrostatic potential Φ (written as A_0). The charge density ρ is coupled to the electrostatic potential A_0. Similarly, the current $\vec{J}$ is coupled to the vector potential $\vec{A}$ via $\int d^3x\, e\vec{J}(x) \cdot \vec{A}(x)$. These terms are then packaged together using the relativistic notation already introduced into $\int d^3x\, eJ^\mu A_\mu$.

After quantization, the electromagnetic potential A_μ is then written in terms of creation and annihilation operators. It is capable of creating a photon (corresponding to emission) and of annihilating a photon (corresponding to absorption). (This is entirely analogous to the fact that the position operator in the harmonic oscillator is proportional to the sum of a creation operator and an annihilation operator.)

In short, the electromagnetic potential A_μ is treated as a quantum field, does what a quantum field does, and commands our respect.

In stark contrast, the poor electron is described by its wave function ψ, just as in good old basic quantum mechanics. Suppose the atom in question starts out with seven electrons, say; then no matter how we shake and bake the Schrödinger equation, we will end up with seven electrons. No more, no less. Quite unlike photons, which can appear and disappear at will. The electron, unlike the photon, is not described by a quantum field.

This unfair and unequal[10] treatment, typically taught in the last few weeks of an extended course on quantum mechanics stopping just short of quantum field theory, is known in academic slang as "half-assed." This is but one of several motivations[11] to have a fully quantum field theoretic description of the world, with all elementary particles described by quantum fields.

And thus Dirac and other pioneers of quantum field theory proposed introducing an electron field $\psi(x)$ in analogy with the electromagnetic field $A_\mu(x)$. (That the electron field and the electron wave function are denoted by the same Greek letter ψ has perhaps confused some beginning students of quantum field theory, but we are not concerned about that here.) Also, for reasons[12] I don't have space to go into here, the conjugate of $\psi(x)$ is written as $\overline{\psi}(x)$, rather than as $\psi^\dagger(x)$.

In quantum field theory, the electromagnetic coupling between the electron and the photon is described by the Lagrangian density $\mathcal{L} = e\overline{\psi}\gamma^\lambda\psi A_\lambda$. As remarked earlier, this represents the familiar interaction $e\psi^\dagger\psi A_0$ in quantum mechanics promoted and generalized.

It should be understood that when I show a Lagrangian density (for example, here), I am writing down, for clarity and simplicity, only the piece of the total Lagrangian density relevant for the discussion. In other words, $\mathcal{L} = e\overline{\psi}\gamma^\lambda\psi A_\lambda$ describes only the interaction between the electron and the photon, and not the motion of the electron nor the dynamics of the electromagnetic field. (The total $\mathcal{L}$ is a sum of terms, including, for example, $-\frac{1}{4}F_{\mu\nu}F^{\mu\nu}$.)

Some readers may be wondering about the symbol γ^λ (which happens to be four matrices, known as gamma matrices,[13] with entries equal variously to 0, ± 1, or $\pm i$) in $\mathcal{L}$. My strategy here is not to explain any details not directly relevant to the task at hand. Instead, I will focus on the important physical points.

Feynman diagrams

First, let us understand what $\mathcal{L} = e\overline{\psi}\gamma^\lambda\psi A_\lambda$ describes when applied to the absorption of a photon by an electron. As mentioned earlier, the electromagnetic field A_λ can annihilate a photon. Analogously, the electron field ψ

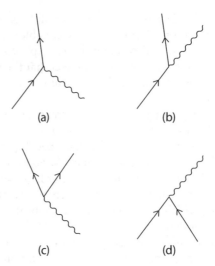

FIGURE 1. The coupling vertex of an electron to a photon can be read four ways: (a) the absorption of a photon by an electron, (b) the emission of a photon by an electron, (c) the production of an electron positron pair by a photon, (d) the annihilation of an electron positron pair into a photon. The wavy line represents the photon, the solid line the electron.

annihilates the electron. Then the conjugate field $\bar{\psi}$ creates an electron. Finally, the number e fixes the probability amplitude for this process to occur.

The absorption of a photon by an electron is described in quantum field theory in what seems to be a somewhat laborious fashion. Reading $e\bar{\psi}\gamma^\lambda\psi A_\lambda$ from right to left: a photon and an electron disappear (are annihilated by A_λ and ψ, respectively, in the jargon) and then an electron appears (is created by $\bar{\psi}$).

Feynman famously invented diagrams to describe what happens in quantum field theory. As a simple example, the process just described is represented pictorially in the Feynman diagram shown in figure 1(a).

Crossing

How about the emission of a photon by an electron? The interaction $e\bar{\psi}\gamma^\lambda\psi A_\lambda$ is capable of describing that also. The field A_λ can also create a photon.

This is described by the Feynman diagram in figure 1(b). We see that this can be obtained from figure 1(a) by bending the photon line "to go forward in time."

This possibility of bending the lines in Feynman diagrams is a deep property of quantum field theory, known as "crossing symmetry."

If the electromagnetic field can create a photon and annihilate a photon, how about the electron field ψ? Can ψ also create as well as annihilate an electron?

The answer is no, due to electric charge conservation. The act of creating an electron adds a unit of (negative) charge to the universe, while the act of annihilating an electron subtracts a unit of (negative) charge. Thus, the two acts cannot be superposed and appear together in a single field ψ. This is why we need the conjugate field $\overline{\psi}$ to create an electron. In contrast, the electromagnetic field A_λ does not carry charge and is its own conjugate field.

In essence, A_λ originates in a real classical field. In contrast, there is no classical analog of ψ. As the reader might know, this is deeply connected to the necessity for complex numbers in quantum physics.

This line of thought suggests that if ψ can annihilate an electron, it should be able to create a particle with electric charge opposite to that of the electron. Indeed, that guess is correct, as is celebrated in song and dance around our communal campfire, and undergirds Dirac's brilliant insight about antimatter. The electron field ψ annihilates an electron and creates a positron, the antiparticle corresponding to the electron. This implies that the conjugate electron field $\overline{\psi}$ creates an electron and annihilates a positron. This table "summarizes" quantum electrodynamics:

	annihilates	creates
ψ	electron	positron
$\overline{\psi}$	positron	electron
A	photon	photon

So go ahead and bend (or more correctly, cross) the incoming electron line in figure 1(a) to transform the Feynman diagram into the one shown in figure 1(c) describing the production of an electron positron pair by a photon. In other words, we now read $\mathcal{L} = e\overline{\psi}\gamma^\lambda\psi A_\lambda$ as follows: A_λ annihilates a photon, while ψ creates a positron and the conjugate field $\overline{\psi}$ creates an electron.

Now that you have learned some quantum field theory, you can also play the game. Cross the outgoing electron line in figure 1(b) to obtain the process in figure 1(d). Figure out what it describes! Yes, the annihilation of an electron positron pair into a photon. We now read $\mathcal{L} = e\overline{\psi}\gamma^\lambda\psi A_\lambda$ as follows: A_λ creates a photon, while ψ annihilates an electron and $\overline{\psi}$ annihilates a positron.

Take home message: This string of symbols $e\overline{\psi}\gamma^\lambda\psi A_\lambda$ describes four apparently different physical processes.

Incidentally, you can now see how Feynman's somewhat cryptic remark[14] that an antiparticle is a particle traveling backward in time comes about.

Needless to say, all this is incredibly sketchy and vague. But my goal here is not to teach you quantum field theory in a peanut shell,[15] which is manifestly impossible, but to give you a flavor of some very deep physics.

Electron electron scattering

Using the photon-electron-electron vertex in figure 1 as a building block, we can construct more and more involved Feynman diagrams, such as the one in figure 2 describing electron electron scattering.[16] An electron emits a photon, which is subsequently absorbed by another electron. Evidently, two

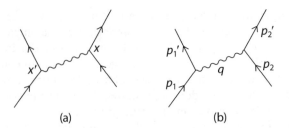

FIGURE 2. Feynman diagrams showing electron electron scattering.

electrons exert an influence on each other through the electromagnetic field. The Coulomb potential $V(r) = e^2/r$ pops out!

You might have noticed that there are two identical Feynman diagrams in figure 2. Figure 2(a) shows the scattering in spacetime: as indicated, the photon propagates from x' to x. (Because we are working with a relativistic quantum theory, the notation is understood to be 4-dimensional. For example, $x = (x^0, \vec{x}) = (t, \vec{x})$.) Figure 2(b), in contrast, shows the same process in momentum space. The lines are labeled by the momentum they carry. For example, the photon carries momentum $q = p_1 - p_1' = p_2' - p_2$, since momentum is conserved. We are talking about 4-momentum here, of course: $q = (q^0, \vec{q})$. The two figures are related by a Fourier transform via e^{iqx}.

Indeed, the Feynman diagrams for Compton scattering we encountered way back in chapter II.3 are just constructed by putting together figures 1(a) and 1(b) in the appropriate order.

Feynman rules

Feynman of course did much more than draw diagrams to describe what is happening. He derived the rules, known as Feynman rules, that enable any trained bozo* to calculate the probability amplitude associated with any process described by these diagrams. This prompted Julian Schwinger to grumble, "Feynman brought quantum field theory to the masses."

Associated with each vertex is a coupling constant in a Feynman diagram (such as e in electromagnetism), associated with each internal line is an expression called a "propagator," and so on and so forth. At each vertex, momentum is conserved, a fact that was used in figure 2(b) to determine the photon momentum q. These rules are derived and listed in quantum field theory textbooks.[17] All you need to know at this point is that once you master the Feynman rules,[18] you can calculate probability amplitudes in quantum field theory.

I would like to tell you something about the propagator for use in chapter IX.3. Surprise surprise, you already know what the propagator is, at least in spacetime, whether you realize it or not!

---

*I am willfully misquoting Feynman's dictum "What one fool can do, another can." See *QFT Nut*, p. 522.

Well, in appendix Gr on Green functions for use in our discussion of electromagnetic waves, I derived

$$G(t - t', \vec{x} - \vec{x}') = \frac{\delta(t - t' - \frac{1}{c}|\vec{x} - \vec{x}'|)}{|\vec{x} - \vec{x}'|} \tag{6}$$

telling us how a disturbance in the electromagnetic field at the spacetime point x is felt at the spacetime point x'. But that is exactly what a propagator does. So G is essentially the propagator to be used in spacetime Feynman diagrams.

But what is the propagator $\tilde{G}(q)$ in momentum space? We simply Fourier transform: $\tilde{G}(q) = \int d^4 x e^{iqx} G(x)$. (We will use the more compact 4-dimensional notation.) A fly by day physicist would dutifully evaluate the Fourier integral, but we fly by night physicists are going to take the easy path: Do dimensional analysis. Since $\int dt \delta(t) = 1$, $\delta(t)$ has dimension of $1/T = 1/L$ and hence from (6), $[G(x)] = 1/L^2$ (in particle physics units with $c = 1$). Thus, $[\tilde{G}(q)] = L^4/L^2 = L^2$, and thus $\tilde{G}(q) \sim 1/q^2$.

Here is another way. In appendix Gr, $G(x)$ was obtained as the solution of the equation $\Box G(x) = \delta^{(4)}(x)$, where $\Box \equiv (\nabla^2 - \frac{1}{c^2} \frac{\partial^2}{\partial t^2})$ is the generalized Laplacian for spacetime. Fourier transforming this equation, we have

$$\int d^4 x e^{iqx} \Box G(x) = \int d^4 x e^{iqx} \delta^{(4)}(x) = 1 = q^2 \int d^4 x e^{iqx} G(x) = q^2 \tilde{G}(q) \tag{7}$$

(The third equality follows on integration by parts.) Once again, we obtain $\tilde{G}(q) \sim 1/q^2$, all consistent, of course. This is the propagator we should associate with the photon line in figure 2(b).

Step by step, we can build up, more or less in this way, the magnificent edifice of quantum electrodynamics, through which electromagnetism is derived and explained.

Incidentally, some individuals well established in other areas of physics have told me that they could not quite grasp the Feynman propagator. I am confused by their confusion. The notion of a "disturbance" propagating from here and now to there and then is basic in many areas of physics. When we talk about electromagnetic radiation, we are talking about propagators. Similarly with water waves, and so on and so forth.

Preparing to do some dimensional analysis

In chapter IX.2, when we apply dimensional analysis to the weak interaction, we will need to know the dimension of the electron field ψ.

All set up and ready! We already know that, in particle physics units, $[\mathcal{L}] = M^4$, $[A] = M$, and $[e] = 1$. Just plug these into $\mathcal{L} = e\bar{\psi} \gamma^\lambda \psi A_\lambda$ to obtain $M^4 = [\mathcal{L}] = [e][\bar{\psi}][\psi][A] = M[\psi]^2$. By high school algebra, we find the important result

$$[\psi] = M^{\frac{3}{2}} \tag{8}$$

The electron field has dimension of mass to a half integer value, namely, $\frac{3}{2}$.

Exercise

(1) Quantum electrodynamics goes nonlinear! Maxwell electromagnetism is linear and is described by the Lagrangian density $\mathcal{L} \sim F_{\mu\nu}F^{\mu\nu} \sim F^2$. But quantum fluctuations generate nonlinear effects, so that two photons can scatter off each other. With the vertex shown in figure 1(c) one photon can morph into an electron-positron pair, off which the other photon can scatter. Then the electron positron pair can reconstitute itself into a photon through the vertex shown in figure 1(d). This process was first calculated[19] in 1936 by W. Heisenberg and H. Euler. The relevant Feynman diagram is shown in figure 3. Write down the corresponding Lagrangian density.

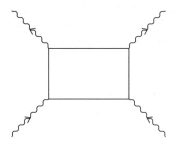

FIGURE 3. Feynman diagram for photon photon scattering.

Notes

[1] Even the proverbial guys and gals (at least the educated ones) on the street know about light years and all that.

[2] In certain high powered areas of modern cutting edge condensed matter physics, in particular, critical phenomena, the opposite choice of using a length is standard.

[3] In contrast to say, viscosity.

[4] In my humble opinion, a good place to start might be chapter II.3 of *GNut*.

[5] That is, either minimizes or maximizes.

[6] One advantage of this formulation is that it renders explicit and clear how classical physics emerges out of quantum physics. Classical physics takes over if $S \gg \hbar$, quantum physics if $S \lesssim \hbar$.

[7] When I said this, some of my students actually intoned "Amen."

[8] If the variations of $\vec{E}^2$ and of $\vec{B}^2$ cancel in the combination $(\vec{E}^2 - \vec{B}^2)$, then they cannot possibly cancel in the combination $(\vec{E}^2 + \vec{B}^2)$.

[9] This is mentioned in appendix Eg.

[10] Einstein, troubled by the electron treated as a particle and light as a wave, was motivated to introduce the photon. Inspired by Einstein's work, de Broglie in turn proposed that the electron is also described by a wave, thus completing the circle, so to speak. ·

[11] For a full description of the inadequacies of quantum mechanics and of the necessity of having quantum field theory, see, for example, *QFT Nut*.

[12] Because time and space have opposite signs. See *QFT Nut*, p. 97.

[13] For the sake of definiteness, I display for the curious reader one of these matrices:

$$\gamma^3 = \begin{pmatrix} 0 & 0 & 1 & 0 \\ 0 & 0 & 0 & -1 \\ -1 & 0 & 0 & 0 \\ 0 & 1 & 0 & 0 \end{pmatrix}$$

[14] See *QFT Nut*, p. 113.

[15] So go read a book on quantum field theory. The one I like is *QFT Nut*.

[16] This diagram is worked out in great detail in, for example, *QFT Nut*, p. 134.

[17] For example, *QFT Nut*, Appendix C, p. 534.

[18] They are actually not difficult to derive, with the clarity of hindsight. See, for example, *QFT Nut*, chapter I.7.

[19] The calculation was extremely arduous before the advent of Feynman diagrams, but I can now assign it as a homework problem in a quantum field theory course. See D. Kaiser, *Drawing Theories Apart: The Dispersion of Feynman Diagrams in Postwar Physics*, University of Chicago Press, 2005. Incidentally, this paper is often cited incorrectly as Euler-Heisenberg.

Weak interaction

a few basic facts

Some basic facts about the weak interaction

I start our discussion of the weak interaction[1] with the sketchiest of all sketches, mentioning only a few relevant facts.[2] Incidentally, we have already mentioned the weak interaction in chapter III.1 and in chapter VII.1 in connection with stellar burning.

The history of the weak interaction begins with the discovery of nuclear β decay: $(Z, A) \rightarrow (Z + 1, A) + e^- + \bar{\nu}$. (Due to charge conservation, the number Z of protons in the nucleus has to increase by one, while the total number A of nucleons remains the same.) Later, this process was understood in terms of the more elementary process $n \rightarrow p + e^- + \bar{\nu}$. A neutron[3] inside the nucleus transmutes itself into a proton while emitting an electron and an antineutrino. Then much later, this process in turn was understood in terms of the even more elementary process $d \rightarrow u + e^- + \bar{\nu}$. A down quark* d inside the neutron transmutes itself into an up quark u while emitting an electron and an antineutrino.

I assume that the reader is at least somewhat familiar with the story of the neutrino. Measuring the energy of the electron in nuclear β decay, $(Z, A) \rightarrow (Z + 1, A) + e^- + \bar{\nu}$, and knowing the mass difference between the mother nucleus (Z, A) and the daughter nucleus $(Z + 1, A)$, experimentalists found that energy was missing. Pauli had the insight that the missing energy was carried away by an invisible particle, which Fermi called the "neutrino" (that is, "little neutron" in Italian), to distinguish it from the neutron. (Within the standardized naming convention used nowadays, this particle emitted in beta decay is actually an antineutrino, and I have written it as such, at the

---

*The proton is now known to be made of two up quarks and a down quark, thus $P = (uud)$; and the neutron is made of two down quarks and an up quark, thus $N = (ddu)$.

risk of being slightly anachronistic.) The actual detection of neutrinos and antineutrinos came decades later.

Recall that while orbital angular momentum takes on integer values, the particles n, p, and e^- all have spin $\frac{1}{2}$. From the quantum mechanical rule on the addition of angular momentum, we then know from neutron decay, $n \rightarrow p + e^- + \bar{\nu}$, that the neutrino must have half integer spin, and spin $\frac{1}{2}$ is the simplest default option.

Meanwhile, the muon μ^-, which looks and acts like a more massive version of the electron e^-, was discovered. The muon μ^- decays into the electron e^- plus some missing energy. Comparing the decay rate with the rate of neutron beta decay, it was deduced that the weak interaction was responsible.[4] Assuming that the neutrino is again involved in carrying off the missing energy, we see that a single neutrino does not suffice: The angular momentum does not add up right. Thus, the missing energy has to be carried off by two invisible particles. The muon is now known to decay ($\mu^- \rightarrow e^- + \bar{\nu} + \nu$) into three spin-$\frac{1}{2}$ particles. Again, conforming to contemporary convention, we call one of the novel particles a neutrino, the other an antineutrino.

For pedagogical reasons, I have written $\bar{\nu}$ and ν in the decay products of the muon, as if one were the antiparticle of the other. In fact, we now know—and that was a big surprise in the history of particle physics—that there are distinct neutrino species! The $\bar{\nu}$ and ν in muon decay belong to different species and are not antiparticles of each other.

How this fascinating discovery came about is one of the main subjects of our subsequent discussion.

To understand physics, it is advisable to place yourself in historical contexts from time to time. I truly detest particle physics textbooks that start with the standard model, as if that were something that fell into our laps from the sky. But alas, the scope of this book certainly does not allow us to delve deeply into the tortuous history.

Fermi's theory of the weak interaction

In 1933, Enrico Fermi proposed his celebrated[5] theory of the weak interaction by writing down the interaction Lagrangian $\mathcal{L} = G \, (\bar{e} \, \Gamma^\lambda \, \nu) \, (\bar{p} \, \Gamma_\lambda \, n)$ to describe neutron beta decay $n \rightarrow p + e^- + \bar{\nu}$. Here n, p, e, ν denote, respectively, the neutron field, the proton field, the electron field, and the neutrino field. (Note that, evidently, particle physicists often use the same letter for a field and the particle associated with it. Also, note that the index λ is summed over, but never mind such details.) Now that you have almost mastered quantum field theory, you are ready to be initiated into reading this secret handwriting: in $\mathcal{L}$ from right to left, n annihilates a neutron, $\bar{p}$ creates a proton, ν creates an antineutrino, and $\bar{e}$ creates* an electron, thus describing the process

---

*I write this so casually, and you will read it even more casually. Fermi's conceptual breakthrough, which many at the time had difficulty accepting, was that the observed outgoing electron in β decay, $(Z, A) \rightarrow (Z+1, A) + e^- + \bar{\nu}$, is actually created during the decay process. More on this later.

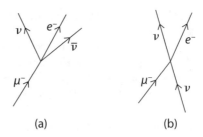

(a)   (b)

FIGURE 1.   (a) The Feynman diagram describing muon decay $\mu^- \to e^- + \bar{\nu} + \nu$; (b) the Feynman diagram describing neutrino scattering $\nu + \mu^- \to \nu + e^-$. The two processes are related by crossing.

$n \to p + e^- + \bar{\nu}$. The Fermi constant G measures the probability amplitude for the process to occur. (I have simplified by suppressing various numerical factors, which can be absorbed into the matrices Γ^λ, which I have not even specified. By doing this, I have glossed over several Nobel Prizes, including the one for parity violation.)

For various reasons, it is somewhat clearer (and for my purposes, preferable) to discuss the weak decay of the muon, $\mu^- \to e^- + \bar{\nu} + \nu$, instead of neutron beta decay. Having become almost an expert on quantum field theory, you can probably now follow in Fermi's footsteps and write down the interaction Lagrangian:

$$\mathcal{L} = G \, (\bar{e} \, \Gamma^\lambda \, \nu) \, (\bar{\nu} \, \Gamma_\lambda \, \mu) \tag{1}$$

You are already a fluent reader of this quantum field theory code: once again, the field μ annihilates a muon, $\bar{e}$ creates an electron, ν creates the antineutrino $\bar{\nu}$, and $\bar{\nu}$ creates the neutrino ν.

The Feynman diagram describing muon decay is shown in figure 1(a).

But you know more than how to write down interaction Lagrangians and to draw Feynman diagrams. You also know crossing symmetry! You could cross the outgoing antineutrino $\bar{\nu}$ and turn it into an incoming neutrino ν, as shown in figure 1(b). Crossing turns muon decay into the scattering process $\nu + \mu^- \to \nu + e^-$: A neutrino scatters off a muon, transforming itself into a neutrino and the muon into an electron, a process also described by the same Lagrangian $\mathcal{L} = G \, (\bar{e} \, \Gamma^\lambda \, \nu) \, (\bar{\nu} \, \Gamma_\lambda \, \mu)$ we just encountered.

The value of the Fermi coupling constant

Historically, the value of the Fermi coupling constant G had to be laboriously deduced from measurements of nuclear and neutron β decay rates and the muon decay rate. Muon decay $\mu^- \to e^- + \bar{\nu} + \nu$ offers the cleanest result, since none of the particles involved participates in the strong interaction, and we do not have to disentangle various nasty nuclear and strong interaction effects.

Since the mass of the electron is much less than the mass of the muon m_μ, we can effectively treat the muon as decaying into three massless particles. In the next section, we will determine the dimension of G to be an inverse mass squared. Given this, you can work out the decay rate (or inverse lifetime) to be $\Gamma \sim G^2 m_\mu^5$. See exercise (1) and especially the comment embedded therein. Notably, the rate varies as the 5th power of m_μ.

Decades of measurements and theoretical calculations have yielded the value $G \simeq 10^{-5}/m_p^2$. Talking about using appropriate units (see chapter I.3 from way back), I much prefer to relate G to the proton mass m_p rather than to express it in terms of ergs, as some are wont to do.

Dimension of the Fermi coupling constant

The key to obtaining some rudimentary understanding of the weak interaction is to first determine the dimension of its coupling strength G. I will give you two ways of obtaining $[G]$. First, a quick and handwaving way, and then later, a more involved but absolutely correct way.

Experimentalists eventually realized that nuclear beta decay[6] occurs entirely inside the nucleus, which (as you may recall) is teeny, essentially a point compared to the size of the atom. Thus, the range or distance scale over which the weak interaction operates is infinitesimal. The weak interaction is said to be short ranged, in contrast to the long ranged electromagnetic interaction (or the gravitational interaction). Instead of the electromagnetic potential $V(\vec{x}) = e^2/r$, the corresponding weak interaction potential has the form $V(\vec{x}) = G\delta^3(\vec{x})$: The delta function (see appendix Del) indicates that it all happens at a point.

Since $\delta^3(\vec{x})$ has dimension $1/L^3$, evaluating the dimension of $V(\vec{x})$, we obtain $E = [G]/L^3$, and thus in the particle physics units we are using,

$$[G] = ML^3 = \frac{1}{M^2} \tag{2}$$

The Fermi constant has the dimension of an inverse mass squared! In contrast, the coupling strength[7] of electromagnetism $\alpha \sim e^2 \simeq 1/137$ is dimensionless.

Shocked? What, you are not shocked?

Scattering amplitude blows up

Consider some scattering process governed by the weak interaction, for example, $\nu + \mu^- \to \nu + e^-$, depicted in figure 1(b). Even better, to not muddy the issue with the presence of the muon (which perhaps not every reader is on speaking terms with), consider a neutrino scattering off an electron, $\nu + e^- \to \nu + e^-$, a process that has actually been observed in modern times. Let the scattering occur at an energy E much larger than the mass of the electron, so that the electron can be treated as effectively massless.[8]

Imagine calculating the probability amplitude $\mathcal{M}$ by dimensional analysis. To lowest order, $\mathcal{M} \propto G$, since the scattering would not occur if G were to vanish.

Let us try to write down the amplitude to the next order: $\mathcal{M} \sim G + G^2 X$, where we will try to guess what X is. Matching the dimensions of the two terms, $[G^2 X] = [G]$, we see from (2) that $[X] = 1/[G] = M^2$: X has dimension of M^2. Since the energy E is the only quantity around with the dimension of a mass, we conclude that

$$\mathcal{M} \sim G(1 + aGE^2 + \cdots) \tag{3}$$

with a some numerical factor.

As we crank up the energy past $\sim 1/\sqrt{G}$, the scattering amplitude threatens to blow up.

The reader with a long memory (and who has read this book from cover to cover) would realize that this is precisely the same argument given in chapter IV.4 that quantum gravity would go haywire at the Planck energy. Indeed, recall that in quantum physics, the scattering amplitude is limited by a unitarity bound, since probability cannot exceed 1 by definition.

Another way of saying this is that if the correction term GE^2 is comparable to the leading term 1, then every subsequent term, with ever increasing powers of G, contained in the dots in (3), is of order 1 also.[9] Verify this.

Yes, in contrast to electromagnetism with its elegantly dimensionless coupling α, the weak interaction and gravity are afflicted by the same "disease." Almost everybody knows the dimension of Newton's constant, fewer know the dimension of Fermi's constant, but the instant you realize that they both have dimension of $1/M^2$ in fundamental units, you know that the corresponding quantum theory is in trouble.

Field theoretic determination of the dimension of the Fermi coupling constant

Historically, this realization that Fermi's theory of weak interaction behaves nastily at high energy $(E \sim 1/\sqrt{G})$ shook the foundation of particle physics and set off a search for a more sensible theory, culminating in the unification of the weak interaction with the better behaved electromagnetism. Marry the nasty guy to a nice girl was the traditional practice in many ancient cultures, often with disastrous consequences. In this case, it turned out to be a happy marriage: The weak interaction is now tamed as a piece of the electroweak interaction.

Since the dimension of the Fermi coupling constant G is such an important issue, you might not trust[10] the quick and dirty determination given in (2). Let us now move on to a more sophisticated field theoretic determination, as promised.

We have already established, in chapter IX.1, that the Lagrangian density $\mathcal{L}$ has dimension M^4 and that a spin $\frac{1}{2}$ field has dimension* $M^{\frac{3}{2}}$.

For the Fermi theory of weak interaction, the Lagrangian density in (1) has the schematic form $\mathcal{L} = G(\overline{\psi}\psi)(\overline{\psi}\psi)$, with ψ denoting generically a spin $\frac{1}{2}$ field, each with dimension $M^{\frac{3}{2}}$. Doing one of the most important calculations in particle physics,

$$4 = -2 + 4\left(\frac{3}{2}\right) = -2 + 6 \tag{4}$$

(got that?) we deduce that $[G] = M^{-2}$, in agreement with the quick and dirty determination in (2).

Fermi's theory cried out

The sound we hear is that of Fermi's theory crying out[11] that something dramatic has to happen at the energy scale[12]

$$E_{\text{weak}} \sim \left(\frac{1}{G}\right)^{\frac{1}{2}} \tag{5}$$

No reason to keep the suspense. We now all know what that something was: The particle responsible for the weak interaction, known as the W boson, was produced around the energy E_{weak} in 1983. The weak interaction was cracked open, so to speak.

The boson responsible for the weak interaction

But we are getting ahead of ourselves. Let us return to our narrative: Something was wrong with the Fermi theory. The search was on for a cure. After several dead ends and numerous puzzles, the avenue for progress came from comparing figure 2 in chapter IX.1 depicting $e^- + e^- \to e^- + e^-$ and figure 1(b) in this chapter depicting $\nu + \mu^- \to \nu + e^-$. We see that if the point vertex in figure 1(b) could be pulled apart, the figure could be made to look like figure 2 in chapter IX.1, provided that we invent[13] a particle (now called the W boson, as was just noted) to play the same role for the weak interaction as that played by the photon for electromagnetism. The pulled-apart process is shown in figure 2.

---

*Strictly speaking, we have only made this plausible for the electron field. However, all fundamental spin $\frac{1}{2}$ fields are described by the Dirac equation, and hence by the Dirac action, which we will briefly touch on in chapter IX.4. As we saw in chapter XI.1, the action, being always dimensionless, determines the dimension of the fields composing it. Hence all fundamental spin $\frac{1}{2}$ fields have the same dimension of $M^{\frac{3}{2}}$. By necessity, I am avoiding discussing the Dirac equation in this book. Readers who desire more details can find them easily in *QFT Nut*.

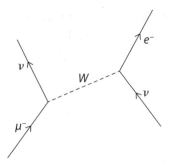

FIGURE 2. The scattering $\nu + \mu^- \to \nu + e^-$ occurs via the emission and subsequent absorption of a W boson. Compare this with figure 2 in chapter IX.1.

The scattering $\nu + \mu^- \to \nu + e^-$ is now understood to occur in two steps. First, the muon μ^- emits a W boson and turns itself into an outgoing neutrino ν. The W boson then travels on, encounters the incoming neutrino ν, and transforms it into an electron e^-.

Then the 4-fermion vertex can be built out of more elementary vertices describing the coupling of the W to fermions. In one vertex, a muon μ^- transforms itself into a neutrino ν by emitting a W. By electric charge conservation, we see that the W must carry negative charge, and so we denote it by W^-. In the other vertex, a neutrino ν absorbs a W^- and transforms itself into an electron e^-.

What do we mean by pulling apart the point vertex in figure 1(b) at which four fermions interact?

Consider the scattering $\nu + \mu^- \to \nu + e^-$ at low energy, so that the incoming muon does not have enough energy to produce an actual W: The W in figure 2 is known as a virtual particle. But by the Heisenberg uncertainty principle, its energy can be uncertain by the amount ΔE over a short time interval of order $\hbar/\Delta E$, during which the W can travel no farther than $\hbar/M$. (Keep in mind that we are using units with $c = 1$.)

Thus, if the mass M of the W boson were large enough, the weak interaction would be sufficiently short ranged. We now know that the range of the weak interaction is about a thousand times shorter than the range of the strong interaction, and thus it appears to act at a point.

In contrast, the photon is massless, thus implying that electromagnetism is infinitely long ranged on the scale of the weak and strong interactions. (In fact, it falls off like a power $1/r$ rather than exponentially.) Since it is massless, the photon is readily emitted, at the slightest provocation. You are in all likelihood exploiting this basic fact[14] of the universe in order to read this book.

Pulling the delta function apart: how bold is that?

Mathematically, replacing an interaction at a point by a short ranged interaction would be like pulling apart the delta function in the weak interaction potential $V(\vec{x}) = G\delta^3(\vec{x})$. You might wonder how that is possible.

To render this plausible, I will show you a function that, as a parameter varies, morphs from the electrostatic potential $\propto 1/r$ to the weak interaction potential $\propto \delta^3(\vec{x})$.

Consider the integral[15] $J = \int d^3q \frac{e^{i\vec{q}\cdot\vec{x}}}{\vec{q}^2+M^2}$. For $M=0$, the integral has dimension $[q^3]/[q^2] = [q]$ and thus by dimensional analysis, $J \sim 1/|\vec{x}| = 1/r$, the Coulomb potential. For $M = \infty$ (that is, for M much larger than the typical values of q that contribute to the integral), we have $J = \frac{1}{M^2} \int d^3q\, e^{i\vec{q}\cdot\vec{x}} \propto \delta^3(\vec{x})$ (see appendix Del), namely, the desired weak interaction potential just mentioned. For intermediate values of M, we have[16] $J \propto e^{-Mr}/r$. (All this is worked out in an exercise worked out in appendix Gr.) So mathematically, it is possible to "pull the delta function apart." Indeed, this operation reflects the physical reasoning about range and mass dictated by the uncertainty principle.

Remember that at the time when this kind of theorizing was going on, the neutrino was a theoretical entity that may or may not exist, and so proposing the hypothetical W boson, with a mass far exceeding what was known, was bold indeed. The great attraction to theorists was that this made the structure of the weak interaction look analogous to the structure of the electromagnetic interaction. As we now know, this turned out to be the prime consideration that swept aside numerous other objections.

The W propagator

To wrap up this chapter, and for use in chapter IX.3, I need to tell you about the propagator to be associated with the W in figure 2. In chapter IX.1, I told you about the photon propagator $1/q^2$. Recall that we obtain this by Fourier transforming the equation $\Box G(x) = \delta^{(4)}(x)$, an equation that follows from the equation of motion for the electromagnetic potential $\Box A_\mu = 0$. See appendix Gr.

So, let us find the equation of motion for the W. The energy E and momentum $\vec{p}$ of a particle with mass M satisfies $E^2 = \vec{p}^2 + M^2$, or $p^2 = M^2$ in terms of the 4-momentum $p = (p^0, \vec{p}) = (E, \vec{p})$. Denote the field associated with W by[17] $\Phi(x)$. If we want $\Phi(x) \sim e^{ipx}$ with $p^2 = M^2$ to be a solution of the field equation, then that equation should have the form $(\Box + M^2)\Phi = 0$ (since $\Box e^{ipx} = -p^2 e^{ipx}$).

Proceeding by blind and naive analogy to the electromagnetic case, we guess that the Green function for the W should satisfy the equation $(\Box + M^2)G(x) = \delta^{(4)}(x)$. Multiplying this equation by e^{iqx} and integrating (in other words, Fourier transforming this equation), and following the same steps as in chapter IX.1, the fly by day physicist finds $(-q^2 + M^2)\tilde{G}(q) = 1$.

The fly by night physicist, in contrast, simply follows dimensional analysis: $[\Box] = 1/L^2$, and so $\Box \to q^2$ in momentum space.

Therefore, both kinds of physicists agree that the propagator of a massive particle has the form[18] $\sim 1/(q^2 - M^2)$. Indeed, for $M=0$, we recover the propagator of the massless photon.

Incidentally, you can now also sense where the integral in the preceding section might have come from.

Creating the electron: a scandal bringing complete disorder into physics?

> [I was] criticized very strongly for this assumption by extremely good physicists. I got one letter saying that it was really a scandal to assume that there were no electrons in the nucleus because one could see them coming out; I would bring a complete disorder into physics by such unreasonable assumptions ... it is really difficult to go away from something which seems so natural and so obvious that everybody had always accepted it. I think the greatest effort in the developments of theoretical physics is always necessary at those points where one has to abandon old concepts. [W. Heisenberg][19]

You could literally see them coming out. So they must have been inside the nuclei all along!

Nowadays, students of quantum field theory routinely accept that electrons can be created by an electron field. But at one time, it was totally natural to suppose, as I have already mentioned earlier in this chapter, that the electron emitted in nuclear β decay, $(Z, A) \to (Z + 1, A) + e^- + \bar{\nu}$, had been sitting inside the mother nucleus (Z, A) all along.

To say that the electron shooting out of the nucleus was created out of thin air required a fantastic leap of faith. We revere the greats of physics, such as Fermi and Heisenberg, for leaps such as these.

Exercises

(1) Using dimensional analysis, show that the decay rate of the muon $\Gamma \sim G^2 m_\mu^5$, as mentioned in the text.

Comment: A rather tedious but straightforward calculation[20] determines the overall numerical constant to be $(192\pi^3)^{-1} \simeq 1/6,000$, which is certainly not of order 1. Some grouches would hold this up as a failure of dimensional analysis, but I disagree. Sometimes the 2πs simply go against you en masse, but they can be understood or even anticipated to some extent with experience. In this case, those (like me!) who have suffered through endless calculations of the decay rates and cross sections might even expect the pileup of πs. In a 3-body decay, such as $\mu^- \to e^- + \bar{\nu} + \nu$ (in contrast to a 2-body decay, such as $\mu \to e + \gamma$, to be discussed in chapter IX.3), the direction and magnitude of the momenta of the two neutrinos can take on continuously many possibilities (with the direction and magnitude of the momentum of the electron then determined by energy-momentum conservation). Indeed, back in chapter III.5, we counted the number of states in the momentum interval $(dp)^3$ to be $d^3 p/(2\pi)^3$ for each particle in the final state in the continuum limit. Thus, the formula[21] for the decay rate Γ starts off with $(\pi^3)^3$ in the denominator. Some of these πs, but not all, are canceled off. Perhaps the reader realizes that these 2πs can be traced back to the pendulum problem discussed in chapter I.1!

(2) A historical oddity and a caution: Not all apparently attractive theoretical ideas turn out to be right. Two particles that participate in the weak interaction (two electrons, say) could exchange two massless neutrinos (as shown in figure 3) and hence feel a force between them. Estimate the dependence of the corresponding interaction on the distance r between the two particles. It was hoped that this would produce the gravitational attraction! Well, you just killed this attractive idea. (While this idea failed to produce gravity, it allegedly led Yukawa to the notion of pion exchange generating the strong interaction.)

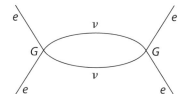

FIGURE 3. With the weak interaction acting twice, two electrons can exchange two neutrinos, as described by this Feynman diagram. Can the resulting force generate gravity?

(3) Use the uncertainty principle to argue that the electron is created in beta decay and not just released from captivity.

Notes

[1] The weak interaction is so weird that Nambu, in his book mentioned in endnote 3 in the prologue to part IX, referred to it as "God's mistake?" I would not go that far, but its history is intriguingly convoluted.

[2] You are urged to consult one of the texts on the subject, for example, E. Commins and P. Bucksbaum, *Weak Interactions of Leptons and Quarks*, Cambridge University Press, 1983.

[3] The neutron was discovered by James Chadwick in 1932. Up until then, it was generally believed that the nucleus was made up of protons and electrons. For a while after the discovery, some continued to believe that the neutron was a bound state of the proton and the electron, until its mass was measured to clearly exceed the sum of the proton mass and the electron mass.

[4] Of course, you understand that in this kind of breathless discussion, I have to suppress years of confusion, debate, etc. Due to phase space factors, the two experimental rates in fact differ by many orders of magnitude.

[5] Perhaps not surprisingly, his paper was rejected by the journal *Nature*. For an English translation of his 1934 paper, see http://microboone-docdb.fnal.gov/cgi -bin/RetrieveFile?docid=953;filename=FermiBetaDecay1934.pdf;version=1. The generally negative reaction to his paper apparently prompted Fermi to turn to experimental work, during which he discovered the activation of certain nuclear processes by slow neutrons, crucial to later developments.

[6] This phrase, used so cavalierly by a theorist, is of course meant to summarize years of experimental work. I am being highly impressionistic here: Of course, alpha and gamma decay also occur in the nucleus.

[7] I am intentionally sloppy here about a factor of 4π, irrelevant for our discussion.

[8] Indeed, it is now known that neutrinos have tiny masses, but we are treating them as effectively massless. If you prefer, you can also consider neutrino neutrino scattering, even though that has not yet been (and may not be for a long time) realized experimentally.

[9] This is analogous to saying that the series $1 + x + x^2 + \cdots = (1 - x)^{-1}$ goes haywire as x approaches 1 from below.

[10] This argument is given on p. 534 of R. B. Leighton, *Principles of Modern Physics*, McGraw-Hill, 1959, the assigned textbook of a sophomore-level course I took. The argument smelled somewhat fishy to me even at that tender age.

[11] I am glossing over all kinds of stuff laced with words like "cutoff" and "nonrenormalizability." See, for example, *QFT Nut*.

[12] I hesitate to call this energy the "Fermi energy" (in analogy with the Planck energy) because that term is already used in connection with the Fermi gas.

[13] H. Yukawa, when he suggested the meson theory for the nuclear forces, also proposed that an intermediate vector boson could account for the Fermi theory of the weak interaction. (In the 1930s the distinction between the strong and the weak interactions was far from clear.)

[14] Examples of the ready emission of photons: by electrons rushing about in the sun, jostling along with the crowd down a wire filament in an incandescent bulb, or hopping about as carbon atoms combine with oxygen atoms in a cradle flame.

[15] In field theory, it takes but a few pages to show that the exchange of a particle of mass M between two "external" particles generates an interaction potential equal to $J = \int \frac{d^3 q}{(2\pi)^3} \frac{e^{i\vec{q}\cdot\vec{x}}}{\vec{q}^2 + M^2} = \frac{e^{-Mr}}{4\pi r}$ (up to overall coupling constants). See QFT Nut, p. 28. (The factor $(2\pi)^3$ comes from counting states, as shown in chapter III.5.) I won't do the integral here (it is done on p. 31 in QFT Nut), but I can tell you how the πs go: The integral over the azimuthal angle in $d^3 q$ produces a 2π, and the integral over the magnitude q via Cauchy's theorem (picking up the pole at $q = iM$) gives another 2π. The remaining 4π is precisely the difference between Lorentz-Heaviside and Gauss; see appendix M.

[16] The Yukawa potential mentioned in appendix Gr on Green functions.

[17] As the better informed might have realized by now, I am ignoring the spin of the W boson.

[18] All this is explained in more detail in QFT Nut, pp. 21–24.

[19] *From a Life of Physics*, World Scientific, 1989, p. 48.

[20] See, for example, E. Commins and P. Bucksbaum, pp. 92–98.

[21] This is given in almost all relevant books, for example, QFT Nut, (38) on p. 141.

Private neutrinos

Radiative muon decay

By electric charge conservation, W must be electrically charged, as was pointed out in chapter IX.2. This realization leads us back to the issue of whether the two missing neutrinos involved in muon decay are related or not, thanks to some clever reasoning due to G. Feinberg[1] in 1958.

Go back to figure 2 in chapter IX.2, showing the W boson mediating the scattering process $\nu + \mu^- \to \nu + e^-$. The electrically charged W, by virtue of its electric charge, knows about the electromagnetic field and is thus capable of emitting a photon while it is en route between the muon and the electron. The Feynman diagram showing this process, $\nu + \mu^- \to \nu + e^- + \gamma$, is given in figure 1(a). From left to right, we see the W emitted by the muon subsequently emitting a photon γ, before combining with the incoming neutrino to form the outgoing electron. Note that the Feynman diagram in figure 1(a) is just the Feynman diagram in figure 2 in chapter IX.2 with a γ attached to the W propagator.

Looking at figure 1(a), Feinberg realized that he could bend the neutrino line emerging from the muon and connect it to the incoming neutrino. In other words, the incoming neutrino shown in figure 1(a) is actually the neutrino emitted by the muon. This neutrino then combines with the W to form the outgoing electron, as shown in figure 1(b). The net result is that the muon decays into an electron plus a photon: $\mu^- \to e^- + \gamma$.

Students often ask how the lines in a Feynman diagram can be bent. Don't relativistic particles travel in straight lines? The point is that we are talking about the relativistic quantum world. These lines actually represent quantum waves!

Another way of seeing that radiative muon decay does occur is to break it up into a three-step quantum process (as was listed[2] by Feinberg): (1) $\mu^- \to W^- + \nu$, (2) $W^- \to W^- + \gamma$, and (3) $W^- + \nu \to e^-$. Steps 1 and 3 proceed by

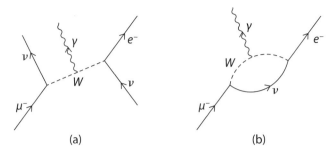

FIGURE 1. (a) The W boson involved in the process $\nu + \mu^- \to \nu + e^-$ emits a photon; and (b) the muon decays into an electron plus a photon.

the weak interaction, and step 2 by the electromagnetic interaction. Putting these steps together, we obtain $\mu^- \to e^- + \gamma$. Make sure you follow this; it reminds me of high school chemistry class.

Thus Feinberg argued that if the W boson existed, then some of the time, the muon μ^-, instead of decaying into $e^- + \nu + \bar{\nu}$, would decay into $e^- + \gamma$ (known as a radiative decay, "radiative" being a generic word for electromagnetic radiation).

How to calculate without a theory

If the W boson exists, then the muon ought to decay into an electron and a photon some of the time. Now that is a striking prediction. None of the usual wishy washy either this or that maybe yes maybe not stuff that characterizes contemporary particle theories!

But experimentalists want a definite number for the branching ratio r, defined as the ratio of the probability for $\mu^- \to e^- + \gamma$ to occur to the probability for $\mu^- \to e^- + \nu + \bar{\nu}$ to occur. How often should they expect to see the muon decaying into an electron and a photon?

But in 1958, with a working theory of the weak interaction nowhere in sight, how could Feinberg come up with a number?

Now is when fly by night physics gets to shine. The branching ratio r is by definition dimensionless. The amplitude for $\mu^- \to e^- + \nu + \bar{\nu}$ is $\propto G$, while the amplitude for $\mu^- \to e^- + \gamma$ is $\propto Ge$, since e measures the coupling of the photon to the W. So the dimensionless r, being equal to the ratio of probabilities, is given by the absolute square of the ratio of amplitudes, namely, $(Ge/G)^2 = e^2 \sim \alpha$.

In the last step, I argue that since we are working in fundamental units with $\hbar$ and c set to 1 rather than to some arbitrary human-made units, the various factors of $\hbar$ and c better gather themselves so as to produce the fine structure constant $\alpha \simeq 1/137$.

You might worry that there is also the muon mass m_μ and the electron mass m_e, so that one could include an unknown function of m_e/m_μ. But no need to worry: Compared to the muon, the electron is effectively massless, so that

$(m_e/m_\mu) \simeq 0$. In other words, we are invoking the dull function hypothesis of chapter I.4: r should also be proportional to some function $f(m_e/m_\mu)$, presumed to be dull, so that $f(0) \sim 1$.

Thus, Feinberg was able to predict that the branching ratio[3] should be of order $\alpha \sim 10^{-2}$. Meanwhile, experimentalists did search for radiative muon decay.[4] They didn't see any and concluded that $r < 2 \times 10^{-5}$.

A main message of this book for would-be theorists

Here is one of the main messages of this book. Many undergraduate physics majors think that theoretical physicists do exact calculations all day long, much as students solve homework problems. Yes, some theorists do that. But the truly major advances are often made, almost by definition, in the absence of an established theory during a time when confusion reigns. By now of course, with electroweak theory firmly established, anybody who can read a textbook can learn to calculate exactly the rate for muon radiative decay. But I have news for the students: nobody will be impressed. The challenge is to have the right idea of what to calculate and then to muddle through some 20 years before people figure out how to do the calculation correctly.

Two neutrinos and the beginning of the family problem

Faced with the experimental upper bound clashing with Feinberg's theoretical estimate, particle theorists had two choices.

The natural but wrong choice, favored by many theorists at the time, is to deny the existence of the W. The correct choice, which would seem to be much more far fetched, is to say that the W exists, but that the neutrino associated with the muon (now called the "muon neutrino" ν_μ) and the neutrino associated with the electron (now called the "electron neutrino" ν_e) are distinct.[5] The ν_μ cannot simply metamorphose into ν_e. Then the Feynman diagram in figure 1(b) is forbidden, and all is well with the experimental upper bound.

The muon actually decays into the electron plus a muon neutrino and an antielectron neutrino: $\mu^- \rightarrow e^- + \nu_\mu + \bar{\nu}_e$. You should go back to chapter IX.2 and add subscripts e and μ to the neutrinos in the text and in the figures (and of course also in figure 1).

The electron and the muon each have their own private neutrinos, much like wealthy people who each have their own chauffeurs. Neutrinos are not to be shared. This possibility, which in 1958 would have seemed extravagant and unlikely, is now firmly established.[6]

When the muon was discovered in 1936, I. I. Rabi famously exclaimed in annoyance, "Who ordered that?" The universe apparently could function quite well without the electron having a much more massive cousin. But now this seemingly redundant muon even has its own neutrino! This marks the beginning of one of the deepest puzzles of particle physics, the family problem. We now know that the quarks and leptons come in three copies.[7] Nature appears to be unreasonably extravagant. Surely there is a deep reason that we have not yet grasped!

Conservation laws: electron number and muon number

The fact that experimentalists did not observe the radiative decay $\mu^- \to e^- + \gamma$ could also be accommodated by introducing two conservation laws, for electron number L_e and for muon number L_μ separately. We simply define $L_e = 1$ for the electron and for the electron neutrino (and also $L_e = -1$ for their antiparticles) and $L_e = 0$ for all other particles (such as the muon). Similarly, define $L_\mu = 1$ for the muon and for the muon neutrino (and also $L_\mu = -1$ for their antiparticles) and $L_\mu = 0$ for all other particles (such as the electron). Then these two conservation laws forbid $\mu^- \to e^- + \gamma$ (since $L_e = 0$ and $L_\mu = 1$ in the initial state, but $L_e = 1$ and $L_\mu = 0$ in the final state) while allowing $\mu^- \to e^- + \nu_\mu + \bar{\nu}_e$.

The astute reader will recognize that these conservation laws are just convenient bookkeeping devices and do not explain anything. Why is Nature so extravagant, not only with particles, but also with conservation laws? Were conservation laws on sale somewhere?

Coupling of the W boson

Let us go back to the period around 1960, during which confidence in the existence of the W boson steadily increased. In analogy with the interaction Lagrangian $\mathcal{L} = e\bar{\psi}\gamma^\lambda\psi A_\lambda$, coupling the photon to charged spin-$\frac{1}{2}$ fields, we can write down the interaction Lagrangian:

$$\mathcal{L} = g\, \bar{\nu}_\mu \gamma^\lambda \mu\ W_\lambda + \text{h.c.} \tag{1}$$

Here we use the standard suggestive notation with the field associated with the muon μ^- denoted by μ and the field associated with the muon neutrino by ν_μ.

Again reading from right to left, we see that $\mathcal{L}$ describes the emission of a W and the annihilation of a muon followed by the creation of a muon neutrino, with the probability amplitude for the process to occur denoted by g.

The abbreviation h.c. in (1) stands for hermitian conjugate. In quantum physics, since the Hamiltonian has to be hermitian to conserve probability, so does the Lagrangian. We thus have to add to the term shown explicitly in (1) its hermitian conjugate. This added term in $\mathcal{L}$ (namely, $g\, \bar{\mu}\gamma^\lambda\nu_\mu\ W_\lambda^\dagger$) describes the absorption of a W and the annihilation of a muon neutrino followed by the creation of a muon, with the probability amplitude for the process to occur denoted by g; in short, the conjugate or the reverse of the process just described.

Note that in the case of electromagnetism, it is not necessary to add the hermitian conjugate of $e\bar{\psi}\gamma^\lambda\psi A_\lambda$ to $\mathcal{L}$, since A_λ is already hermitian (in light of its origin in classical physics, where it is in fact real), and since $\bar{\psi}$ is the conjugate of ψ.

Thus far, we have described the vertex on the left side of the Feynman diagram in figure 2 in chapter IX.2 and in figure 1(a). To describe the vertex on the right side, we have to couple the W to the electron and to its neutrino as well. Thus, we amend (1) to be

$$\mathcal{L} = g\left(\bar{\nu}_e\gamma^\lambda e\ W_\lambda + \bar{\nu}_\mu\gamma^\lambda\mu\ W_\lambda\right) + \text{h.c.} \tag{2}$$

The vertex on the right side of the Feynman diagram in figure 2 in chapter IX.2 and in figure 1(a) is then generated by the term conjugate to the term actually shown in (2), namely, $g\,\bar{e}\gamma^\lambda\nu_e\ W_\lambda^\dagger$, describing the absorption of a W and the annihilation of an electron neutrino followed by the creation of an electron.[8]

But wait, the astute reader cries. Why did you use the same g in (2) for the muon and for the electron? In our discussion, this merely represents a reasonable guess: The W plays fair and treats the electron and its more massive cousin the muon equally. This hypothesis is known as electron-muon universality, now firmly established through a long series of experiments.[9] So the answer to the question is: empirical fact.

Hints of electroweak unification

But the really exciting thing about all this talk about the W boson is the hint that electromagnetism and the weak interaction might be unified. Just compare (2) with the coupling of the electromagnetic field to the electron and to the muon:

$$\mathcal{L} = e\left(\bar{e}\gamma^\lambda e\ A_\lambda + \bar{\mu}\gamma^\lambda\mu\ A_\lambda\right) \tag{3}$$

The similarity between (2) and (3) is striking: The W boson and the photon sure appear to be related. Indeed, electron-muon universality was inspired by the universal coupling of the photon to charged particles. The photon couples equally to the electron and to the muon. The deep puzzle in the late 1950s, and the major stumbling block to unification, is how the W can be so massive, while the photon is strictly massless.[10]

The mass of the W boson

A wild fly by night guess even led to an estimate of how massive the W boson might be. All right, guess a reasonable value for g.

If you said $g \sim e$, you might have done well in the late 1950s!

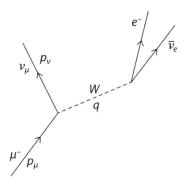

FIGURE 2. Muon decay generated by the W boson

Now we go back to the Feynman diagram for muon decay, $\mu^- \to e^- + \bar{\nu}_e + \nu_\mu$, given in figure 1(a) in chapter IX.2. In a theory with the W, the vertex in this figure is pulled apart (or you might say, magnified to expose its inner workings), resulting in the Feynman diagram shown in figure 2.

Denote the momentum of the muon by p_μ and the momentum of the muon neutrino by p_ν. By momentum conservation, the momentum q carried by the W equals $(p_\mu - p_\nu)$. But you need hardly worry about these momenta, because in muon decay, they are all tiny (order of 100 MeV, comparable to the muon mass) compared to M (order of 100 GeV; see below). We learned in chapter IX.2 that the propagator associated with the dotted line representing the W is given by $\sim 1/(q^2 - M^2)$, with M the W mass. With $q^2 \ll M^2$, the propagator becomes[11] $1/M^2$.

Each vertex in figure 2 is associated with a factor of g, and thus the amplitude associated with the Feynman diagram is just g^2/M^2. Equating this to the amplitude for muon decay in the Fermi theory, we determine the Fermi constant G in terms of the coupling and mass of the W boson:

$$G \sim \frac{g^2}{M^2} \qquad (4)$$

Thus, the unknown mass of the W should be around $M \sim g/\sqrt{G}$.

With our guess that $g \sim e$, we can even predict that the W has mass $M \sim e/\sqrt{G} \sim 10^2$ GeV. Indeed, that turned out to be the case!

What the Fermi theory was crying about

But now we can even see what the Fermi theory was crying about. Recall that it was crying about its foretold failure at the energy $\sim 1/\sqrt{G} \sim M/e > M$. Indeed, this energy is more than enough to produce the W boson. And that is exactly what happens in the electroweak theory unifying electromagnetism and the weak interaction. The Fermi theory was telling us that new physics was going to appear!

I find it sobering that theories in physics have the ability to announce their own eventual failure and hence their domains of validity, in contrast to theories in some other areas of human thought.

Even now, Einstein's theory is still crying out, as was explained in chapter IV.4. The theory tells us that at the Planck energy* $1/\sqrt{G_N} \sim M_{Planck}$, new physics must appear. Fermi's theory cried out, and the new physics turned out to be the electroweak theory. Einstein's theory is now crying out. Will the new physics turn out to be string theory?[12] Or something else?

Quarks and leptons

Thus far, we have discussed the weak interaction involving e, ν_e, μ, and ν_μ, collectively known as leptons. Leptons do not interact strongly. As mentioned earlier, strongly interacting particles are now known to be made of quarks. Quarks and leptons have spin $\frac{1}{2}$ and behave as fermions.

I also mentioned in chapter IX.2 that neutron and nuclear beta decays can be accounted for by the weak decay of the down quark $d \to u + e^- + \bar{\nu}$. This necessitates introducing the interaction Lagrangian

$$\mathcal{L} = g_1 \, \bar{u}\gamma^\lambda d \, W_\lambda + \text{ h.c.} \tag{5}$$

The relevant Feynman diagram is as in figure 2 but with the vertex $\mu^- \to \nu_\mu + W^-$ replaced by $d \to u + W^-$. In the interest of being concise and getting quickly to my main points, the presentation here is necessarily a bit anachronistic; quarks were of course unknown in the late 1950s, and people would have written p and n for proton and neutron in (5) instead.

You might have noticed that I wrote g_1 in (5), not g as in (2). In fact, for a long time, it was thought that electron-muon universality could be extended so that $g_1 = g$, but ever more accurate measurements showed that g_1 is slightly less than g. With this sentence, I am also glossing over a lot of arduous experimental and theoretical work. Since the neutron, unlike the muon, is a strongly interacting bound state of three quarks, extracting g_1 from experiments was not a simple task. I will come back to the issue of why $g_1 \lesssim g$.

---

*Evidently, Newton's constant is denoted by G_N to avoid confusion with the Fermi constant G.

Notes

[1] G. Feinberg, *Physical Review* (1958), p. 1482 (letter to the editor).

[2] This was in 1958; Feynman diagrams are so well known by now that nobody would feel that they need to break down a Feynman diagram into steps for the readers of a research journal.

[3] I am cutting a minor corner here. Otherwise I would have to digress and tell you about phase space. See exercise (1) in chapter IX.2. A historical note. Actually, Feinberg did do a fly by day calculation of the diagram in figure 1(b) using the by-then well-known Feynman rules. He obtained $r = \alpha X/24\pi$, with X given by a divergent integral. He then argued that X should be set equal to 1. That X came out formally infinite offered another symptom that Fermi theory was sick. Quantum field theory was also not fully understood at that time. So in some sense, the fly by day calculation that Feinberg did, while precise, is less accurate than the fly by night calculation. If you are using a theory not yet in final form, your calculation could well be precise but not accurate. In some cases, precision in calculation may be overrated, or at least unwarranted.

[4] S. Lokanathan and J. Steinberger, *Physical Review* 98 (1955), p. 240.

[5] Another historical note. Feinberg never mentioned this possibility in his paper. Instead, he referenced earlier work, by E. J. Mahmoud and H. M. Konopinski and by J. Schwinger, which proposed that muon number and electron number are separately conserved. To modern eyes, with the benefit of hindsight, these proposals seem to be based on rather shaky ground. They appear to be driven by the desire to have a 4-component field for the neutrino.

[6] The convoluted history of particle physics lies far outside the scope of this book, particularly since I am not a historian. My colleague S. Pakvasa informs me that already in 1943, S. Sakata and T. Inoue proposed that the muon decays into an electron and two distinct neutrinos, but their paper was not published until 1946 due to wartime conditions. In 1962, Sakata returned to the topic of two neutrinos, and in a remarkable paper with Z. Maki and M. Nakagawa, extended the footnote of the Gell-Mann–Lèvy paper of 1959 (which I mention in chapter IX.4) to leptons and thus suggested the possibility of neutrino flavor oscillations. S. Sakata and T. Inoue, *Progress in Theoretical Physics* (Kyoto) 1 (1946), p. 143; Z. Maki, M. Nakagawa, and S. Sakata, *Progress in Theoretical Physics* (Kyoto) 28 (1962), p. 870.

[7] Indeed, the τ lepton, yet another cousin of the muon and electron, discovered decades later, also comes with its own private neutrino.

[8] With the proliferation of particles and fields, we have abandoned the generic symbol ψ in favor of e for the electron field. I need hardly say this except to appease the nitpickers.

[9] For instance, by comparing the decays of the charged pion $\pi^+ \to \mu^+ + \nu_\mu$ and $\pi^+ \to e^+ + \nu_e$.

[10] As you probably know, this was later solved through the so-called Higgs mechanism.

[11] By the way, this point was already hinted at in the discussion of the integral $J = \int d^3q e^{i\vec{q}\cdot\vec{x}}/(\vec{q}^2 + M^2)$ in chapter IX.2.

[12] J. Polchinski, *String Theory*, Cambridge University Press, 1998.

Strangeness and charm

Strange particles are strange

In the early 1950s, a whole zoo of strange particles was discovered. They are called "strange," because they acted strangely, duh. "Strangely" means contrary to the theoretical expectations of that time.

Later, it was understood that besides the up and down quarks u and d, there was a strange quark s, which carries the same electric charge as d. Strange particles contain the strange quark s. (What delightfully childish nomenclature particle physicists have!) For example, the strange meson $K^+ = \{\bar{s}u\}$. The bracket notation means that the positively charged K meson is a bound state of an antistrange quark $\bar{s}$ and an up quark u.

Strange particles decay via the weak interaction. Just as the weak decay of the down quark $d \to u + e^- + \bar{\nu}$ is responsible for neutron decay, the weak decay of the strange quark $s \to u + e^- + \bar{\nu}_e$ (and its charge conjugate $\bar{s} \to \bar{u} + e^+ + \nu_e$) is responsible for strange particle decays, for example, the decay $K^+ \to \pi^0 + e^+ + \nu_e$, as depicted in figure 1. (Here π^0 denotes the electrically neutral π meson, known as the pion.)

The observed decay can then be explained in terms of the decay of the strange quark as follows. While the $\bar{s}$ quark in $K^+ = \{\bar{s}u\}$ decays into $\bar{u} + e^+ + \nu_e$, the u quark just hangs around (and hence is known as a spectator[1] quark). The $\bar{u}$ then forms a state $\{\bar{u}u\}$ with the spectator u, thus producing a π^0 (since the neutral pion consists partly of $\{\bar{u}u\}$). Altogether, these processes yield the K^+ decay indicated above. As usual in part IX, I have to brush over decades of work in a few sentences.

By now, you are practically an expert on how to extend weak interaction theory to accommodate strange particle weak decays. Simply add to the interaction Lagrangian the term

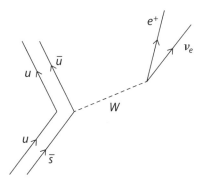

FIGURE 1. The $\bar{s}$ quark in K^+ decays into $\bar{u} + e^+ + \nu_e$, while the u quark goes along for the ride.

$$\mathcal{L} = g_2 \, \bar{u}\gamma^\lambda s \, W_\lambda + \text{h.c.} \tag{1}$$

and let the W do its thing.

Just as the interaction Lagrangian $\mathcal{L} = g_1 \, \bar{u}\gamma^\lambda d \, W_\lambda$ in (IX.3.5) generates the vertex $d \to u + W^-$, the interaction Lagrangian in (1) generates the vertex $s \to u + W^-$.

But please compare the two interaction Lagrangians carefully. Here I wrote g_2, not g_1. Historically, some physicists expected electron-muon universality might manifest analogously in the quark sector as a *d-s* universality, so that $g_2 = g_1$. But no, experiments showed g_2 to be significantly smaller, about $0.23g_1$. (Again, the extraction of g_2 from the measured weak decay rates of the strange particles was a highly involved and difficult task best left to the fly by day professionals.)

We can combine the two interaction Lagrangians and write

$$\mathcal{L} = \left(g_1 \, \bar{u}\gamma^\lambda d + g_2 \, \bar{u}\gamma^\lambda s\right) W_\lambda + \text{h.c.} \tag{2}$$

Orienting the weak interaction

The next important step was taken in 1960 by Murray Gell-Mann and Maurice Lèvy, who suggested, in a celebrated footnote in a paper[2] they published on a different though somewhat related subject, that in (2), we should write $g_1 = g' \cos\theta$ and $g_2 = g' \sin\theta$.

Well, any physics undergraduate could tell them that this doesn't make any difference: They are merely trading g_1 and g_2 for g' and the angle θ, now known as the Cabibbo angle. But MGM, as Gell-Mann is sometimes referred to, did not dominate particle physics for decades for nothing!

First, by defining the "rotated" quark field $d(\theta) = \cos\theta \, d + \sin\theta \, s$, we can write (2) as

$$\mathcal{L} = g' \left(\cos\theta \, \bar{u}\gamma^\lambda d + \sin\theta \, \bar{u}\gamma^\lambda s\right) W_\lambda + \text{h.c.} = g' \, \bar{u}\gamma^\lambda d(\theta) \, W_\lambda + \text{h.c.} \tag{3}$$

Still, you say, this is just a rewrite. No new physics.

The physics comes when Gell-Mann and Lèvy guess that electron-muon universality can be extended to say that g', not g_1, is equal to the g in (IX.3.2) describing the coupling of the W to the electron and to the muon. So

$$g' = g, \ g_1 = g\cos\theta, \ \text{and} \ g_2 = g\sin\theta \qquad (4)$$

We should erase the prime in (3).

Given that $g_2 \simeq 0.23 g_1$ and so $\tan\theta \simeq 0.23$, we obtain $g_1 = g\cos\theta \simeq 0.97g$. Recall the longstanding puzzle I mentioned in chapter IX.3 of why the W coupling extracted from neutron beta decay is a tiny bit smaller than that extracted from muon decay. Remarkably, the puzzle was resolved just like that!

Best example ever of an attractive notation suggesting a new question

With the Cabibbo angle $\theta \simeq 0.2$, we see that the combination that the W couples to, $d(\theta) = \cos\theta \ d + \sin\theta \ s$, is mostly d with a bit of s mixed in.

To summarize, the observational fact that $g(\mu \to \nu) \gtrsim g(d \to u) > g(s \to u)$ is now translated into the mathematical statement $g \gtrsim g\cos\theta > g\sin\theta$, with $\theta \simeq 0.2$.

The resolution of why the amplitude for neutron decay was a bit less than that for muon decay was nice, but even more importantly, the "mere notation" $d(\theta) = \cos\theta \ d + \sin\theta \ s$ raised the question of what happened to the orthogonal combination $s(\theta) = -\sin\theta \ d + \cos\theta \ s$. Should we look for $s(\theta)$ as the long-lost sibling of $d(\theta)$ somehow separated "at birth"?

The answer to this question led to a significant breakthrough in particle physics, as we will see shortly. To me, the Gell–Mann–Lèvy rewrite offers the best example ever of an attractive notation suggesting a new question and then leading to previously unsuspected physics.

The Cabibbo angle indicates that the weak interaction is not quite aligned with the strong and electromagnetic interactions. Why such a screwball arrangement exists between the three interactions poses a mystery[3] not yet fully understood.

K_L decay into a pair of muons

Heeding Gell-Mann's famous dictum that in the quantum world, whatever is not forbidden is mandatory, strange particles decay every which way. Among these decays is $K_L \to \mu^+ + \mu^-$: the so-called long-lived (hence the subscript L) neutral K meson decays[4] into a muon μ^- and an antimuon μ^+.

The $K_L = \{\frac{1}{\sqrt{2}}(\bar{s}d - \bar{d}s)\}$ is a linear combination of two bound states, an anti s quark with a d quark, and an anti d quark with an s quark. Why that is so, while an extremely interesting issue in particle physics, also does not concern us in this book; our narrow interest here is in setting up to do an important fly by night calculation.

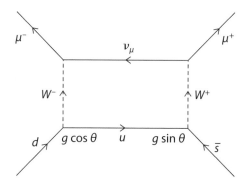

FIGURE 2. The box diagram for the decay $K_L \to \mu^+ + \mu^-$. The anti s quark and the d quark can convert with each other only by going through the u quark.

Focus on the bound state $\{\bar{s}d\}$. Somehow the $\bar{s}$ and the d disappear and turn into μ^+ and μ^-. The first important point is that in (3), d and s do not couple directly via the weak interaction but only "know" about each other through u. Hence the W has to act twice to generate the decay $K_L \to \mu^+ + \mu^-$. This is shown in the Feynman diagram in figure 2, known as a box diagram.

Let us read. The d quark transforms itself into a u quark by emitting a W^-, with an amplitude $g \cos \theta$, as indicated by (3). The u quark then wanders over to the anti s quark, whereupon they annihilate each other and turn into a W^+, with an amplitude $g \sin \theta$, as also indicated by (3). This W^+ then transforms itself into a μ^+ and a muon neutrino ν_μ, with amplitude g, as indicated by (IX.3.2). Finally, the muon neutrino ν_μ absorbs the W^- and changes into a μ^-, with amplitude g, as also indicated by (IX.3.2).

Feynman diagrams allow us to see these quantum processes occurring step by step, so to speak; hence their utility. Once you become proficient, you can dispense with all those words that I just labored to write. To facilitate reading Feynman diagrams, keep in mind that electric charge (as well as momentum) must be conserved at each vertex.

Another way of reading this box diagram is to chop it into two halves by cutting[5] the two W lines. The decay $K_L \to \mu^+ + \mu^-$ can then be thought of as occurring in two stages: d quark and $\bar{s}$ antiquark collide to produce a $W^+ W^-$ pair; this pair then collides to produce a $\mu^+ \mu^-$ pair.[6]

A famous infamous box diagram

Since the W boson has to go to work twice (that is, since the weak interaction has to act twice), you would expect the amplitude for $K_L \to \mu^+ + \mu^-$ to be of order G^2. (If it helps, think about second order perturbation in quantum mechanics.) Indeed, multiplying the coupling strength at the four vertices gives us $g^4 \cos \theta \sin \theta \propto G^2$. But surprise! A simple calculation yields a significantly larger amplitude, as we will see presently.

We have never calculated a Feynman diagram with a closed loop* before. What do we do with the momentum circulating in the loop?

The easiest way to see this circulating momentum is to set the external momenta, p_d, $p_{\bar{s}}$, p_{μ^+}, p_{μ^-}, of the d, $\bar{s}$, μ^+, μ^-, respectively, all to[7] 0. Anyway, they are small compared to the mass M of the W. Let the W^- carry momentum[†] $(-q)$. Then by momentum conservation at each of the four vertices, the u quark and the muon neutrino must both have momentum q, and then also the W^+. In other words, momentum conservation asserts that a momentum q circulates around the loop.

What is q equal to? Since it is not fixed by momentum conservation, it could be anything. So, what do we do about it?

Feynman tells us to integrate over all values of q. While this rule can be derived[8] mathematically, the idea of integrating over q essentially expresses the quantum dictum that anything not forbidden by conservation laws is allowed. All values of q are allowed. Make sense?

But large values of q are suppressed by the propagators. We already know the propagator $\sim 1/(q^2 - M^2)$ for the W. What about the propagators for the u quark and the muon neutrino, both spin-$\frac{1}{2}$ fermions? I state here without proof that their propagators are both $\sim 1/q$ in order not to break the narrative flow. I promise to give a fly by night derivation soon.

Fly by night evaluation of a Feynman integral

Fly by day physicists would now study a quantum field theory textbook, learn the Feynman rules precisely, write down the integral with all the 2s and πs, and integrate using well-developed tricks, one of which was mentioned in math medley 3. I highly recommend that the interested reader do the same. But we will presently do a fly by night calculation instead. Actually, we are not being wild men and women here: This kind of rough and tumble calculation was widely practiced in particle physics. As I have said, after one understands what is going on, one is then free to do the integral exactly.

Putting everything together, we obtain the amplitude

$$\mathcal{M} \sim g^4 \cos\theta \sin\theta \int d^4q \left(\frac{1}{q^2 - M^2}\right)^2 \frac{1}{q}\frac{1}{q} \tag{5}$$

The integral can be evaluated instantly by dimensional analysis: four powers of q in the numerator, six powers in the denominator, thus dimension of $1/q^2$, and since M is the only guy around, the integral $\sim 1/M^2$.

We could also do it somewhat more carefully by arguing that the integrand[9] falls off rapidly for $q \gg M$, and thus the important contribution comes from

---

*We merely applied dimensional analysis to the diagram for radiative muon decay in figure 1 of chapter IX.3.

†The sign choice is purely for convenience.

the region $q \lesssim M$, so that the integral can be replaced by

$$\int^{q \lesssim M} d^4 q \left(\frac{1}{q^2 - M^2}\right)^2 \frac{1}{q^2} \sim \frac{1}{M^4} \int^{q \lesssim M} d^4 q \frac{1}{q^2} \sim \frac{1}{M^4} M^2 \sim \frac{1}{M^2} \quad (6)$$

In the first step, we approximated the W propagator by $1/(q^2 - M^2) \sim -1/M^2$, since $q \lesssim M$. We arrive at the same answer as before, of course.

Punchline: The amplitude for $K_L \to \mu^+ + \mu^-$ comes out to be

$$\mathcal{M} \sim g^4 \cos\theta \sin\theta / M^2 \sim (g^2/M^2)g^2 \sim Gg^2 \sim Ge^2 \sim G\alpha \quad (7)$$

In other words, the amplitude $\mathcal{M}$ is actually of order $G\alpha$, namely, first order in the weak and electromagnetic interactions, instead of the naive expectation of Fermi constant squared, G^2. To say it in words, the many quantum possibilities for q produce a factor of M^2, which overwhelms one factor of G, turning it into the electromagnetic $\alpha \simeq 10^{-2}$. The decay $K_L \to \mu^+ + \mu^-$, instead of being a second order weak process, is down by only a factor of α compared to a first order weak decay, such as $K^+ \to \pi^0 + e^+ + \nu_e$.

Once again, a theoretical estimate far exceeds what experimentalists see. What to do?

Fermion propagator: $5 \neq 4$

To maintain the suspense, let me now digress to give you a fly by night derivation, as promised, that spin-$\frac{1}{2}$ fermions (such as the electron and the neutrino) have propagator $\sim 1/q$, in contrast to the photon propagator $1/q^2$, or the W propagator $1/(q^2 - M^2)$.

Recall from way back that the energy density in an electromagnetic field equals something like $\vec{E}^2 + \vec{B}^2$, and the Lagrangian density is like $\mathcal{L} = \frac{1}{2}(\vec{E}^2 - \vec{B}^2)$. Recall also that $\vec{E}$ and $\vec{B}$ can be packaged into the field strength tensor $F_{\mu\nu} = \partial_\mu A_\nu - \partial_\nu A_\mu$, with A_μ the electromagnetic vector potential. Most recently, we encountered all this in chapter IX.1. But I don't even expect you to know all this intimately, just that the electromagnetic fields $\vec{E}$ and $\vec{B}$ are obtained by taking a partial derivative of A, and thus $F \sim \partial A$. It follows that $\mathcal{L} = -\frac{1}{4}F_{\mu\nu}F^{\mu\nu} \sim (\partial A)^2$ involves two powers of derivative, and this is why the equations of electromagnetism $\Box A_\mu = J_\mu$ and the equation for the Green function $\Box G(x) = \delta^{(4)}(x)$ all involve two derivatives. (See appendix Gr on Green functions.) As we discussed in chapter VII.3, derivatives can neither vanish into thin air nor drop down from the sky. Derivatives are conserved.[10]

We learned in chapter IX.2 that the propagator is just the Fourier transform of the Green function $G(x)$. Since the Fourier transform of $\Box$ is $\sim q^2$, the propagator of the photon is just* $\sim 1/q^2$.

We also learned in chapter IX.1 that the action $S = \int d^4 x \mathcal{L}$ is dimensionless in particle physics units, and hence $[\mathcal{L}] = M^4$. Let us check this: Since $[A] = M$, we have $[\mathcal{L}] = [\partial]^2 [A]^2 = (1/L)^2 M^2 = M^4$ indeed.

---

*Because the Fourier transform of $\Box G(x) = \delta^{(4)}(x)$ reads $q^2 \tilde{G}(q) \sim 1$.

To summarize, electromagnetism, the Lagrangian density, and the equations for the field and for the Green function all involve two spacetime derivatives,[11] and these lead to the quadratic dependence $\sim 1/q^2$ of the photon propagator.

With all this under our belt, we are now ready to give a fly by night derivation of the electron propagator. We already argued in chapter IX.1 that the electron field ψ has dimension $[\psi] = M^{\frac{3}{2}}$. It follows immediately that the Lagrangian density $\mathcal{L}$ for the electron field, in contrast to the Lagrangian density for the photon field, cannot have two powers of ∂, since $[\mathcal{L}] = M^4$ while $[\bar{\psi}\partial\partial\psi] = M^2(M^{\frac{3}{2}})^2 = M^{2+3} = M^5$. Well, $5 \neq 4$: There is room for only one power of ∂. The Lagrangian density is forced to have the schematic form $\mathcal{L} \sim \bar{\psi}\partial\psi$. As a result, the Dirac equation for the field and the equation defining the Green function involve only one spacetime derivative. Thus, the electron propagator has the form[12] $\sim 1/q$, as indicated in (5).

Actually, this fly by night derivation is not too far off. Dirac, in deriving his famous equation[13] in 1928, was motivated by his desire (for reasons of only historical interest now) to have an equation with only one spacetime derivative. A deeper derivation is based on the representation theory of the Lorentz group.[14]

We can even include a mass m for the field, again by dimensional analysis. Since $[\bar{\psi}\psi] = M^3$ already, the only possibility would be $m\bar{\psi}\psi$, if the Lagrangian density $\mathcal{L}$ is to have dimension M^4.

In fact, for the record, the Dirac Lagrangian density for a spin-$\frac{1}{2}$ field is

$$\mathcal{L} = \bar{\psi}(i\gamma^\mu \partial_\mu - m)\psi \tag{8}$$

Once again, you see the gamma matrices γ^μ mentioned but not explained in chapter IX.1. The Lorentz index μ carried by ∂_μ has to be contracted with something. Thus, the electron propagator is actually a matrix inverse defined by $\sim 1/(\gamma^\mu q_\mu - m)$. (By the way, Feynman got tired of writing $\gamma^\mu q_\mu$ and so invented his famous slash notation $\displaystyle{\not{q}} \equiv \gamma^\mu q_\mu$.)

This also clears up a confusion that you might have had. What does $1/q$ mean if q is actually a momentum vector? Well, it means the inverse of the matrix $\displaystyle{\not{q}}$.

Charm

Back to the puzzle of the theoretical estimate $K_L \to \mu^+ + \mu^-$ exceeding the observed rate! Remarkably, the resolution calls on the other puzzle mentioned earlier of why the W "sees" the combination $d(\theta) = \cos\theta \ d + \sin\theta \ s$, connecting it to the u quark, but ignores the orthogonal combination $s(\theta) = -\sin\theta \ d + \cos\theta \ s$.

Glashow, Iliopoulos, and Maiani[15] solved both problems in 1970 by proposing that, to the contrary, the W does know about $s(\theta)$ and transforms it into a hitherto unknown quark c they called the "charm quark." They postulated that the charm quark c is significantly more massive than the three quarks, u, d, s,

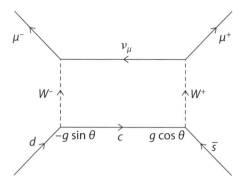

FIGURE 3. Compare this with figure 2. Find the crucial differences. The up quark u has been replaced by the charm quark c, two coupling constants have been interchanged, and an all-important minus sign introduced. The invention of this diagram contributed to Glashow's Nobel Prize.

known at the time, and hence was unknown. Particles containing c would have been too massive to be produced at accelerators.

Thus, the coupling of the W to quarks given in (3) is to be amended (recall that we already set $g' = g$) by adding the term

$$\mathcal{L} = g \left(- \sin\theta \; \bar{c}\gamma^\lambda d + \cos\theta \; \bar{c}\gamma^\lambda s \right) W_\lambda + \text{h.c.} = g \, \bar{c}\gamma^\lambda s(\theta) \, W_\lambda + \text{h.c.} \quad (9)$$

What about the decay $K_L \to \mu^+ + \mu^-$? Now, in addition to the Feynman diagram in figure 2 describing the decay, we have another diagram (see figure 3) obtained by replacing the up u quark by the charm c quark, but with a crucial difference: The coupling $g\cos\theta$ in figure 2 is replaced in the new diagram by $-g\sin\theta$, and the coupling $g\sin\theta$ is replaced by $g\cos\theta$, as we see by comparing (9) with (3). The charm quark contributes to the decay amplitude for $K_L \to \mu^+ + \mu^-$ an additional amount given by (5) with the combination $\cos\theta\sin\theta$ replaced by $-\sin\theta\cos\theta$. In other words, the additional amount is obtained, duh, by flipping the overall sign of (5).

Hence, while the diagram with the u quark contributes $g^4 \cos\theta \sin\theta/M^2$ to the decay amplitude $\mathcal{M}$, the diagram with the c quark contributes $-g^4 \sin\theta \cos\theta/M^2$. Cancellation! These two diagrams cancel each other out completely. For convenience, let us write $\mathcal{M} = \mathcal{M}_u + \mathcal{M}_c$, evidently with $\mathcal{M}_u$ and $\mathcal{M}_c$ denoting the contribution of the u quark and the c quark to $\mathcal{M}$, respectively. We just found that $\mathcal{M}_c = -\mathcal{M}_u$ and thus $\mathcal{M} = 0$: The decay rate from $K_L \to \mu^+ + \mu^-$ just went from too large to 0.

This is bad. Experimentalists did observe this decay, albeit with a rate much smaller than predicted. But then we remember that we did not include the mass m_u of the u quark in (5). We have to put m_u into $\mathcal{M}_u$, and similarly the mass m_c of the c quark into $\mathcal{M}_c$. Since quark masses were assumed to be small compared to the W mass M (which of course was unknown at the time), we can expand to leading order in m/M. Furthermore, since $\mathcal{M} = \mathcal{M}_u + \mathcal{M}_c$ would vanish if $m_c = m_u$, we expect $\mathcal{M} \propto (m_c - m_u)$, the difference in the masses of the c quark and the u quark. Actually, for a simple quantum field theoretic reason,[16]

$\mathcal{M} \propto (m_c^2 - m_u^2) = (m_c - m_u)(m_c + m_u)$. Since by assumption, m_u is negligible compared to m_c, we have $(m_c^2 - m_u^2) \simeq m_c^2$.

Thus, we expect the residue after cancellation to be given by $g^4 \cos\theta \sin\theta / M^2$ multiplied by the factor $(m_c^2 - m_u^2)/M^2 \simeq m_c^2/M^2$. The net value of the amplitude is reduced by this almost-cancellation to

$$\mathcal{M} \sim (g^4 \cos\theta \sin\theta / M^2)(m_c^2/M^2) \sim G^2 m_c^2 \cos\theta \sin\theta \qquad (10)$$

in line with experimental observation. Everything worked like a charm, Glashow et al said.

In summary, and at the risk of repetition, the contributions of the u and c quarks would have added up to zero were their masses neglected. Including their small masses thus means that the net contribution is suppressed by a factor $(m_c^2 - m_u^2)/M^2 \simeq m_c^2/M^2 \ll 1$, much less than naive expectation.

The particularly attractive feature of this bold proposal is that the unknown charm quark cannot be too massive, since that would make $\mathcal{M}$ large, thus defeating its purpose in life. The excitement over a possible fourth quark culminated in the discovery in 1974 of particles containing the charm quark.

Another attractive feature, based more on aesthetics, is that with four quarks, the lepton and the quark sectors would have a pleasing symmetry between them. Let us introduce a doublet notation to indicate that the $W^\pm$ could transform the top entry into the bottom entry of a doublet and vice versa. Then much of the weak interaction can be summarized by writing

$$\begin{pmatrix} \nu_e \\ e \end{pmatrix} \quad \begin{pmatrix} \nu_\mu \\ \mu \end{pmatrix} \quad \Big| \quad \begin{pmatrix} u \\ d(\theta) \end{pmatrix} \quad \begin{pmatrix} c \\ s(\theta) \end{pmatrix} \qquad (11)$$

In other words, the W boson couples to leptons and quarks according to

$$\mathcal{L} = g\left(\bar{\nu}_e \gamma^\lambda e + \bar{\nu}_\mu \gamma^\lambda \mu + \bar{u}\gamma^\lambda d(\theta) + \bar{c}\gamma^\lambda s(\theta) \right) W_\lambda + \text{h.c.} \qquad (12)$$

In fact, after the realization that $\nu_e \neq \nu_\mu$ but some years before the resolution of the $K_L \to \mu^+ + \mu^-$ decay rate puzzle, theorists[17] had already proposed the existence of four quarks by appealing to aesthetics. A sense of balance, I guess: four leptons and four quarks. These days, we have six leptons and six quarks, but still no deep understanding of the family problem.[18] Why does Nature choose to repeat Herself?

Notes

[1] This was already implicit when we talked about neutron beta decay. While one of the d quarks decays, the other d quark and the u quark spectate.

[2] M. Gell-Mann and M. Lèvy, *Il Nuovo Cimento* 16 (1960), pp. 705–726. See also the mention in an endnote in chapter IX.3.

[3] For an early attempt to solve this, see F. Wilczek and A. Zee, *Physical Review* D15 (1977), p. 3701; and *Physical Review Letter* 42 (1979), p. 421; and references cited therein.

[4] Most of the time, this meson decays into three pions: $K_L \to \pi^+ + \pi^0 + \pi^-$, but this does not concern us here.

[5] Known as a Cutkosky cut. See *QFT Nut*, p. 215.

[6] Authors of popular books sometimes feel obliged to provide a cartoon for their readers. See, for example, *Fearful*, p. 238. But now that you know Feynman diagrams, you are way past that.

[7] If we don't, we would reach the same conclusion somewhat more laboriously. Assign a momentum to each of the four internal lines. At each of the four vertices, there is a 4-dimensional δ function enforcing momentum conservation. One of these reduces to a δ function specifying overall momentum conservation $p_d + p_{\bar{s}} = p_{\mu^+} + p_{\mu^-}$. The other three fix 3 of the 4 internal momenta. The one that is left free could be denoted by q.

[8] See, for example, *QFT Nut*, p. 57. It's easy.

[9] Some readers might be worried by the pole at $q^2 = M^2$. The answer is that we should Wick rotate to Euclidean spacetime. See any quantum field theory textbook.

[10] As mentioned in chapter VII.3, in more advanced discussions, such as the one regarding the form of $\mathcal{L} \sim (\partial A)^2$, the derivative ∂ remains unexecuted.

[11] For a few more words on this, see appendix Eg.

[12] For a really handwaving argument, you could try saying that, since the electron has spin $\frac{1}{2}$ and the photon spin 1, the electron propagator should go like the square root of the photon propagator $\sim 1/q^2$, and hence like $\sqrt{1/q^2} \sim 1/q$.

[13] See, for example, *QFT Nut*, p. 93.

[14] See *Group Nut*, chapter VII.4.

[15] S. L. Glashow, J. Iliopoulos, and L. Maiani, *Physical Review D* (1970), p. 1285.

[16] We learned in the preceding section that the propagator for a massive spin-$\frac{1}{2}$ fermion is given by $\frac{1}{q-m}$. Since $m \ll q \sim M$, we can expand $\frac{1}{q-m} = \frac{1}{q} + \frac{1}{q}m\frac{1}{q} + \frac{1}{q}m\frac{1}{q}m\frac{1}{q} + \cdots$. The terms odd in q vanish upon integration, since the rest of the integrand is even in q. (We live in 4-dimensional spacetime!) When we combine the contribution of the u and c quarks, the leading term $\frac{1}{q}$ drops out, since it does not know about the quark masses. The term $\frac{1}{q}m\frac{1}{q}m\frac{1}{q} \propto m^2$ gives the contribution proportional to $(m_c^2 - m_u^2)$.

[17] See papers by Z. Maki and Y. Hara, and by J. Bjorken and S. Glashow.

[18] Shucks, the classic Hollywood movie *Seven Brides for Seven Brothers* (1954) does not shed any light here.

Appendix Cp

Critical points

As promised, we now determine the critical quantities T_c, P_c, and v_c of a van der Waals gas. (Refer back to chapter V.2 for notation, etc.)

First, a quick fly by night determination using dimensional analysis. Since $[a] = EL^3$ and $[b] = L^3$, we obtain, almost instantaneously, $T_c \sim a/b$, $P_c \sim a/b^2$, and $v_c \sim b$.

Next, we put on our fly by day hat. Start by recalling

$$P = \frac{T}{v - b} - \frac{a}{v^2} \tag{1}$$

and

$$\left. \frac{\partial P}{\partial v} \right|_T = -\frac{T}{(v - b)^2} + \frac{2a}{v^3} \tag{2}$$

Physically, v must exceed b.

Here is a quick review of the physics behind the phase transition, as discussed in chapter V.2. Look at figure 2 there. For sufficiently high temperature T, the slope $\left. \frac{\partial P}{\partial v} \right|_T$ is negative. But for sufficiently low temperature, as v increases past b, the slope $\left. \frac{\partial P}{\partial v} \right|_T$ goes from negative to positive and back to negative, crossing zero twice: $P(v)$ has a minimum and a maximum, as you can see in the figure. As T increases, the two extrema move closer to each other, then merge and disappear at some critical temperature T_c. The values of P and v at which the merging occurs are denoted by P_c and v_c, respectively.

Newton and Leibniz taught us that the minimum and maximum of $P(v)$ occur when $\frac{\partial P}{\partial v}|_T = 0$, namely, when

$$Tv^3 = 2a(v - b)^2 \tag{3}$$

For ease of exposition, define $f(v) \equiv Tv^3$ and $g(v) \equiv 2a(v-b)^2$. A potential confusion is that the cubic $f(v) = g(v)$ should have three roots, so that $P(v)$ would have three extrema, not two. But a moment's reflection shows that one root occurs for $v < b$, which is unphysical. To see this immediately, plot $f(v)$ and $g(v)$ for low T. As v increases, $f(v)$ starts out small compared to $g(v)$. But $g(v)$ decreases and then vanishes at $v = b$: manifestly, $f(b) > g(b) = 0$. As v increases past b, both functions increase, but $f(v)$ increases more slowly, since by assumption, T is small. But eventually, for large v, $f(v) > g(v)$ again. Thus, $f(v)$ must have intersected $g(v)$ twice in the region $v > b$. A picture is worth a thousand words: just draw a picture, and all is clear!

Hence, for $v > b$, the two curves $f(v)$ and $g(v)$ intersect twice. But as T increases toward the critical T_c, the intersection points move closer together, until the two curves just touch at some $v_c > b$.

The critical values are thus determined by equating the two functions and their slopes, that is, by solving $f(v) = g(v)$ and $f'(v) = g'(v)$ simultaneously. The critical pressure P_c is then determined by plugging into (1). I leave it to you to carry this calculation to the bitter end, and verify that the results are consistent[1] with our fly by night expectations, of course.

Note

[1] But numerically far off: for example, $P_c = a/27b^2$.

Appendix Del

Delta function

The Dirac[1] delta function, or simply delta function, is almost indispensable in theoretical physics. This brief account is intended for those readers who might need a brush up.

Consider a function $f(x)$ that rises sharply just before $x = 0$, rapidly reaches its maximum at $x = 0$, and then drops rapidly to 0. Except for a tiny interval around $x = 0$, $f(x)$ is practically 0 everywhere.

The precise form of the function does not matter. For example, we could take $f(x)$ to rise linearly from 0 at $x = -a$, reaching a peak value of $1/a$ at $x = 0$, and then to fall linearly to 0 at $x = a$. For $x < -a$ and for $x > a$, the function $f(x)$ is defined to be zero. As another example, consider the scaled Gaussian

$$f(x) = \frac{1}{(2\pi)^{\frac{1}{2}} a} e^{-\frac{1}{2} x^2 / a^2} \tag{1}$$

In the limit $a \to 0$, both functions are (infinitely) sharply peaked, with width $\sim a$ going to 0 and peak value $\sim 1/a$ going to infinity. The scaled Gaussian is practically 0 everywhere except for the interval $-a \lesssim x \lesssim a$. The limit of such sharply peaked functions is known to physicists as the delta function[2] $\delta(x)$. The delta function is traditionally normalized by

$$\int dx \, \delta(x) = 1 \tag{2}$$

We have chosen both of our examples such that $\int dx \, f(x) = 1$.

Speaking loosely, physicists think of $\delta(x)$ as a function whose juice is all concentrated at the point $x = 0$. This implies that, for a sufficiently smooth function $g(x)$, $\int_{-\infty}^{\infty} dx \, g(x) \delta(x - c) = g(c)$. In other words, the integral picks out the value of the function $g(x)$ at an arbitrary point $x = c$.

The delta function goes all the way back to Joseph Fourier, who gave the immensely useful[3] representation

$$\delta(x) = \int_{-\infty}^{\infty} \frac{dk}{2\pi} e^{ikx} \qquad (3)$$

(To verify this, integrate from $-K$ to K, and study the result as a function of x in the limit $K \to \infty$.)

An important identity in 3-dimensional space

Since much of physics occurs in 3-dimensional space, it is natural to define the 3-dimensional delta function $\delta^3(\vec{x}) \equiv \delta(x)\delta(y)\delta(z)$. For example, the charge density $\rho(\vec{x})$ of a point particle carrying charge e and sitting at the point $\vec{c}$ in space would be given by $\rho(\vec{x}) = e\delta^3(\vec{x} - \vec{c})$.

We now introduce an identity of great importance:

$$\nabla^2 \left(\frac{1}{r} \right) = -4\pi \delta^3(\vec{x}) \qquad (4)$$

with $r^2 = x^2 + y^2 + z^2$, as usual.

Also as usual, it is convenient to introduce indices, with $\vec{x} = (x^1, x^2, x^3) = (x, y, z)$. Differentiating $r^2 = \sum_i (x^i)^2$, we obtain $r dr = \sum_i x^i dx^i$, and hence $\frac{\partial r}{\partial x^i} = \frac{x^i}{r}$. In other words,

$$\vec{\nabla} r = \frac{\vec{x}}{r} \qquad (5)$$

Therefore, in 3-dimensional space,

$$\nabla^2 \frac{1}{r} = \vec{\nabla} \cdot \vec{\nabla} \frac{1}{r} = \sum_i \frac{\partial}{\partial x^i} \left(-\frac{1}{r^2} \frac{x^i}{r} \right) = \sum_i \left(\frac{3}{r^4} \frac{x^i}{r} x^i - \frac{1}{r^3} \frac{\partial x^i}{\partial x^i} \right) = \frac{3}{r^3} - \frac{3}{r^3} = 0 \quad (6)$$

The left side of (4) indeed vanishes, except possibly at $\vec{x} = 0$. At this point, we already know that $\nabla^2(\frac{1}{r})$ equals some constant times the delta function.

To obtain the (-4π) in (4), simply integrate over a small ball B of radius a centered at $\vec{x} = 0$. Using Gauss's theorem to turn the volume integral into a surface integral over the sphere S that bounds B, we obtain

$$\int_B d^3x \nabla^2 \frac{1}{r} = \int_B d^3x \vec{\nabla} \cdot \vec{\nabla} \frac{1}{r} = \int_S d\vec{S} \cdot \left(-\frac{1}{r^2} \frac{\vec{x}}{r} \right) \bigg|_{r=a} = -4\pi \qquad (7)$$

We have thus verified (4) and understood that the 4π comes from the surface area of a unit sphere. This identity is basic to electromagnetism, for example, and hence will appear in appendices Gr and M (and also in chapter IX.2).

Exercises

(1) Generalize Fourier's representation to 3-dimensional space.

(2) Verify this useful representation:

$$\frac{e^{-mr}}{4\pi r} = \int_{-\infty}^{\infty} \frac{d^3 k}{(2\pi)^3} \frac{e^{i\vec{k}\cdot\vec{x}}}{\vec{k}^2 + m^2} \tag{8}$$

In particular, the Coulomb potential can be represented by setting m to 0. We will need the result of this exercise in chapter IX.2.

Notes

[1] Also introduced by Cauchy, Poisson, Hermite, Kirchhoff, Kelvin, Helmholtz, and Heaviside. See J. D. Jackson, *American Journal of Physics* 76 (2008), pp. 707–709.

[2] Rigorous mathematicians go berserk at physicists' use of the word "function" here; they prefer to call it a distribution, defined as the limit of a function. But working physicists do not give a flying barf about such niceties. In any case, I do not personally know a theoretical physicist suffering any harm by calling $\delta(x)$ a function.

[3] Exploited almost constantly in quantum field theory.

Appendix Eg

Einstein gravity: a lightning review

A whiff of a profound theory

Evidently, in a few pages I can do no more than give you a flavor of Einstein gravity.[1] I will just tell you what I need for our fly by night trips. If you are puzzled by any sentence below, you could almost surely find it expanded, elaborated on, and explained in detail in my, or any other, textbook on Einstein gravity.

For those readers who know, this appendix serves as a reminder. For those who do not, it is meant to whet their appetites.

The parable of the curious flyer

Start with a parable.[2] Imagine flying from Los Angeles to Taipei. Looking idly at the flight map, you might notice that the plane follows a curved path arcing toward the Bering Strait. Is the Bering Strait exerting a mysterious attractive force on the plane? See figure 1.

On your next trip, you try another airline. This pilot follows exactly the same curved path. Don't these pilots have any sense of personality or originality? Why don't they sometimes, just for the heck of it, swing south and fly over Hawaii? They seem to prefer flying over[3] grim and unsuspecting Inuit hunters rather than cheerful Polynesian maidens.

Not only is the mysterious force attractive, it is universal, independent of the make of the airplane. Should you seek enlightenment from the guy sitting next to you? Dear reader, surely you are chuckling. You know perfectly well that the Mercator projection distorts the earth, and pilots follow scrupulously

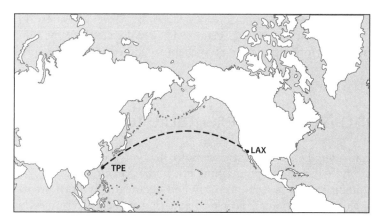

FIGURE 1. Is the Bering Strait exerting a mysterious attractive force on airplanes flying from Los Angeles to Taipei? From Zee, A. *On Gravity: A Brief Tour of a Weighty Subject*, Princeton University Press, 2018.

the shortest possible path between Los Angeles and Taipei. The answer to the universality of the mystery force is to be sought, not in the physics, but in the economics, department.

Gravity, too, acts universally. We all learned in school that Galileo dropped cannonballs off the Leaning Tower of Pisa to see whether they would all hit the ground at the same time. Only a small fraction of school children now grown up, but assuredly including the readers of this book, now remember why he did this. The rest of our fellow citizens would guess that Galileo was either crazy or high.

But you and I know that the gravitational force exerted by the earth, of mass M and radius R, on a cannonball of mass m, is given by $F = GMm/R^2$, and hence, according to Newton for the second time in a single sentence, the acceleration of the cannonball equals $a = F/m = GM/R^2$. This profound but elementary bit of math, that something divided by itself gives 1 ($m/m = 1$), indicating that all falling objects on the surface of the earth accelerate toward the ground at the same rate,[4] is known as the universality of gravity.

No gravity, only the curvature of spacetime

To Einstein, that an apple and a stone fall in exactly the same way in a gravitational field is no more amazing than different airlines, regardless of national or political affiliation, choosing exactly the same path getting from Los Angeles to Taipei. We might see an "obvious" connection, but hindsight[5] is, of course, too easy.

For three hundred years, the universality of gravity had been whispering "curved spacetime" to us. Finally, Einstein heard it.

Just as there is no mysterious force emanating from the Bering Strait, one could say that there is no gravity, only the curvature of spacetime. The gravity we observe is due to the curvature of spacetime. More accurately, gravity is equivalent to the curvature of spacetime: They really are the same thing.

We did not go looking for curved spacetime; curved spacetime came looking for us!

Making Newton presentable to Lorentz

Imagine, in a wildly anachronistic scenario, that you are to present[6] Newton to Lorentz, who will not accept anything not Lorentz invariant.

Newtonian gravity is ostentatiously and aggravatingly not Lorentz invariant. Consider, in particular, Poisson's equation, which determines Newton's gravitation potential ϕ given the mass density ρ:

$$\vec{\nabla}^2 \phi \sim G\rho \tag{1}$$

It rudely clashes with Lorentz invariance.

Electromagnetism illuminates the way out. Compare this equation with the equation $\vec{\nabla}^2 \phi \sim e\rho$ determining Coulomb's potential ϕ, given the electric charge density ρ. These two equations are structurally the same, with G replaced by e. Physicists even use the same letters.

Well, you know full well (or, if you don't know, you would see in appendix M) how electromagnetism got reconciled* with Lorentz invariance. This equation of electrostatics was replaced by

$$\left(\vec{\nabla}^2 - \frac{\partial^2}{\partial t^2}\right)A^\mu \sim eJ^\mu \tag{2}$$

with the index $\mu = 0, 1, 2, 3$ and the spacetime coordinates $x^\mu = (x^0, x^i) = (t, \vec{x})$.

To start with, the charge density ρ is promoted in (2) to the time component J^0 of a 4-vector J^μ, whose spatial components J^i form the usual current density 3-vector $\vec{J}$.

This much would have been obvious even to Galileo, if he had known about charges and currents. Consider a box of electric charges. To an observer, call him Mr. Prime,[7] in motion relative to the box, the charges are moving and constitute an electric current. Evidently, ρ and $\vec{J}$ come together.

What would not be at all obvious to Galileo are the Lorentz factors like $1/\sqrt{1 - \frac{v^2}{c^2}}$ characteristic of special relativity. But you, a 21st century reader, could explain to Galileo that, physically, the box is Lorentz contracted. Volume goes down, and so density goes up. Mr. Prime sees the charge density enhanced by $\left(1/\sqrt{1 - \frac{v^2}{c^2}}\right) > 1$.

To match the 4-vectorial nature of the right side, the electrostatic potential ϕ on the left side has to be promoted to the time component A^0 of a 4-vector potential $A^\mu = (A^0, A^i) = (\phi, \vec{A})$. This 4-vector potential's spatial components

---

*Indeed, historically, electromagnetism begat Lorentz invariance.

A^i form the vector potential $\vec{A}$ that some late 19th century theoretical physicists thundered against.[8]

Finally, a notational "triviality": With the coordinates $x^\mu = (x^0, x^i) = (t, \vec{x})$, it is convenient to introduce the corresponding partial derivatives $\partial_\mu \equiv \frac{\partial}{\partial x^\mu} = (\frac{\partial}{\partial x^0}, \frac{\partial}{\partial x^i}) = (\frac{\partial}{\partial t}, \frac{\partial}{\partial x}, \frac{\partial}{\partial y}, \frac{\partial}{\partial z}) = (\frac{\partial}{\partial t}, \vec{\nabla})$. We will use these derivatives below.

Gravity versus electromagnetism: a key difference

Now that we have electromagnetism solidly under our belt, we can sketch the same doodle for gravity. Back to the equation (1), $\vec{\nabla}^2 \phi \sim G\rho$, for Newtonian gravity. First off, in special relativity, the mass density ρ must be promoted[9] to the energy density ε.

We see here the crucial difference between gravity and electromagnetism that makes gravity so much more difficult. Now consider a box containing some massive particles. Previously, to Mr. Prime, the charge density is enhanced because the box is Lorentz contracted. Here, not only is the box Lorentz contracted, but the masses in the box constituting ρ are also moving. The energy of a mass in motion is enhanced from m to $m/\sqrt{1 - \frac{v^2}{c^2}}$.

Moving masses carry kinetic energy. Compared to the charge density, which is enhanced once, mass density is enhanced doubly, by two factors of $\left(1/\sqrt{1 - \frac{v^2}{c^2}}\right)$!

Mathematically, this means that, in contrast to charge density, which is promoted to the time component J^0 of a 4-vector J^μ, mass density is promoted to the time-time component of a 4-tensor $T^{\mu\nu}$, known as the energy momentum tensor. Specifically, T^{00} is the energy density ε. (Those readers who have studied fluid dynamics or elasticity are already acquainted with T^{ij}, the space-space components of $T^{\mu\nu}$, known as the stress tensor in those subjects.)

Enhanced doubly, hence a tensor instead of a vector. Two indices instead of one index: get it? Math and physics work hand in hand.

Indeed, the attentive reader might have realized that we already encountered $T^{\mu\nu}$ in chapter VII.2, together with the physical argument presented here.

Tensor on the right, then tensor on the left

But if the right side of Poisson's equation (1) transforms like a tensor, then the left side, namely $\vec{\nabla}^2 \phi$, must also be replaced by something that transforms like a tensor. In particular, the gradient operator $\frac{\partial}{\partial x^i} \equiv \nabla_i$ in $\vec{\nabla}^2$ must bring its friend the time derivative $\frac{\partial}{\partial t}$ also into play: In special relativity, time and space derivatives are packaged into a 4-vector $\partial_\mu \equiv \frac{\partial}{\partial x^\mu}$, as was just noted and already used[10] in (2).

Another clue is that the tensor we are searching for to put on the left side has to reduce to $\vec{\nabla}^2\phi$ in the Newtonian limit. Since derivatives cannot pop in and out of thin air,[11] this tensor must contain two powers of ∂_λ. Furthermore, we must figure out what ϕ is promoted to.

The bottom line is that we expect Newton's equation to generalize to something of the form

$$(\cdots\partial\cdots\partial\cdots)^{\mu\nu} = GT^{\mu\nu} \tag{3}$$

It is only a slight exaggeration to say that it took Einstein ten years of arduous work to figure out what this mystery expression $(\cdots\partial\cdots\partial\cdots)^{\mu\nu}$ must be! So there ain't noway nohow that you could grasp it in two seconds.

This point about two powers of ∂ figures in part IX and in some sense permeates physics.

Why did the flat earth theory last for so long?

At this point, we make contact with the guy on the airplane wondering about the attractive force exerted by the Bering Strait. So, gravity has something to do with the curvature of spacetime?

To understand the curvature of spacetime, we first have to understand the curvature of space. Let's start with a question. Do you know why it took humans so long to realize that the world was round?

Of course you know that answer. Because the world is locally flat, and humans can't walk very fast. In everyday life, we have no need to know that the world is actually round, as long as the distance of interest is small compared to the earth's radius. If you focus on a small enough region on any smooth surface (in fact any space in any dimension for that matter), it's going to look flat, and you could simply study the small deviation from flat space. That is the core idea in Riemannian geometry, which we will now quantify.

First, we have to grasp the concept of a metric. On Euclid's plane, Pythagoras tells us that the infinitesimal distance ds between two nearby points with Cartesian coordinates (x,y) and $(x+dx, y+dy)$ is given by $ds^2 = dx^2 + dy^2$. Next, consider a sphere of unit radius. The points on the sphere are located by latitude[12] θ and longitude φ. The distance ds between two neighboring points with coordinates (θ, φ) and $(\theta+d\theta, \varphi+d\varphi)$ is given by the familiar $ds^2 = d\theta^2 + \sin^2\theta d\varphi^2$. See figure 2.

For a sufficiently smooth surface, with (x,y) two suitable coordinates, not necessarily Cartesian, this expression is generalized to $ds^2 = g_{xx}dx^2 + g_{xy}dxdy + g_{yx}dydx + g_{yy}dy^2$. (Evidently, $g_{xy} = g_{yx}$, since by definition, $dxdy = dydx$.)

For a D-dimensional manifold, this expression is further generalized to $ds^2 = g_{\mu\nu}(x)dx^\mu dx^\nu$, with the repeated indices μ, ν summed over $1, \ldots, D$. For example, for the unit sphere, $g_{\theta\theta} = 1$, $g_{\theta\varphi} = 0$, and $g_{\varphi\varphi} = \sin^2\theta$; the metric is

North pole

FIGURE 2. The metric on a sphere: The distance between two nearby points with the same latitude but with longitudes differing slightly by $d\varphi$ is given by $\sin\theta\, d\varphi$. Note that $\sin\theta$ is equal to 1 at the equator, decreases steadily as we move north, and vanishes at the north pole. From Zee, A. *On Gravity: A Brief Tour of a Weighty Subject*, Princeton University Press, 2018.

diagonal but not constant (of course). In general, the metric $g_{\mu\nu}(x)$ can be thought of as a real symmetric* D by D matrix.

In 1828, Gauss the Great discovered that the curvature of a surface is entirely determined by its metric. He was so struck by this fact that he named it the Theorema Egregium,[13] the "outstanding" or "extraordinary" theorem. Bernhard Riemann, who was two years old when this theorem was discovered and who as a student attended Gauss's lectures, took the profound step of extending this remarkable fact to arbitrary dimensions. Given a metric $g_{\mu\nu}$, Riemann can tell us what the curvature is. He found an expression for what is now known as the Riemann curvature tensor, constructed out of $g_{\mu\nu}$.

Needless to say, I am hardly going to explain this monument of mathematics to you in a page or less.[14] Instead, I will motivate, as concisely as I am able, what we need for our purposes here.

Since you would agree that the curvature of a sphere is the same everywhere, we can choose to look at the point where the calculation is particularly simple, namely, a point P on the equator, say, $\theta = \frac{1}{2}\pi + \epsilon$, $\varphi = 0$. Then $g_{\varphi\varphi} \simeq (1 - \frac{1}{2}\epsilon^2)^2 \simeq 1 - \epsilon^2$, and thus $\frac{\partial^2 g_{\varphi\varphi}}{\partial\theta^2} = -2$. Up to some sign and normalization, this is the curvature of the sphere.[15] Of course, this proves nothing, since the expression $\frac{\partial^2 g_{\varphi\varphi}}{\partial\theta^2}$ depends on θ. It can't be that easy! Otherwise, Gauss would hardly call it the Theorema Egregium. What this handwaving discussion in the

---

*That is, with $g_{\mu\nu}(x) = g_{\nu\mu}(x)$.

fly by night spirit of this book does show is that a more complicated formula involving two partial derivatives acting on the metric $g_{\mu\nu}$ and its matrix inverse might work.

The crucial insight, as previewed just now, is that in a small enough neighborhood of the point P, the world looks flat, the deviation from flat appearing only in second order (namely, ϵ^2 in this example).

It is instructive to see how this works in Cartesian coordinates. Consider a sphere of radius L defined by $x^2 + y^2 + z^2 = L^2$. Since all points on it are equivalent, focus on the plane touching the north pole and perpendicular to the z-axis joining the north and south poles. The northern hemisphere is defined by $z = +\sqrt{L^2 - x^2 - y^2}$. Near the north pole, $z \simeq L(1 - \frac{1}{2L^2}(x^2 + y^2) + \cdots)$: The sphere is locally an inverted parabolic bowl and is well approximated by the tangent plane. The surface is locally flat, and the deviation from flatness is second order and hence has gone unnoticed by Inuit hunters for ages.

In general, around a point P on a Riemannian manifold[16] let us, for writing convenience, first shift our coordinates x so that the point P is labeled by $x = 0$. Expand the metric around P out to second order:[17] $g_{\mu\nu}(x) = g_{\mu\nu}(0) + A_{\mu\nu,\lambda}x^\lambda + B_{\mu\nu,\lambda\sigma}x^\lambda x^\sigma + \cdots$.

A basic theorem in linear algebra assures us that we can diagonalize and scale the matrix $g_{\mu\nu}(0)$ so that it can be replaced by the unit matrix[18] $\delta_{\mu\nu}$. Next, a simple counting argument shows that the linear deviation denoted by A above can be transformed away.[19] Around the point P, we end up with the metric $g_{\mu\nu}(x) = \delta_{\mu\nu}(0) + B_{\mu\nu,\lambda\sigma}x^\lambda x^\sigma + O(x^3)$.

The Riemann curvature tensor, written as $R_{\mu\nu\lambda\sigma}$, is given[20] by the expansion coefficients B, which we can extract by differentiating the metric with respect to x twice. Thus, as we anticipated, curvature is given by an expression of the form $(\cdots \partial \cdots \partial \cdots)$. Our discussion here is for curved space, but it generalizes immediately to curved spacetime.

Now comes the moment of excitement: Physics and mathematics come together.

Tying three strands together

That was a lot to absorb, so let me summarize. The narrative consists of three strands:

(1) Objects affected by gravity follow universal paths in spacetime, which suggests that the curvature of spacetime is responsible.
(2) Newton and Lorentz tell us that the left side of the field equations for gravity must involve two partial derivatives with respect to spacetime coordinates, something like $(\cdots \partial \cdots \partial \cdots)^{\mu\nu}$.
(3) Gauss and Riemann tell us that the curvature tensor has the form $(\cdots \partial \cdots \partial \cdots)$.

Einstein ties these three strands together and writes down his celebrated field equation

$$E^{\mu\nu} = GT^{\mu\nu} \qquad (4)$$

where the Einstein tensor $E^{\mu\nu} \equiv R^{\mu\nu} - \frac{1}{2}g^{\mu\nu}R$ is given in terms of $R^{\mu\nu}$ and R, known as the Ricci tensor and the scalar curvature, respectively, and constructed out of the Riemann curvature tensor $R_{\mu\nu\lambda\sigma}$. Meanwhile, ϕ has been promoted to the metric tensor $g_{\mu\nu}$.

Compare this equation with (1). Einstein's equation, while it looks a lot more involved, logically and naturally extends Newton's equation. In a suitable limit, Einstein's equation reduces to Newton's equation, as it must.

Energy momentum tensor

We already know that the time-time component T^{00} of the energy momentum tensor $T^{\mu\nu}$ corresponds to the energy density ϵ. What about the other components?

Consider a gas of point particles zipping around with 4-momentum p_a^μ, with a labeling the particles. By Lorentz symmetry, the tensor $T^{\mu\nu}$ must be given by $p_a^\mu p_a^\nu$ summed over all the particles found in a small region around $\vec{x}$ and averaged over some time interval centered at t. Then by isotropy, $T^{0i} = 0$; on average, no special direction is favored. T^{ij} measures the amount of $p_a^i p_a^j$, that is, the amount of zipping around in the gas. This can only correspond to the pressure P at that point in spacetime.[21] Furthermore, again by isotropy, $T^{xx} = T^{yy} = T^{zz}$. Or, in a more compact notation, $T^{ij} \propto \delta^{ij}$. I have used this result in chapters IV.2 and VII.3.

Notes

[1] After all, my textbook *GNut* on Einstein gravity runs to 866 pages.

[2] Adapted in part from my book *On Gravity*.

[3] I am abusing geography slightly.

[4] As the reader surely knows, I am talking about the equality of inertial mass and gravitational mass, experimentally verified to an impressive degree of accuracy.

[5] Staircase wit, l'esprit d'escalier, Treppenwitz, firing the cannon after the cavalry had already charged by you.

[6] Perhaps you would like Lorentz to consider Newton as a potential son-in-law, even though the historical Newton was a lifelong bachelor opposed to marriage.

[7] He first appeared on p. 18 of *GNut*, together with Ms. Unprime.

[8] Saying that only the measurable quantities $\vec{E}$ and $\vec{B}$ should be admitted into physics.

[9] Indeed, this is the physics behind $E = mc^2$.

[10] Indeed, see our discussion of electromagnetism in chapter II.1.

[11] See endnote 10 in chapter IX.4 about the conservation of derivatives. This argument shows that Einstein's equation must have at least some terms containing two powers of ∂. There could be additional terms that disappear in the Newtonian limit. For instance, various spacetime derivatives could act on expressions that tend to a constant in that limit.

[12] Differing from the latitude of everyday life, physicists' latitude assigns 0 to the north pole and $\pi/2$ to the equator.

[13] The meaning of this Latin word has been much distorted in the English word "egregious." Here is to hoping that you find your Theorema Egregium some day.

[14] Consult any number of books. For a particularly gentle introduction, see *GNut*. In particular, consult p. 83.

[15] You might be expecting the curvature to depend on the inverse square of the radius, but recall that we already scale the radius out by setting it to 1.

[16] For the purpose of this book, we define a Riemannian manifold as a space whose metric is smooth enough to be differentiated an appropriate number of times. This may require finding an appropriate set of coordinates.

[17] Let me assure the abecedarians that nothing profound is going on. We are merely expanding $g_{\mu\nu}(x)$ in a power series, with the coefficients given the names $A_{\mu\nu,\lambda}$ and $B_{\mu\nu,\lambda\sigma}$. The commas in the subscripts carried by A and B are purely for notational clarity, to separate two sets of indices.

[18] Here δ refers to the Kronecker delta.

[19] See *GNut*, p. 88.

[20] If you want to see the formula, it appears in *GNut* on p. 344, equation (14).

[21] This is worked out precisely and in detail in *GNut*, pp. 226–231.

Appendix ENS

From Euler to Navier and Stokes

Euler from Newton

The Euler equation

$$\frac{\partial \vec{v}}{\partial t} + (\vec{v} \cdot \vec{\nabla})\vec{v} = -\frac{1}{\rho}\vec{\nabla}P + \vec{f} \qquad (1)$$

describing fluid flow is readily derived from Newton's law of motion $\vec{a} = \vec{F}/m$. We sketch the derivation here.

Denote the velocity field of the fluid by $\vec{v}(t,\vec{x})$. Then the acceleration of an infinitesimal volume of fluid in an infinitesimal time lapse δt is determined by

$$\vec{v}(t+\delta t, \vec{x}+\vec{v}\delta t) - \vec{v}(t,\vec{x}) = \delta t \frac{\partial \vec{v}}{\partial t} + (\vec{v}\delta t \cdot \vec{\nabla})\vec{v} = \delta t\left(\frac{\partial \vec{v}}{\partial t} + (\vec{v}\cdot\vec{\nabla})\vec{v}\right) \qquad (2)$$

The convective or fluid derivative $\frac{D\vec{v}}{Dt} \equiv \frac{\partial \vec{v}}{\partial t} + (\vec{v}\cdot\vec{\nabla})\vec{v}$ appearing on the left side of (1) pops out naturally. That $\frac{D\vec{v}}{Dt}$ is not linear in $\vec{v}$ is the root cause of all the difficulties in fluid dynamics. Evidently, the "extra" term $(\vec{v}\cdot\vec{\nabla})\vec{v}$ appears because the fluid element has moved from $\vec{x}$ to $\vec{x}+\vec{v}\delta t$ in time δt.

The right side of (1) consists of the sum of the forces per unit mass acting on the fluid. Consider, in particular, the pressure P acting on an infinitesimal volume of length δx and cross sectional area A. See figure 1. The force $\left(P(t,x,y,z) - P(t,x+\delta x,y,z)\right)A = -\frac{\partial P}{\partial x}\delta x A$ is then determined by the pressure gradient $-\frac{\partial P}{\partial x}$.

In specific situations, various external forces also enter. The most common example: for fluid flow in the earth's gravitational field, we have to include in $\vec{f}$ the acceleration $\vec{g}$ due to gravity.

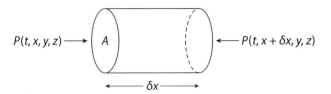

FIGURE 1. A pressure gradient produces a net force acting on a fluid element.

The conservation of mass is indicated by the equation of continuity

$$\frac{\partial \rho}{\partial t} + \vec{\nabla} \cdot (\rho \vec{v}) = 0 \tag{3}$$

Clearly, we also have to inform (1) and (3) how ρ varies with P, that is, we have to specify the equation of state $\rho(P)$ that characterizes the fluid.

Note in passing that if the fluid is incompressible (which holds for water under ordinary conditions), that is, if ρ is constant, then (3) implies that

$$\vec{\nabla} \cdot \vec{v} = 0 \tag{4}$$

If we add the viscosity term $\nu \nabla^2 \vec{v}$ to Euler's equation, we obtain the Navier-Stokes equation

$$\frac{\partial \vec{v}}{\partial t} + (\vec{v} \cdot \vec{\nabla})\vec{v} = -\frac{1}{\rho}\vec{\nabla}P + \nu\nabla^2\vec{v} + \vec{f} \tag{5}$$

We discuss viscosity in chapters VI.2 and VIII.5.

When can we linearize?

As the reader surely knows, back of the envelope calculations are routinely used in physics to determine when we are allowed to make various approximations.

Often (that is, in introductory textbooks on fluid dynamics), we can drop the nasty nonlinear term $(\vec{v} \cdot \vec{\nabla})\vec{v}$ in the Navier-Stokes equation (5). For example, consider[1] a water wave with wavelength λ, frequency ω, and amplitude a. The fluid elements oscillate over a characteristic time period $\tau \sim 1/\omega$, so that $v \sim a/\tau$. Thus, the nasty term is of order $(\vec{v} \cdot \vec{\nabla})\vec{v} \sim (a/\tau)(a/\tau)/\lambda \sim a^2/(\tau^2\lambda)$, since the wave varies over the characteristic distance λ. In contrast, the leading term on the left side of (5) is of order $\frac{\partial \vec{v}}{\partial t} \sim (a/\tau)/\tau \sim a/\tau^2$. Hence, requiring that the nasty term be negligible compared to the leading term leads to the eminently reasonable condition $a \ll \lambda$, requiring the amplitude to be much smaller than the wavelength.

Note that this also implies, rather pleasingly, that the fluid velocity $v \sim a/\tau$ is much smaller than the phase or group velocity $v_{\text{wave}} \sim \omega/k \sim \lambda/\tau$.

Should ρ be outside or inside?

This section answers a potential question that some readers might have. The rank beginner may skip over it.

You are of course free to multiply the Euler equation (or the Navier-Stokes equation) by ρ so as to obtain an equation for $\rho \frac{\partial \vec{v}}{\partial t}$. When I was a student, I was confused by why ρ is outside the partial time derivative $\frac{\partial}{\partial t}$. Shouldn't it be inside, so as to give the rate of change of momentum density $\frac{\partial(\rho \vec{v})}{\partial t}$? Which form is correct? Think for a minute before you read on.

Both forms are correct! It is a bit clearer if we use index rather than vector notation, so that (1) multiplied by ρ reads $\rho \frac{\partial v_i}{\partial t} + \rho v_j \nabla_j v_i = -\nabla_i P$, where $i = 1, 2, 3$ or x, y, z. For ease of writing, we have dropped the external force; it is just going along for the ride. So, put ρ inside and plug ahead:

$$\frac{\partial \rho v_i}{\partial t} = \rho \frac{\partial v_i}{\partial t} + \frac{\partial \rho}{\partial t} v_i = -\rho v_j \nabla_j v_i - \nabla_i P - v_i \nabla_j (\rho v_j)$$
$$= -\nabla_j (\rho v_i v_j + \delta_{ij} P) \equiv -\nabla_j T_{ij} \qquad (6)$$

We used the equation of continuity (3) in the second equality, and the Kronecker delta δ_{ij} in the third equality. The energy momentum tensor T_{ij}, as defined here, emerges naturally. The relativistic version of this object appears in chapter IV.2 and appendix Eg.

Note

[1] We follow Landau and Lifshitz, *Fluid Mechanics*, p. 37.

Appendix FSW
Finite square well

A brief review of the ground state wave function for the finite square well

As promised, here is an exceedingly brief review of the ground state wave function for the finite square well $V(x) = -W < 0$ for $L > x > -L$ and $V(x) = 0$ everywhere else. The solution is of course well known and may be found in almost all introductory texts on quantum mechanics. I need to go through it here quickly for a result needed in chapter III.3, and in order to impart a couple of lessons to the budding theoretical physicist.

For $L > x > -L$, we have the ground state wave function[1] $\psi = \cos kx$, with $k^2 = \frac{2m}{\hbar^2}(E + W) = \frac{2m}{\hbar^2}(W - |E|)$ since the ground state energy E is negative. In contrast, for $x > L$, $\psi = Ae^{-\kappa x}$, with $\kappa^2 = \frac{2m}{\hbar^2}|E| = \frac{2m}{\hbar^2}W - k^2$ and A a normalization constant. Defining $a^2 \equiv \frac{2m}{\hbar^2}W > 0$ for convenience, we write this as

$$k^2 + \kappa^2 = a^2 \tag{1}$$

Since we have to match ψ and $\psi' \equiv \frac{d\psi}{dx}$ at $x = L$, we might as well match $\psi'/\psi = (\log \psi)'$. For $L > x > -L$, $\log \psi = \log \cos kx$, while for $x > L$, $\log \psi = -\kappa x + \log A$. Matching $(\log \psi)'$ at $x = L$, we watch A drop out, and thus obtain

$$\kappa = k \tan kL \tag{2}$$

The bound state energies $|E| = \frac{\hbar^2}{2m}\kappa^2$ are determined by solving (1) and (2). Note that a^2 measures the depth of the potential, and L its width.

As is well known, the resulting transcendental equation can be solved numerically by plotting the left and right sides of (2) as functions of k and looking for intersections. The left side traces out a quarter circle (note that $\kappa > 0$ by definition) with radius a, while the right side describes a forest of quasi-vertical "tangent" trees reaching for the sky at $k = \pi/2L, 3\pi/2L, \ldots$. Do sketch if this sentence is not quite clear.

Shallow and deep wells

All this is straightforward, and the precise numerical solution is not particularly interesting. But after you have solved a physics problem without breaking a sweat, it is always instructive to take some limits. For $W \to \infty$, the quarter circle, with radius $a = (\frac{2m}{\hbar^2} W)^{\frac{1}{2}}$, is very large and slices through "the tips of the trees."[2]

More interestingly, for W small, no matter how small, we see geometrically that there is always one solution. The quarter circle always slices through one tree.

The physics thus revealed is much more intriguing than all this transcendental stuff: No matter how shallow the well, it can still trap the particle.

How well is your intuition humming today? Would the result still hold in 3-dimensional space? Well, maybe not; the particle could leak out sideways. The answer will be revealed in due course.

As $W \to 0$, how does the bound state energy $|E|$ behave?

Then $|E|$, κ, and k all $\to 0$. The tangent in (2) becomes approximately linear, and thus $\kappa \simeq k^2 L$. Hence, (1) becomes $k^2 + k^4 L^2 \simeq k^2 = a^2 = \frac{2m}{\hbar^2} W$, and $\kappa^2 \simeq (k^2 L)^2 \simeq (\frac{2m}{\hbar^2} WL)^2$.

We obtain

$$|E| = \frac{\hbar^2}{2m} \kappa^2 \simeq \frac{2m}{\hbar^2} (WL)^2 \tag{3}$$

The bound state energy vanishes like W^2.

Yet another lesson. The attentive reader might notice the appearance of WL, sort of the "area" of the well but not quite, since W is an energy and L a length. As long as you keep WL the same, the energy $|E|$ does not care whether you make the well narrower and deeper, or wider and shallower. Is there any significance to this result?

For the answer to this question, see chapter III.3.

Easier or harder to escape in 3 dimensions

We saw that an attractive potential well in 1 dimension, no matter how shallow, will hold a bound state. I asked you to exercise your intuition to decide whether the same is true in general in 3 dimensions.

Consider the attractive spherically symmetric potential well $V(r) = -W < 0$ for $r < L$ and $V(r) = 0$ for $r > L$. We learn in appendix L that if we define $\psi(r) =$

$u(r)/r$, the Schrödinger equation for zero angular momentum states reduces to, for $r < L$,

$$\frac{d^2u}{dr^2} = -\frac{2m}{\hbar^2}(E - V(r))u = -\frac{2m}{\hbar^2}(W - |E|)u = -k^2u \qquad (4)$$

This looks very much like the 1-dimensional Schrödinger equation, but with the crucial difference that $u(r)$ must vanish at $r = 0$.

So, instead of a cosine, $u(r) = \sin kr$. Thus, at $r = 0$, $u(r)$ rises up, and the potential $V(r)$ must have enough oomph to bend it down to meet $e^{-\kappa r}$ at $r = L$. No bound state exists unless the potential is attractive enough!

Indeed, matching u'/u at $r = L$, we now obtain, instead of (2),

$$\kappa = k \cot kL \qquad (5)$$

Amusingly, the tangent tan has turned into a cotangent cot. For small enough W, the quarter circle is too small to intersect the cotangent tree.

Notes

[1] According to the theorems of chapter III.1, since $V(x) = V(-x)$, the wave function is either even $\psi(x) = \psi(-x)$, or odd $\psi(x) = -\psi(-x)$. Further, the ground state wave function cannot have a node.

[2] And we recover half of the spectrum of the infinite square well.

Appendix Gal

Galilean invariance and fluid flow

Navier-Stokes equation is Galilean invariant

Let us check that the Navier-Stokes equation is Galilean invariant, as promised in chapter VI.2.

First, we have to work out how the various partial derivatives in the Navier-Stokes equation transform. To do this, we need to invert the Galilean transformation given in chapter VI.2, thus $t = t'$, $x = x' - ut'$, $y = y'$, $z = z'$. Next, the important thing is to be clear about what is being held fixed as we differentiate (just as in thermodynamics, as mentioned in chapter V.1). So,

$$\left.\frac{\partial}{\partial t'}\right|_{x'} = \left.\frac{\partial t}{\partial t'}\right|_{x'} \left.\frac{\partial}{\partial t}\right|_{x} + \left.\frac{\partial x}{\partial t'}\right|_{x'} \left.\frac{\partial}{\partial x}\right|_{t} = \left.\frac{\partial}{\partial t}\right|_{x} - u\left.\frac{\partial}{\partial x}\right|_{t} \tag{1}$$

and

$$\left.\frac{\partial}{\partial x'}\right|_{t'} = \left.\frac{\partial x}{\partial x'}\right|_{t'} \left.\frac{\partial}{\partial x}\right|_{t} + \left.\frac{\partial t}{\partial x'}\right|_{t'} \left.\frac{\partial}{\partial t}\right|_{x} = \left.\frac{\partial}{\partial x}\right|_{t} \tag{2}$$

More trivially, $\left.\frac{\partial}{\partial y'}\right|_{t'} = \left.\frac{\partial}{\partial y}\right|_{t}$ and $\left.\frac{\partial}{\partial z'}\right|_{t'} = \left.\frac{\partial}{\partial z}\right|_{t}$.

The convective derivative $\frac{\partial \vec{v}}{\partial t} + (\vec{v} \cdot \vec{\nabla})\vec{v}$ is derived in appendix ENS. We now show that, as mentioned in chapter VI.2, this particular form is also mandated by Galilean invariance, not surprisingly.

To keep the arithmetic simple, let us for a first pass restrict ourselves to 1-dimensional flow, with $\vec{v}(t,x) = (v(t,x), 0, 0)$ pointing in the x-direction and depending on only t, x, and not on y, z. Then $v'(t', x') = v(t,x) + u$. Applying (1), we obtain, since u is a constant, $\left.\frac{\partial v'}{\partial t'}\right|_{x'} = \left.\frac{\partial v}{\partial t}\right|_{x} - u\left.\frac{\partial v}{\partial x}\right|_{t}$. Even more simply, applying (2), we obtain $\left.\frac{\partial v'}{\partial x'}\right|_{t'} = \left.\frac{\partial v}{\partial x}\right|_{t}$.

To lessen clutter, we now drop the vertical bars; it should be evident by now what is being held fixed for each partial derivative. We see that, while $\frac{\partial v'}{\partial x'} = \frac{\partial v}{\partial x}$ and so is invariant, $\frac{\partial v'}{\partial t'} \neq \frac{\partial v}{\partial t}$ is not invariant.

But we also have $v' \frac{\partial v'}{\partial x'} = (v + u) \frac{\partial v}{\partial x} \neq \frac{\partial v}{\partial x}$, and thus

$$\frac{\partial v'}{\partial t'} + v' \frac{\partial v'}{\partial x'} = \frac{\partial v}{\partial t} - u \frac{\partial v}{\partial x} + v \frac{\partial v}{\partial x} + u \frac{\partial v}{\partial x} = \frac{\partial v}{\partial t} + v \frac{\partial v}{\partial x} \tag{3}$$

is Galilean invariant. Lo and behold, the left side of the Navier-Stokes equation is invariant. Physics works!

Galilean and rotational invariances

I will let you have the fun of checking Galilean invariance for the general case[1] $\vec{v}(t, x, y, z) = (v_x, v_y, v_z)$. In other words, both the magnitude and direction of the flow velocity $\vec{v}$ vary with space and time. Note that Galilean invariance fixes the relative coefficient between the two terms on the left side of the Navier-Stokes equation to be 1, as mentioned in chapter VI.2.

Suppose you were to figure out what the left side of the fluid equation could contain besides $\frac{\partial \vec{v}}{\partial t}$. You would realize that, as far as P, T, and rotational invariance are concerned, you could add two terms, $(\vec{v} \cdot \vec{\nabla})\vec{v}$ and $\vec{\nabla}(\vec{v} \cdot \vec{v})$. Why didn't those smart guys Euler, Navier, and Stokes include the second term? Because it is not Galilean invariant, as you should check.

Finally, I leave you to check that the viscosity term $\nabla^2 \vec{v}$ added by Navier and Stokes is Galilean invariant.

Note

[1] Again, to lessen clutter, I have not indicated the dependence of v_x, v_y, v_z on t, x, y, z.

Appendix Gr

Green functions

Please, why so much verbiage?

When I was a student, once I understood the Green[1] function, I was puzzled by how much verbiage some standard textbooks devoted to such a simple (and exceedingly beautiful) idea. Granted, the textbooks have to deal with various complicated boundary conditions and so on. But still, the basic idea is just that, for a linear equation, we can add solutions together.

Consider the equation

$$\nabla^2 \varphi(\vec{x}) = -4\pi \rho(\vec{x}) \tag{1}$$

determining the electrostatic potential $\varphi(\vec{x})$ generated by a given charge distribution $\rho(\vec{x})$. If $\nabla^2 \varphi_1(\vec{x}) = -4\pi \rho_1(\vec{x})$ and $\nabla^2 \varphi_2(\vec{x}) = -4\pi \rho_2(\vec{x})$, then evidently, $\nabla^2 \big(\varphi_1(\vec{x}) + \varphi_2(\vec{x})\big) = -4\pi \big(\rho_1(\vec{x}) + \rho_2(\vec{x})\big)$. The only thing we need is that ∇^2 acts linearly.

Where to put the pesky 4πs in electromagnetism is explained in appendix M. In contrast to the discussion there, here we are more interested in the solutions of Maxwell's equations than in the equations themselves, and so in this appendix, we put the 4π in the equations as in (1). But surely you see that the 4πs are not the issue here.

Linear superposition, symmetry, and dimensional analysis join forces

In particular, let $\rho(\vec{x}) = \delta^{(3)}(\vec{x})$ describe a unit point charge sitting at the origin. You and I know that it generates an electrostatic potential $\varphi(\vec{x}) = 1/r$ with the radial coordinate $r = |\vec{x}|$.

But what if we didn't know, and the equation $\nabla^2 \varphi = -4\pi \delta^{(3)}(\vec{x})$ just fell on our heads?

The fly by night physicist could still solve (1) by symmetry and dimensional analysis!

Since $[\nabla^2] = 1/L^2$ and $[\delta^{(3)}(\vec{x})] = 1/L^3$, we have $[\varphi(\vec{x})] = 1/L$, and thus $\varphi(\vec{x}) \sim 1/r$ (as required by rotational invariance). This of course reproduces the basic results: the electrostatic potential is $\varphi \sim 1/r$ around a point charge, thus giving an electric field $\vec{E} = -\vec{\nabla}\varphi$ that falls off like $1/r^2$. (The overall coefficient can be fixed by integrating. See appendices Del and M.)

By translation invariance, a unit point charge sitting at $\vec{x}_1$ generates an electrostatic potential $\varphi(\vec{x}) = 1/|\vec{x} - \vec{x}_1|$. Similarly, a unit point charge sitting at $\vec{x}_2$ generates an electrostatic potential $\varphi(\vec{x}) = 1/|\vec{x} - \vec{x}_2|$.

So, gentle reader, what is the electrostatic potential generated by two unit charges, one sitting at $\vec{x}_1$ and the other at $\vec{x}_2$?

If you said $\varphi(\vec{x}) = \frac{1}{|\vec{x}-\vec{x}_1|} + \frac{1}{|\vec{x}-\vec{x}_2|}$, advance to square one.

Next, after this baby question, a child question: What is the electrostatic potential generated by N unit charges, sitting at $\vec{x}_a$ for $a = 1, \ldots, N$? The answer is $\varphi(\vec{x}) = \sum_{a=1}^{N} \frac{1}{|\vec{x}-\vec{x}_a|}$. Just keep adding.

In the limit $N \to \infty$, with the charges smeared out to form a continuous charge distribution $\rho(\vec{x})$, the sum, as Newton and Leibniz taught us, merges to an integral:

$$\varphi(\vec{x}) = \int d^3x' \frac{\rho(\vec{x}')}{|\vec{x} - \vec{x}'|} \tag{2}$$

The infinitesimal volume d^3x' located at $\vec{x}'$ contains an amount of charge equal to $d^3x' \rho(\vec{x}')$.

Green function

The result in (2) completely solves the problem of determining the electrostatic potential $\varphi(\vec{x})$ generated by any charge distribution $\rho(\vec{x})$, as stated in (1). So what more is there to say, you might ask? Indeed, essentially nothing, but in physics, it is often highly useful to formulate a language that we can generalize and carry over to other situations. And the Green function is an example.

Go back and examine how we solved the problem. We see that the key is to know that the electrostatic potential generated by a unit point charge sitting at the origin, corresponding to the distribution $\rho(\vec{x}) = \delta^{(3)}(\vec{x})$, is equal to $\varphi(\vec{x}) = 1/r$. So define $G(\vec{x})$, known as a Green function, as the solution of

$$\nabla^2 G(\vec{x}) = -4\pi \delta^{(3)}(\vec{x}) \tag{3}$$

But you exclaim, we already know what $G(\vec{x})$ is; namely, $G(\vec{x}) = 1/r$. We merely gave the electrostatic potential of a point charge another name.

What good is inventing a name? It means that we can rewrite the general solution (2) as

$$\varphi(\vec{x}) = \int d^3x' \, G(\vec{x} - \vec{x}')\rho(\vec{x}') \tag{4}$$

The reader who has never seen this before can verify this equation by applying ∇^2 to both sides and using (3): $\nabla^2 \varphi(\vec{x}) = \int d^3x' \, \nabla^2 G(\vec{x} - \vec{x}')\rho(\vec{x}') = -4\pi \int d^3x' \, \delta^3(\vec{x} - \vec{x}')\rho(\vec{x}') = -4\pi \rho(\vec{x}')$.

That's it.

Green's point is that this procedure works for any linear equation.

Time enters

To discuss radiation, we have to let time into the game. Generalize* (1) to[2]

$$\Box A_\mu(t, \vec{x}) \equiv \left(\nabla^2 - \frac{1}{c^2} \frac{\partial^2}{\partial t^2} \right) A_\mu(t, \vec{x}) = -\frac{4\pi}{c} J_\mu(t, \vec{x}) \tag{5}$$

Note first that φ and ρ have been promoted to two 4-vectors, of which they are the time components: $A_\mu = (A_0, A_i) = (\varphi, A_i)$ and $J_\mu = (J_0, J_i) = (c\rho, J_i)$. Second, the Laplacian ∇^2 has been promoted to the d'Alembertian $\Box \equiv \nabla^2 - \frac{1}{c^2} \frac{\partial^2}{\partial t^2}$, that is, it has been extended to include time variation.

The corresponding Green function is determined by

$$\Box G(t, \vec{x}) = \left(\nabla^2 - \frac{1}{c^2} \frac{\partial^2}{\partial t^2} \right) G(t, \vec{x}) = -4\pi \delta(t) \delta^{(3)}(\vec{x}) \tag{6}$$

Time enters!

Physically, the Green function describes the effect of a point charge at the origin ($\vec{x} = 0$) that pops in and out of existence for an instant at $t = 0$. It is important to emphasize to some unsuspecting students that the point charge is not moving around, but does something not terribly physical. But we are writing down a mathematical equation, not a physical equation.

Determining the time dependent Green function

Again, symmetry and dimensional analysis allow us to determine the time dependent Green function $G(t, \vec{x})$, together with a dose of physical sense. Due to the finite speed of light c, the effect of a charge popping in and out of existence at the origin of space and time cannot be felt at $\vec{x}$ until time $t = |\vec{x}|/c$.

---

*See appendix M.

Thus, $G(t, \vec{x})$ must be proportional to a delta function $\delta(t - |\vec{x}|/c)$. In addition, by integrating (6) over time, we see that $\int dt G(t, \vec{x})$ must be equal to our previous Green function $G(\vec{x}) = 1/|\vec{x}|$.

At this point, the fly-by-night physicist doesn't even need our old friend dimensional analysis: We have already determined that[3]

$$G(t, \vec{x}) = \frac{\delta(t - \frac{r}{c})}{r} \tag{7}$$

with $r = |\vec{x}|$, as before. But let us use dimensional analysis as a check: Since $[\Box] = 1/L^2$ and $[\delta(t) \delta^{(3)}(\vec{x})] = 1/(TL^3)$, we obtain directly from (6) that $[G(t, \vec{x})] = 1/(TL)$, and of course our solution checks out.

For the record, let us write down, by translating in time and in space (that is, by letting $t \to t - t'$, $\vec{x} \to \vec{x} - \vec{x}'$ in (7)),

$$G(t - t', \vec{x} - \vec{x}') = \frac{\delta(t - t' - \frac{1}{c}|\vec{x} - \vec{x}'|)}{|\vec{x} - \vec{x}'|} \tag{8}$$

The value of t' is mandated by the vanishing of the delta function, namely,

$$t'_R(t, |\vec{x} - \vec{x}'|) \equiv t - \frac{1}{c}|\vec{x} - \vec{x}'| \tag{9}$$

known as the retarded time. It enforces causality: The effect of a signal that originated at the place $\vec{x}'$, if felt at time t at the place $\vec{x}$, must have left the place $\vec{x}'$ at the earlier time $t'_R(t, |\vec{x} - \vec{x}'|)$.

Electromagnetism in all its glory

With the Green function in hand, we can solve (5) in all generality:

$$
\begin{aligned}
A_\mu(t, \vec{x}) &= \frac{1}{c} \int d^3x' \int dt' G(t - t', \vec{x} - \vec{x}') J_\mu(t', \vec{x}') \\
&= \frac{1}{c} \int d^3x' \int dt' \frac{\delta(t - t' - \frac{1}{c}|\vec{x} - \vec{x}'|)}{|\vec{x} - \vec{x}'|} J_\mu(t', \vec{x}') \\
&= \frac{1}{c} \int \frac{d^3x'}{|\vec{x} - \vec{x}'|} J_\mu(t'_R, \vec{x}')
\end{aligned}
\tag{10}
$$

with t'_R a function of t and $|\vec{x} - \vec{x}'|$ as given in (9).

This result summarizes electromagnetism in all its glory. Given any distribution of charge and current $(\rho, \vec{J})$, we merely have to evaluate this integral to determine the electromagnetic potentials $(\varphi, \vec{A})$.

Some students might think this expression formidable looking, but in fact, it is as simple as it can be.

The notation for the argument of J_μ makes the result look clunky, but we are simply integrating over the source current $J_\mu(t', \vec{x}')$ when t' is equal to $t'_R(t, \frac{1}{c}|\vec{x} - \vec{x}'|)$.

To repeat and to summarize: the electromagnetic potentials at the observation point $\vec{x}$ at time t are generated by the currents J_μ at $\vec{x}'$ at a retarded time t'_R to allow the electromagnetic wave enough time to propagate from $\vec{x}'$ to $\vec{x}$ at the speed of light c. The factor $1/|\vec{x} - \vec{x}'|$ must be included for (10) to reproduce the elementary result (4) for a charge distribution just sitting there, that is, with $J_i = 0$ and $J_0(t', \vec{x}') = c\rho(\vec{x}')$ time independent.

Green functions for quantum mechanics

From the Green function in (7), we can generate the Green functions we need for other areas of physics, for instance, scattering in quantum mechanics. Integrate (6) in time to obtain

$$\int dt e^{ickt} \left(\nabla^2 - \frac{1}{c^2} \frac{\partial^2}{\partial t^2} \right) G(t, \vec{x}) = (\nabla^2 + k^2) \int dt e^{ickt} G(t, \vec{x}) = \delta^{(3)}(\vec{x}) \quad (11)$$

The first equality follows on integrating by parts. The second equality simply states that $\int dt e^{ickt} \delta(t) = 1$ on integrating the right side of (6).

Thus, with essentially no work, we have found the Green function for the equation

$$(\nabla^2 + k^2) g(\vec{x}) = \delta^{(3)}(\vec{x}) \quad (12)$$

namely,

$$g(\vec{x}) = \int dt e^{ickt} \frac{\delta(t - \frac{r}{c})}{r} = \frac{e^{ikr}}{r} \quad (13)$$

The Yukawa potential

Next, it follows that the solution to

$$(\nabla^2 - m^2) \phi(\vec{x}) = \delta^{(3)}(\vec{x}) \quad (14)$$

is given by the substitution $k \to im$:

$$\phi(\vec{x}) = \frac{e^{-mr}}{r} \quad (15)$$

This is known as the Yukawa potential in nuclear and particle physics, a result of fundamental importance.

The discerning reader may be impressed by how few fingers we have to lift to derive results crucially needed in several areas of physics.

Notes

[1] George Green (1793–1841) was the almost entirely self-taught son of a baker. From emails I have received, etc., I am aware that many autodidacts read my textbooks.

[2] The learned reader would recognize that I have implicitly chosen the Lorenz gauge, in which $\partial_\mu A^\mu = 0$. I remind you that this is not a textbook on electromagnetism, but see appendix M.

[3] For another derivation of (7), see the poor man's approach given on page 573 of *GNut*.

Appendix Grp

Group versus phase velocity

Forming wave packets: a quick review

Here I offer you a quick review of phase velocity versus group velocity. I assume that you are familiar, from either electromagnetism or quantum mechanics, with the possibility of forming a wave packet in a linear theory by superposing plane waves.

For notational simplicity, we will talk about the 1-dimensional case. (You can readily generalize to 3 dimensions.) A wave packet is defined by

$$\psi(t, x) = \int dk e^{i(\omega(k)t - kx)} f(k) \tag{1}$$

with $f(k)$ a function peaked around some k_*. The integral in (1) sums an infinite number of plane waves with wave vectors k close to k_*.

Respectable fly by day physicists would now approximate $f(k)$ by a Gaussian, expand the integrand, and integrate. Indeed, this is exactly what is done in standard textbooks.[1]

Adding instead of integrating

Instead, we fly by night physicists simply add two waves with almost the same k to see what is going on. Start with the trigonometric identity $\cos A' + \cos A = 2 \cos \frac{A'+A}{2} \cos \frac{A'-A}{2}$. Set $A = \omega(k)t - kx$ and $A' = \omega(k')t - k'x$, with $k' = k + \Delta k$

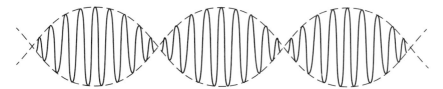

FIGURE 1. The low frequency $\Delta\omega$ envelope moves with the group speed $\frac{\Delta\omega}{\Delta k}$, while the high frequency ω wave it encloses moves with the phase speed $\frac{\omega}{k}$.

and $\omega(k') = \omega(k) + \frac{\Delta\omega}{\Delta k}\Delta k \simeq \omega(k) + \frac{d\omega}{dk}\Delta k$. Thus

$$\cos\left(\omega(k')t - k'x\right) + \cos\left(\omega(k)t - kx\right)$$
$$\simeq 2\cos\left(\omega(k)t - kx\right)\cos\left(\frac{1}{2}\left[x - \frac{d\omega}{dk}t\right]\Delta k\right) \qquad (2)$$

It may be helpful for you to sketch what this looks like. We obtain a rapidly varying wave $\cos(\omega(k)t - kx)$ with large wave number, k, with a correspondingly short wavelength, modulated by a slowly varying envelope

$$\cos\left(\frac{1}{2}\left[x - \frac{d\omega}{dk}t\right]\Delta k\right) \qquad \text{slowly varying envelope} \qquad (3)$$

with small wave number, Δk, and hence a long wavelength. See figure 1.

A point on the rapidly varying wave (for example, where the argument of $\cos(\omega(k)t - kx)$ vanishes) moves in spacetime according to $x = \frac{\omega}{k}t$, while a point on the envelope in (3) moves according to $x = \frac{d\omega}{dk}t$.

This simple example captures the essential physics behind the phase velocity

$$v_p = \frac{\omega}{k} \qquad (4)$$

and the group velocity

$$v_g = \frac{d\omega}{dk} \qquad (5)$$

Note

---

[1] It is not even that difficult to do. Following the steps outlined, we obtain something like $\sim e^{i(\omega_* t - k_* x)} f(k_*) (\int dk\, e^{i(\frac{d\omega}{dk}|_* t - x)(k - k_*)} e^{-\frac{1}{2}a(k - k_*)^2})$. No need to do the integral. Shifting the integration variable by $k \to k + k_*$ just for clarity, you can see that the integral defines a function of $(x - \frac{d\omega}{dk}|_* t)$. A given point on the wave packet moves with group velocity $\frac{d\omega}{dk}|_*$.

Appendix L

Radial part of the Laplacian

Derive from first principles rather than look it up

When I was an undergraduate, a professor told me to always derive any needed expressions from first principles rather than look them up. Good advice! But after working out the Laplacian in spherical coordinates by brute force for what seemed like the hundredth time, I decided I understood how it all worked. You know the drill: $\frac{\partial \psi(r,\theta,\varphi)}{\partial x} = \frac{\partial r}{\partial x}\frac{\partial \psi}{\partial r} + \frac{\partial \theta}{\partial x}\frac{\partial \psi}{\partial \theta} + \frac{\partial \varphi}{\partial x}\frac{\partial \psi}{\partial \varphi}$, evaluating $\frac{\partial r}{\partial x}$, $\frac{\partial \theta}{\partial x}$, $\frac{\partial \varphi}{\partial x}$ explicitly, then acting with $\frac{\partial}{\partial x}$ on these three terms using the same procedure, then repeat with $x \to y$, and repeat again with $x \to z$, and add up the whole mess to get[1] $\vec{\nabla}^2 \psi \equiv \frac{\partial^2 \psi}{\partial x^2} + \frac{\partial^2 \psi}{\partial y^2} + \frac{\partial^2 \psi}{\partial z^2}$ in spherical coordinates. It was certainly tedious.

Later, in learning general relativity, I realized that Riemann's method[2] for dealing with curved spaces works in flat space also (as it must), and it gives an easy way of getting at the Laplacian, especially if we don't care about the angular part. I am surprised that this approach is not widely known.

For those who do not know differential geometry

In fact, you don't even have to know general relativity or differential geometry if you only want the radial part of the Laplacian, which is more or less what we limit ourselves to in this book. The key is to use the integral rather than the differential formulation, as exemplified in (4) of chapter III.1, for example.

The Laplacian appears all over the place in physics. For definiteness, let's focus on, say, the Schrödinger equation: $-\frac{\hbar^2}{2m}\nabla^2\psi + V(x)\psi = E\psi$. Suppressing factors like $\frac{\hbar^2}{2m}$ irrelevant for our purposes here, note that this differential equation follows by varying the energy functional $\int d^3x\, \vec{\nabla}\psi^* \cdot \vec{\nabla}\psi + \cdots$ with respect to ψ^* and then integrating by parts: $\delta \int d^3x\, \vec{\nabla}\psi^* \cdot \vec{\nabla}\psi = \int d^3x\, \vec{\nabla}\delta\psi^* \cdot \vec{\nabla}\psi = -\int d^3x\, \delta\psi^*\nabla^2\psi$.

In spherical coordinates,

$$\int d^3x\, \vec{\nabla}\psi^* \cdot \vec{\nabla}\psi = \int_0^\infty dr\, r^2 \int_{-1}^{+1} d\cos\theta \int_0^{2\pi} d\varphi \left(\frac{\partial\psi^*}{\partial r} \frac{\partial\psi}{\partial r} + \cdots \right) \quad (1)$$

So if we only care about the radial part, we have simply $\int_0^\infty dr\, r^2 \frac{\partial\psi^*}{\partial r} \frac{\partial\psi}{\partial r}$. Varying with respect to ψ^* and integrating by parts, we obtain

$$\int_0^\infty dr\, r^2 \frac{\partial\delta\psi^*}{\partial r} \frac{\partial\psi}{\partial r} = -\int_0^\infty dr\, \delta\psi^* \left(\frac{\partial}{\partial r} r^2 \frac{\partial\psi}{\partial r} \right) = -\int_0^\infty dr\, r^2 \delta\psi^* \left(\frac{1}{r^2} \frac{\partial}{\partial r} r^2 \frac{\partial\psi}{\partial r} \right)$$
$$(2)$$

In the second step, we remembered to put the r^2 back into the integration measure. Hence, ta da:

$$\nabla^2\psi = \frac{1}{r^2} \frac{\partial}{\partial r} r^2 \frac{\partial\psi}{\partial r} + \cdots = \frac{\partial^2\psi}{\partial r^2} + \frac{2}{r} \frac{\partial\psi}{\partial r} + \cdots \quad (3)$$

Of course, even though the derivation here is ostensibly for quantum mechanics, the Laplacian is the Laplacian.

Schrödinger equation in 3 dimensions

As the reader probably knows, there is yet a third form of the radial part of the Laplacian that is highly useful in dealing with the Schrödinger equation in 3 dimensions:

$$\nabla^2\psi = \frac{1}{r} \frac{\partial^2}{\partial r^2} r\psi + \cdots \quad (4)$$

You could of course differentiate away and show that (4) agrees with (3). But again, I prefer to see this via the integral form. Write

$$\int_0^\infty dr\, \frac{\partial(r\psi^*)}{\partial r} \frac{\partial(r\psi)}{\partial r} = \int_0^\infty dr \left(r\frac{\partial\psi^*}{\partial r} + \psi^* \right) \left(r\frac{\partial\psi}{\partial r} + \psi \right)$$

$$= \int_0^\infty dr \left(r^2 \frac{\partial\psi^*}{\partial r} \frac{\partial\psi}{\partial r} + \psi^* r\frac{\partial\psi}{\partial r} + r\frac{\partial\psi^*}{\partial r}\psi + \psi^*\psi \right) \quad (5)$$

Integrating the second term $\psi^* r\frac{\partial\psi}{\partial r}$ on the right side by parts, we see that it knocks out the third and fourth terms. We thus obtain $\int_0^\infty dr\, r^2 \frac{\partial\psi^*}{\partial r} \frac{\partial\psi}{\partial r} = \int_0^\infty dr\, \frac{\partial(r\psi^*)}{\partial r} \frac{\partial(r\psi)}{\partial r}$, which, on integrating by parts, becomes $-\int_0^\infty dr\, r^2 \left(\frac{\psi^*}{r} \right) \frac{\partial^2(r\psi)}{\partial r^2}$. Referring back to (1) and integrating its left side by parts, we see that (4) holds.

The form of the Laplacian in (4) invites us to define

$$\psi \equiv \frac{u}{r} \tag{6}$$

so that $\nabla^2 \psi = \frac{1}{r}\frac{\partial^2 u}{\partial r^2}$. For a spherically symmetric potential, we see that the ground state wave function satisfies a 1-dimensional Schrödinger equation

$$-\frac{\hbar^2}{2m}\frac{d^2 u(r)}{dr^2} + V(r)u(r) = Eu(r) \tag{7}$$

but with one crucial difference, namely, the boundary condition $u(r = 0) = 0$.

Again, when I was an undergrad, I couldn't quite remember whether the nifty substitution was replacing ψ by u/r or ru. We see that it is simply a matter of sharing the measure r^2 between ψ^* and ψ.

The simple mnemonic is that the 3-dimensional measure and the 1-dimensional measure differ by a factor of r^2, thus:

$$\int dr r^2 \psi^* \psi = \int dr u^* u \tag{8}$$

The boundary condition just stated follows from the conservation of probability. Recall that the probability current has the form $\vec{J} \propto (\psi^* \vec{\nabla}\psi - (\vec{\nabla}\psi^*)\psi)$, so that the radial current $J_r \sim \psi \frac{d\psi}{dr} \sim (u/r)(1/r)(u/r) \sim u^2/r^3$. The probability flowing in or out of a small sphere of radius $r \sim 0$ is then $J_r(4\pi r^2) \sim u^2/r$. Thus, as $r \to 0$, $u(r)$ must vanish faster than $r^{\frac{1}{2}}$.

Divergence of a vector field

We now show that this "integration by parts" trick works for the divergence of a vector field also. Consider integrating the functional $\int d^3x \, \vec{\nabla}\phi \cdot \vec{V} = -\int d^3x \, \phi\vec{\nabla}\cdot\vec{V}$ by parts. In spherical coordinates, for a radial vector field $\vec{V} = (V_r, 0, 0)$ independent of θ and φ, we have

$$\int_0^\infty dr \, r^2 \vec{\nabla}\phi \cdot \vec{V} = \int_0^\infty dr \, r^2 \frac{d\phi}{dr} V_r = -\int_0^\infty dr \, \phi \frac{d}{dr}(r^2 V_r) \tag{9}$$

It follows that

$$\vec{\nabla}\cdot\vec{V} = \frac{1}{r^2}\frac{d}{dr}(r^2 V_r) \tag{10}$$

Again, it is "merely" the integration measure. Can you see that the divergence actually has a nice physical interpretation?

Not long ago, I gave a final exam to advanced undergrads with a problem involving the flow of an incompressible fluid in a spherically symmetric situation. Only a few were able to recall the appropriate expression for the divergence (even though the exam was open notebook). But in fact, the incompressible condition $\vec{\nabla}\cdot\vec{V} = 0$ corresponds physically to the conservation of fluid: For spherical incompressible flow, $r^2 V_r$ cannot depend on r. See exercise 2.

A tiny bit of differential geometry

For those who know Riemann's approach to differential geometry, all those operators in various coordinate sytems that bedevil the typical undergraduate (the Laplacian, the divergence, and so on) can be expressed in terms of the metric. In fact, dimensional analysis is basically all you need.

Start with the integral

$$I = \int d^D x \, \vec{\nabla}\varphi \cdot \vec{\nabla}\varphi \tag{11}$$

In curved space or flat space described by curvilinear coordinates x^μ, $\mu = 1, 2, \ldots, D$, this generalizes to[3]

$$I = \int d^D x \, \sqrt{g} g^{\mu\nu} \partial_\mu \varphi \partial_\nu \varphi \tag{12}$$

where g and $g^{\mu\nu}$ respectively denote the determinant and inverse of the metric $g_{\mu\nu}$.

For spherical coordinates in 3 dimensions, for example, the metric is defined by $ds^2 = dr^2 + r^2(d\theta^2 + \sin^2\theta d\phi^2) = g_{\mu\nu}dx^\mu dx^\nu$. Thus, $g_{rr} = 1$, $g_{\theta\theta} = r^2$, and $g_{\phi\phi} = r^2 \sin^2\theta$, so that $g = g_{rr}g_{\theta\theta}g_{\phi\phi} = 1 \cdot r^2 \cdot r^2 \sin^2\theta = r^4 \sin^2\theta$, and $g^{rr} = 1$, etc. If we don't care about the angular stuff, then in (12), we could set $\sqrt{g}$ to $\sqrt{r^4} = r^2$ and keep only the g^{rr} term. (Note to those not fluent in differential geometry: Recall that $d^D x = dr d\theta d\phi$ by definition.) Thus,

$$J = \int_0^\infty dr r^2 \frac{\partial\varphi}{\partial r}\frac{\partial\varphi}{\partial r} \tag{13}$$

Indeed, we don't even need Riemann in his full glory; just appeal to dimensional analysis and compare dimensions with (12) to give us in D-dimensional space

$$J = \int_0^\infty dr r^{D-1} \frac{\partial\varphi}{\partial r}\frac{\partial\varphi}{\partial r} \tag{14}$$

Varying I in (11) gives $\delta I = \int d^D x \, (\vec{\nabla}\delta\varphi) \cdot (\vec{\nabla}\varphi)$. Integrating by parts, we obtain $\delta I = -\int d^D x \, \delta\varphi \vec{\nabla} \cdot \vec{\nabla}\varphi = -\int d^D x \, \delta\varphi \nabla^2\varphi$. On the other hand, applying this procedure to (14) gives

$$\delta J = \int_0^\infty dr r^{D-1} \frac{\partial\delta\varphi}{\partial r}\frac{\partial\varphi}{\partial r} = -\int_0^\infty dr r^{D-1}\delta\varphi \left(\frac{1}{r^{D-1}}\frac{\partial}{\partial r}r^{D-1}\frac{\partial\varphi}{\partial r} \right) \tag{15}$$

Comparing, we obtain

$$\nabla^2\varphi = \frac{1}{r^{D-1}}\frac{\partial}{\partial r}r^{D-1}\frac{\partial\varphi}{\partial r} + \text{angular stuff} = \frac{\partial^2\varphi}{\partial r^2} + \frac{D-1}{r}\frac{\partial\varphi}{\partial r} + \text{angular stuff} \tag{16}$$

For cylindrical coordinates, simply set $D = 2$: The z coordinate just goes along for the ride.

Let me mention for completeness that the Laplacian is given in general[4] by

$$\nabla^2 \varphi = \frac{1}{\sqrt{g}} \partial_\mu \sqrt{g} g^{\mu\nu} \partial_\nu \varphi \tag{17}$$

a result we obtain easily: Simply integrate (12) by parts and follow the same procedure as above. Compare with (16), and note that the expression in (17) includes the angular part.

Exercises

(1) Solve (7) for $V = 0$, and interpret both u and ψ physically.

(2) Show that for spherical incompressible flow, $r^2 V_r$ does not depend on r.

Notes

[1] For those familiar with Guinness stout, the company logo appearing on every bottle uses a Phoenician or Hebrew harp. The symbol ∇ is in fact a Greek word $\nu\alpha\beta\lambda\alpha$ for the Phoenician harp. See https://en.wikipedia.org/wiki/Nablasymbol.

The Hebrew word for harp is "nevel," a word appearing often in the Bible, especially in the Psalms. You do not have to be a linguist to recognize nabla and nevel as having the same Semitic origin. I am grateful to J. Feinberg for pointing out this interesting connection.

[2] See *GNut*, p. 78.

[3] If you are unfamiliar with this stuff or need to brush up, see *GNut*, chapter I.5, for a detailed explanation.

[4] For a more detailed explanation, see *GNut*, pp. 78–79.

Appendix M

Maxwell's equations: a brief review

I need hardly say that this is not a textbook on electromagnetism, and that I have to assume the reader's familiarity with Maxwell's equations. Here I offer merely a quick review, partly to explain units and to establish notation.

Heaviside-Lorentz

The classic text by Jackson contains an appendix listing no less than five systems of units.[1] As mentioned in chapter II.1, everybody is entitled to his or her preference. I favor Heaviside-Lorentz, which is standard for particle theory and quantum field theory. Maxwell's equations in the absence of a medium (no $\vec{D}$ and no $\vec{H}$) read[2]

$$\vec{\nabla} \cdot \vec{E} = \rho \tag{1}$$

$$\vec{\nabla} \times \vec{B} - \frac{1}{c}\frac{\partial \vec{E}}{\partial t} = \frac{1}{c}\vec{J} \tag{2}$$

$$\vec{\nabla} \times \vec{E} + \frac{1}{c}\frac{\partial \vec{B}}{\partial t} = 0 \tag{3}$$

$$\vec{\nabla} \cdot \vec{B} = 0 \tag{4}$$

Note that, since one main purpose of this appendix is to show that, in a well chosen gauge, these equations lead to the more compact $\Box A_\mu = -\frac{1}{c}J_\mu$ (see (17) below), it is convenient to absorb the electromagnetic coupling e into ρ and $\vec{J}$ here, so as not to clutter up the derivation. The numerical value of e, which depends on the system of units used, is discussed in the concluding section of this appendix.

In electrostatics, (3) implies $\vec{\nabla} \times \vec{E} = 0$ and thus $\vec{E} = -\vec{\nabla}\varphi$ for some potential φ. Then (1) becomes $\nabla^2 \varphi = -\rho$, which implies that the electrostatic potential around a point particle sitting at the origin with[3] $\rho = \delta^3(\vec{x})$ equals

$$\varphi = \frac{1}{4\pi r} \quad \text{Heaviside-Lorentz} \tag{5}$$

with $r = |\vec{x}|$ the radial coordinate, as is well known and as was worked out in appendix Del.

Gaussian

The Gaussian system of units differs from the Heaviside-Lorentz system only in the presence of 4π in Maxwell's equations. To go to Gaussian units, let $\rho \rightarrow 4\pi\rho$ and $\vec{J} \rightarrow 4\pi\vec{J}$, so that (1) and (2) are replaced by

$$\vec{\nabla} \cdot \vec{E} = 4\pi\rho \quad \text{Gaussian} \tag{6}$$

and

$$\vec{\nabla} \times \vec{B} - \frac{1}{c}\frac{\partial \vec{E}}{\partial t} = \frac{4\pi}{c}\vec{J} \quad \text{Gaussian} \tag{7}$$

respectively.

It goes without saying that since the other two Maxwell's equations, (3) and (4), do not involve charge and current, they are the same with Gaussian units or with Heaviside-Lorentz units.

Thus, for the same point particle sitting at the origin mentioned above, that is, with $\rho = \delta^3(\vec{x})$, we have

$$\varphi = \frac{1}{r} \quad \text{Gaussian} \tag{8}$$

Where to put the 4π? Sometimes here, sometimes there

From this discussion, we see that the 4πs, which come from the surface area of the unit sphere, are unavoidable. You can either shove the 4πs into Maxwell's equations or into the solutions of those equations. Which do you think is more "sacred": equations or their solutions?

Thus, I vacillate between Heaviside-Lorentz and Gaussian depending on the context. In this appendix, for example, I am working out the theory of electromagnetism in a particular gauge and so naturally prefer not to have 4πs all over the place in the equations. However, in chapter I.3, for example, when we discuss the Bohr atom, of course it would be better to have the electrostatic potential energy between the electron and the proton equal to simply e^2/r without the 4π. (Hence most elementary discussions of quantum physics, such as that of the Bohr atom, are given in Gaussian units.)

Massaging Maxwell's equations: Lorenz gauge

Now that we have taken care of the 4πs, we are ready to solve Maxwell's equations. In the glare of hindsight, we recognize that the two sourceless equations, (3) and (4), stand apart from the other two equations: They are merely constraints on $\vec{B}$ and $\vec{E}$, and hence they urge us to attack them first.

We are instructed by (4) that $\vec{B}$ may be written as the curl of a vector potential $\vec{A}$: $\vec{B} = \vec{\nabla} \times \vec{A}$. It is crucial to note that $\vec{A}$ is not uniquely determined by $\vec{B}$; given $\vec{A}$, we can always, without changing $\vec{B}$, let $\vec{A} \to \vec{A} + \vec{\nabla}\Lambda$, with Λ some scalar function under rotation. This is known as a gauge transformation.

Using $\vec{B} = \vec{\nabla} \times \vec{A}$, we can rewrite (3) as

$$\vec{\nabla} \times \left(\vec{E} + \frac{1}{c}\frac{\partial \vec{A}}{\partial t} \right) = 0 \tag{9}$$

which implies that $\vec{E} + \frac{1}{c}\frac{\partial \vec{A}}{\partial t}$ is the gradient of some function $-\varphi$ (the minus sign is a standard convention), and thus

$$\vec{E} = -\vec{\nabla}\varphi - \frac{1}{c}\frac{\partial \vec{A}}{\partial t} \tag{10}$$

Under a gauge transformation $\vec{A} \to \vec{A} + \vec{\nabla}\Lambda$, we would like to keep the electric field $\vec{E}$, as well as the magnetic field $\vec{B}$, unchanged. This requires us to transform $\varphi \to \varphi - \frac{1}{c}\frac{\partial \Lambda}{\partial t}$ to offset the transformation $\vec{A} \to \vec{A} + \vec{\nabla}\Lambda$.

We now turn to (1) and (2), the equations that do involve charge and current. Plugging $\vec{E} = -\vec{\nabla}\varphi - \frac{1}{c}\frac{\partial \vec{A}}{\partial t}$ into (1), we obtain

$$\nabla^2\varphi + \frac{1}{c}\frac{\partial}{\partial t}\vec{\nabla} \cdot \vec{A} = -\rho \tag{11}$$

The second term in (11) involves a rather nasty looking operator $\frac{\partial}{\partial t}\vec{\nabla}$, mixing time and spatial derivatives. Without this term, (11) would reduce to the electrostatic equation $\nabla^2\varphi = -\rho$. But we don't want to be restricted to electrostatics; we want φ to depend on time as well as on space.

After staring at this, we might come to realize that if we impose the so-called Lorenz* condition[4]

$$\vec{\nabla} \cdot \vec{A} = -\frac{1}{c}\frac{\partial \varphi}{\partial t} \tag{12}$$

then (11) becomes the much nicer looking

$$\left(\nabla^2 - \frac{1}{c^2}\frac{\partial^2}{\partial t^2} \right)\varphi = -\rho \tag{13}$$

---

*Named after Ludvig Lorenz, not to be confused with the more famous Hendrik Lorentz, as people often do!

Defining the d'Alembertian $\Box \equiv \nabla^2 - \frac{1}{c^2}\frac{\partial^2}{\partial t^2}$, we can write this more compactly as

$$\Box \varphi = -\rho \tag{14}$$

This generalizes the electrostatic equation $\nabla^2\varphi = -\rho$.

At this point, you should show that using gauge freedom, we can always impose the Lorenz condition (12) with a judicious choice of Λ.

Putting $\vec{B} = \vec{\nabla} \times \vec{A}$ and $\vec{E} = -\vec{\nabla}\varphi - \frac{1}{c}\frac{\partial \vec{A}}{\partial t}$ into (2), we obtain

$$\vec{\nabla} \times (\vec{\nabla} \times \vec{A}) + \frac{1}{c^2}\frac{\partial^2 \vec{A}}{\partial t^2} + \frac{1}{c}\vec{\nabla}\frac{\partial \varphi}{\partial t} = \frac{1}{c}\vec{J} \tag{15}$$

Using the vector identity (see below) $\vec{\nabla} \times (\vec{\nabla} \times \vec{A}) = -\nabla^2\vec{A} + \vec{\nabla}(\vec{\nabla} \cdot \vec{A})$ on the first term and imposing the Lorenz condition (12) on the third term, we obtain, after a cancellation,

$$\Box \vec{A} = -\frac{1}{c}\vec{J} \tag{16}$$

Packaging the various quantities into two 4-vectors, $A_\mu \equiv (\varphi, \vec{A})$ and $J_\mu \equiv (c\rho, \vec{J})$, we can finally write

$$\Box A_\mu = -\frac{1}{c}J_\mu \tag{17}$$

The appearance of $\Box$, A_μ, and J_μ implies that Maxwell's equations are Lorentz invariant and essentially contain special relativity.

Given J_μ, we can in principle solve (17) to obtain A_μ, from which we can determine $\vec{E}$ and $\vec{B}$.

Moving between $(t, \vec{x})$ and $(\omega, \vec{k})$

An important skill for undergraduates to master is to move with ease between "position space" and[5] "momentum space." Thus, the Maxwell equation $\vec{\nabla} \times \vec{E} + \frac{1}{c}\frac{\partial \vec{B}}{\partial t} = 0$ becomes $\omega\vec{B} = c\vec{k} \times \vec{E}$, which allows us to immediately relate the magnetic field to the electric field in a plane wave. Similarly, the solution to (17) follows immediately in momentum space: $(c^2\vec{k}^2 - \omega^2)A_\mu(\omega, \vec{k}) = cJ_\mu(\omega, \vec{k})$.

A vector identity

For completeness, let us derive the vector identity we used to obtain (16).

I still remember, the first day of my undergraduate course on electromagnetism, the professor writing an identity for $\vec{V} \equiv \vec{A} \times (\vec{B} \times \vec{C})$ on the board and telling us to memorize it. Gentle readers, there is really no need for brute memorization.

Since the vector $\vec{V}$ is perpendicular to $\vec{B} \times \vec{C}$ and since $\vec{B} \times \vec{C}$ is perpendicular to the plane spanned by $\vec{B}$ and $\vec{C}$, the vector $\vec{V}$ must lie in the plane spanned by

$\vec{B}$ and $\vec{C}$. Thus, it can be written as $\vec{V} = b\vec{B} + c\vec{C}$. The coefficients can be fixed by a scaling argument.

Scale $\vec{A} \to \alpha\vec{A}$, $\vec{B} \to \beta\vec{B}$, $\vec{C} \to \gamma\vec{C}$, so that $\vec{V} \to \alpha\beta\gamma\vec{V}$. Then $b \to \alpha\gamma b$, thus implying that $b = \vec{A} \cdot \vec{C}$ with the overall coefficient fixed* by $\alpha \to \infty$, for example. Appealing to antisymmetry under $\vec{B} \leftrightarrow \vec{C}$, we obtain[6]

$$\vec{A} \times (\vec{B} \times \vec{C}) = (\vec{A} \cdot \vec{C})\vec{B} - (\vec{A} \cdot \vec{B})\vec{C} \tag{18}$$

(Note also that the scaling argument is equivalent to dimensional analysis; we simply assign different dimensions to $\vec{A}$, $\vec{B}$, $\vec{C}$.)

The identity $\vec{\nabla} \times (\vec{\nabla} \times \vec{A}) = \vec{\nabla}(\vec{\nabla} \cdot \vec{A}) - \nabla^2\vec{A}$ we needed for Maxwell's equations follows, provided we take due care of the fact that $\vec{\nabla}$ and $\vec{A}$ do not commute.

Fine structure constant

In this appendix, to focus on the gauge structure of electromagnetism, I chose to absorb e, as mentioned in the beginning. Now, as promised, we discuss the numerical value of e.

The Coulomb potential between two electrons measures the strength of the electromagnetic coupling e and implies that e^2 has dimension of energy times length: $[e^2] = EL$, as was emphasized in chapter II.1. Hence the usual freshman physics definition of charge involves the force between two previously specified charged spheres separated by a conventionally chosen distance.

Thus far, no quantum in sight. Now enters the quantum. From Planck's postulate that a photon carries energy $\hbar\omega$, we see that $[\hbar] = ET$ and hence, $[\hbar c] = EL$: $\hbar c$ has the same dimension as e^2.

Ta da! Nature provides us with a dimensionless measure of the strength of the electromagnetic interaction, namely, $e^2/\hbar c$. Happily, in our quantum world, we can measure the strength of electromagnetism without relying on some arbitrary human convention.[†] In quantum physics,[7] the fine structure constant $\alpha \simeq 1/137$ is defined, in the two different systems of units mentioned here, as

$$\alpha = \frac{e^2}{4\pi\hbar c} \qquad \text{Heaviside-Lorentz} \tag{19}$$

and as

$$\alpha = \frac{e^2}{\hbar c} \qquad \text{Gaussian} \tag{20}$$

In quantum field theory and particle physics, Heaviside-Lorentz units are used[8] (as was mentioned) with $\hbar$ and c both set equal to 1. In that case,

$$e = \sqrt{4\pi\alpha} \simeq 0.303 \tag{21}$$

---

*It can also easily be fixed by taking a special case, such as $\vec{B} \perp \vec{C}$ and $\vec{A} = \vec{B}$.

[†]Yet another reason some theoretical physicists feel more comfortable in the quantum world.

Similarly, in Gaussian units,

$$e = \sqrt{\alpha} \simeq 0.0854 \tag{22}$$

Evidently, before blindly plugging e into any given formula, you have to ascertain the system of units that the formula was derived in, whether Heaviside-Lorentz or Gaussian.[9] For instance, when we evaluated the Thomson cross section numerically in chapter II.3, we noted that the relevant formula was derived in Gaussian units.*

---

*Which were certainly more prevalent before the Feynman rules were codified.

Notes

[1] J. Jackson, *Classical Electrodynamics*, p. 618, table 2. In view of all the pain and suffering physics majors go through coping with units in electromagnetism, it may be helpful to recall the historical necessity of these different units. The crucial point is that measuring the force between two current-carrying wires is much easier than measuring the force between two charges. Hence, experimentalists and engineers favor SI, with its amperes and statcoulombs. In contrast, at the level of fundamental physics, a vacuum in quantum field theory is a vacuum is a vacuum, and notions like ϵ_0 and μ_0 are utterly alien. A particularly lucid discussion may be found in Appendix A of the text on quantum mechanics by Sakurai and Napolitano, *Modern Quantum Mechanics*, p. 519.

[2] Strictly speaking, the right side of (3) should be $\vec{0} = (0, 0, 0)$, since these appendices fly by day, in contrast to the main text.

[3] I purposely said this in a somewhat convoluted way to avoid digressing into the dimension of e, which is discussed in chapter II.1.

[4] See *QFT Nut*, chapters II.7 and III.4, particularly the footnote on p. 144. For the analogous condition for gravity, see *GNut*, p. 564.

[5] Called this in certain circles since $(\hbar\omega, \hbar\vec{k})$ denotes the 4-momentum and since $\hbar$ is routinely set to 1.

[6] It is also worth remarking that (18) also follows easily from this identity satisfied by the products of two antisymmetric symbols (which I am fairly sure I learned in the same course): $\varepsilon_{ijk}\varepsilon_{klm} = \delta_{il}\delta_{jm} - \delta_{im}\delta_{jl}$. Thus, the ith component of $\vec{V}$ equals $V_i = \varepsilon_{ijk}A_j(\vec{B} \times \vec{C})_k = \varepsilon_{ijk}A_j\varepsilon_{klm}B_lC_m = (\vec{A} \cdot \vec{C})B_i - (\vec{A} \cdot \vec{B})C_i$. It is a matter of personal taste, but I generally prefer using indices.

[7] See, for example, chapters I.3 and III.5.

[8] Particle theorists do not want the 4π in the Lagrangian, which would then gum up the Feynman rules as in, say, the calculation of the anomalous magnetic moment of the electron. See, for example, *QFT Nut*, p. 198.

[9] Or some other arbitrary system of units, which I either was never taught, or have mercifully forgotten.

Appendix N

Newton's two superb theorems and the
second square root alert

Beyond the usual square root alert

I sketch Newton's two superb theorems, referred to in chapter VI.3.

We all know about the square root alert: Be sure to look at the thingy inside a square root to check the possibility of its going negative. Call this the first square root alert. Here, a more subtle square root alert comes into play, as we will see.

To calculate the gravitational potential at $\vec{r}$ due to a spherically symmetric body, we integrate $1/|\vec{r} - \vec{r}'| = 1/\sqrt{r^2 + r'^2 - 2rr'\cos\theta}$ with $\vec{r}'$ ranging over the body. See figure 1 for the notation.

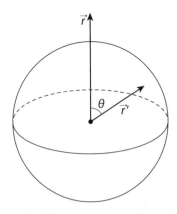

FIGURE 1. Gravitational potential at $\vec{r}$ due to a spherically symmetric body.

The integral over φ is trivial, so

$$\int d^3\vec{r}'\rho(r')\frac{1}{|\vec{r}-\vec{r}'|}=2\pi\int_0^\infty dr'(r')^2\rho(r')\int_{-1}^{+1}d\cos\theta\frac{1}{\sqrt{r^2+r'^2-2rr'\cos\theta}} \tag{1}$$

In fact, the integral over $\vec{r}'$ is also not relevant to our discussion here; in particular, we can set $\rho(r')\propto\delta(r'-s)$, so that the body is actually a shell of radius s, and r can be larger or smaller than s. A general spherically symmetric body can always be built out of shells (which is what the integration over r' in (1) means).

Thus, we need only focus on the integral over θ:

$$\int_{-1}^{+1}d\cos\theta\frac{1}{\sqrt{r^2+s^2-2rs\cos\theta}} \tag{2}$$

Note that at the two limits $\cos\theta=\pm1$, the square root becomes, respectively, $\sqrt{(r+s)^2}=r+s$ and $\sqrt{(r-s)^2}=?$.

The question mark is the cue for you to write down the answer. Here the first square root alert does not come in: in (1), the quantity inside the square root, $(\vec{r}-\vec{r}')^2$, is manifestly positive.

The second square root alert

The subtlety is that the square root itself is manifestly positive, since it started out in life as $|\vec{r}-\vec{r}'|$: The answer equals $r-s$ if $r>s$, but $s-r$ if $r<s$. Clearly, this makes perfect geometric sense. (By the way, the second case here is needed only for the second superb theorem.) Did you get that right?

A second square root alert: A square root can take on different values! Notice that this is not a reminder that a square root can be plus or minus. In this problem, the square root is always positive.

Change the integration variable in (2) to $u=\cos\theta$, and set $a=r^2+s^2$ and $b=2rs$. Then our integral becomes, for $r>s$,

$$\int_{-1}^{+1}\frac{du}{(a-bu)^{\frac{1}{2}}}\propto-\frac{(a-bu)^{\frac{1}{2}}}{b}\Big|_{-1}^{+1}=\frac{(a+b)^{\frac{1}{2}}-(a-b)^{\frac{1}{2}}}{b}=\frac{(r+s)-(r-s)}{2rs}=\frac{1}{r} \tag{3}$$

which is independent of s!

For the case $r<s$, which concerns where hell is, but does not worry about apples, moons, and planets, the integral gives $\big((r+s)-(s-r)\big)/2rs=1/s$. It is independent of the distance r of the observer from the origin!

Outside a spherical shell, the gravitational potential behaves as if the mass of the shell is shrunk to a point at the origin. Inside a spherical shell, the gravitational potential is constant.

Since the gravitational force is given by the derivative of the potential with respect to r, this shows that there is no force. Hence, no leaping flames in hell.

Appendix VdW

Fly by day derivation of van der Waals's law from first principles

We flew by night in chapter V.2 to obtain, using physical arguments and dimensional analysis, the two van der Waals parameters a and b from the intermolecular potential $v(r)$. For the reader who knows statistical mechanics, we now derive the van der Waals law. The reason I include this very fly by day stuff in a fly by night book is because some of the intermediate steps are rather instructive and nice.

Start with the partition function $Z = Tre^{-\beta H}$, where $H = H_0 + \sum_{i<j} v(x_i - x_j)$. Denote by Z_0 the value of Z with the interaction v turned off. Then in semi-classical approximation, we have

$$Z \simeq \frac{1}{h^{3N}} \int d^3p_1 \cdots d^3p_N d^3x_1 \cdots d^3x_N e^{-\beta E(\cdots)}$$

$$= Z_0 \left(\frac{1}{V^N} \int d^3x_1 \cdots d^3x_N e^{-\beta \sum_{i<j} v(x_i - x_j)} \right)$$

$$= Z_0 \left(1 + \frac{1}{V^N} \int d^3x_1 \cdots d^3x_N [e^{-\beta \sum_{i<j} v(x_i - x_j)} - 1] \right) \tag{1}$$

In the second equality, we added and subtracted 1. The point of doing this is that at low densities, we expect that occasionally two molecules will come close to each other, but that three molecules coming together is unlikely. Focus on the square bracket $[\prod_{i<j} e^{-\beta v(x_i - x_j)} - 1]$ in the integrand. When x_i and x_j are far apart, the $v(x_i - x_j)$ vanishes, so that $e^{-\beta v(x_i - x_j)} \simeq 1$. Thus, in the product $\prod_{i<j}$, we can set all but one of the $\simeq N^2/2$ factors to 1.

We might as well call the two molecules that come close $i=1$, $j=2$, and integrate trivially over $d^3x_3 \cdots d^3x_N$ to obtain V^{N-2}. Then the expression in large round parenthesis in (1) becomes

$$(\cdots) = 1 + \frac{V^{N-2}N^2/2}{V^N} \int d^3x_1 d^3x_2 [e^{-\beta v(x_1-x_2)} - 1]$$

$$= 1 + \frac{4\pi N^2}{2V} \int_0^\infty dr r^2 [e^{-\beta v(r)} - 1] \tag{2}$$

with $r = |x_1 - x_2|$.

For the potential sketched in figure 1 of chapter V.2, split the integral $\int_0^\infty [\cdots]$ in (2) into two pieces. First, $\int_0^{r_0} [\cdots] \simeq \int_0^{r_0} [-1] \simeq -r_0^3$, since $v \sim +\infty$ in this integration range. Second, $\int_{r_0}^\infty [\cdots] \simeq \int_{r_0}^\infty dr r^2 [1 - \beta v(r) - 1] \simeq -\frac{1}{T} \int_{r_0}^\infty dr r^2 v(r) \propto -\text{constant}/T$, where we have treated $\beta v(r)$ as small for $r > r_0$.

Thus, $Z = Z_0 \left(1 - \frac{N^2}{V}(B - \frac{A}{T}) + \cdots \right)$, with A, B two constants characteristic of the gas.

Now we merely have to go through the usual thermodynamic steps: from the definition of free energy $-\beta F = \log Z$, we obtain $F \simeq F_0 + \frac{N^2}{V}(BT - A)$, and since $dF = -SdT - PdV$, we determine the pressure:

$$P = -\frac{\partial F}{\partial V}\bigg|_T \simeq nT \left(1 + nB - \frac{nA}{T} + \cdots \right) \tag{3}$$

Comparing with (V.2.3), we see that $a = A$, $b = B$. We have identified the van der Waals parameters with molecular properties.

As a bonus, we find the entropy

$$S = -\frac{\partial F}{\partial T}\bigg|_V \simeq S_0 - NnB + \cdots \tag{4}$$

As expected, the entropy is reduced by the hardcore repulsion. The longer ranged attraction does not do anything to this order.

The moral of the story: adding and subtracting 1 often turns out to be a good move.

Timeline

Galileo Galilei (1564–1642), 78
Johannes Kepler (1571–1630), 59
Henry Power (1623–1668), 45
Robert William Boyle (1627–1691), 64
Robert Hooke (1635–1703), 68
Isaac Newton (1642–1726/27), 85
Edmond Halley (1656–1742), 86
Daniel Gabriel Fahrenheit (1686–1736), 50
Leonhard Euler (1707–1783), 76
John Michell (1724–1793), 69
Henry Cavendish (1731–1810), 79
Charles-Augustin de Coulomb (1736–1806), 70
Joseph-Louis Lagrange (1736–1813), 77
Pierre-Simon, marquis de Laplace (1749–1827), 78
Jean-Baptiste Joseph Fourier (1768–1830), 62
Johann Georg von Soldner (1776–1833), 57
George Green (1793–1841), 48
George Stokes (1819–1903), 84
Ludvig Valentin Lorenz (1829–1891), 62
James Clerk Maxwell (1831–1879), 48
Johannes Diderik van der Waals (1837–1923), 86
Osborne Reynolds (1842–1912), 70
John William Strutt, Lord Rayleigh (1842–1919), 77
Ludwig Eduard Boltzmann (1844–1906), 62
Baron Loránd Eötvös de Vásárosnamény (1848–1919), 71
John Henry Poynting (1852–1914), 62
Hendrik Antoon Lorentz (1853–1928), 75
Johannes Rydberg (1854–1919), 65
Joseph Larmor (1857–1942), 85
Max Karl Ernst Ludwig Planck (1858–1947), 89
Alfred-Marie Liénard (1869–1958), 89
Vesto Melvin Slipher (1875–1969), 94

James Jeans (1877–1946), 69
Albert Einstein (1879–1955), 76
Gunnar Nordström (1881–1923), 42
Max Born (1882–1970), 88
Niels Henrik David Bohr (1885–1962), 77
Erwin Rudolf Josef Alexander Schrödinger (1887–1961), 74
Milton La Salle Humason (1891–1972), 81
Louis Victor Pierre Raymond de Broglie (1892–1987), 95
Satyendra Nath Bose (1894–1974), 80
Wolfgang Ernst Pauli (1900–1958), 58
Enrico Fermi (1901–1954), 53
Werner Karl Heisenberg (1901–1976), 75
Paul Adrien Maurice Dirac (1902–1984), 82
Eugene Wigner (1902–1995), 93
Llewellyn Hilleth Thomas (1903–1992), 89
Matvei Petrovich Bronstein (1906–1938), 32
Lev Davidovich Landau (1908–1968), 60
Hans Heinrich Euler (1909–1941), 32
John Archibald Wheeler (1911–2008), 97
Edward Mills Purcell (1912–1997), 85
Richard Feynman (1918–1988), 70
Julian Schwinger (1918–1994), 76
Murray Gell-Mann (1929–2019), 90
Stephen William Hawking (1942–2018), 76
Jacob David Bekenstein (1947–2015), 68

Note: The final number for each entry lists the approximate age at death. The age at death does not take into account the months of birth and of death.

Solutions to selected exercises

Solutions for Chapter I.1

(1) Since $[a] = L/T^2$, the extra distance covered due to acceleration must be $\sim at^2$ by dimensional analysis. That there are two terms could be argued for by considering the two extreme cases, $v_0 = 0$ or $a = 0$. The factor of $\frac{1}{2}$ is trickier. It follows from the average velocity being half of the initial velocity v_0 and final velocity $v_f = v_0 + at$. In other words, $\Delta x = \frac{1}{2}(v_0 + v_f)t$.

(4) $[k] = [F]/L = (ML/T^2)/L = M/T^2$. Hence the (circular) frequency $\omega \sim \sqrt{k/m}$. As is well known, the coefficient actually equals 1.

Solutions for Chapter I.3

(2) $v \sim e^2/\hbar \sim (e^2/\hbar c)c \sim \alpha c$, so special relativity may be neglected.

(3) In (1), replace e^2 by Ze^2, so that $v \sim Z\alpha c$. Start to worry for $Z \gtrsim 10$ (but depending on who you are).

Solutions for Chapter I.5

(1) $\simeq 50{,}000$ years. See D. Maoz, *Astrophysics in a Nutshell*, Princeton University Press, 2007, p. 39.

Solutions for Chapter I.6

(1) First, note that pressure has dimension of force per unit area: $[P] = (ML/T^2)/L^2 = M/(LT^2)$. Then write $P \sim E^a \rho_0^b t^c$. Matching powers of M, L, and T, we have three

equations for three unknowns: $1 = a + b$, $-1 = 2a - 3b$, $-2 = -2a + c$. We obtain $a = \frac{2}{5}$, $b = \frac{3}{5}$, $c = -\frac{6}{5}$ and hence the stated result.

Interlude: Math medley 1

(3) R and A are both cyclic invariants. Note that R does not vanish as $a + b - c \to 0$, but in fact tends to ∞. Start with abc. Looking at a triangle with $b = c$, $a \to 0$ so that $R \to b/2$, we see that the denominator must vanish like a. This also fixes the coefficient.

Solutions for Chapter II.2

(1) The ω^4 has to be folded into the solar spectrum (see chapter III.5) and the photo sensitivity of the human eye.

Solutions for Chapter II.3

(1) In the Gaussian units used here, $e^2 = \hbar c \alpha \simeq \hbar c / 137$ (see appendix M). From the table of fundamental numbers, $\hbar / m_e c^2 \simeq 3.85 \times 10^{-11}$ cm. Thus, $\sigma_{\text{Thomson}} \simeq (8\pi/3)(3.85 \times 10^{-11}/137)^2$ cm$^2 \simeq 6.6 \times 10^{-25}$ cm^2.

(2) The Klein-Nishina cross section looks like

$$\sigma_{\text{KN}} \sim \sigma_{\text{Thomson}} \left(\frac{mc^2}{\hbar\omega} \right) \left(\log \frac{\hbar\omega}{mc^2} + \cdots \right)$$

The decrease at high frequencies $\frac{mc^2}{\hbar\omega} \to 0$ is expected, since such processes as $\gamma + e^- \to \gamma + e^- + e^+ + e^-$, as mentioned in the text, start to kick in. Readers with a glancing familiarity with quantum field theory would know that the logarithm pops up all over the place. Thus, the high frequency behavior $f(x) \sim x \log x$ as $x \to 0$ would have indeed been my first guess. Numerically, the logarithm does not count for much, and I would have given any student who got the $\frac{mc^2}{\hbar\omega}$ factor full credit. As I've said elsewhere in this book, we are after understanding here, not agreement to n significant digits.

Solutions for Chapter III.1

(2) Define $f(E) \equiv \sqrt{\frac{E_G}{E}} + \frac{E}{T}$. The minimum of $f(E)$ is given by $E_*^{\frac{3}{2}} = \frac{1}{2} E_G^{\frac{1}{2}} T$, thus yielding the stated dependence of E_* on T. The second derivative of $f(E)$ evaluated at E_* gives $1/\Gamma^2$, thus yielding the stated dependence of Γ on T.

(4) We have

$$\frac{d^2u}{dr^2} \left(\frac{m}{\hbar^2}\right)\left(\frac{e^2}{r}\right)u = \varepsilon u$$

Thus, $[\varepsilon] = 1/L^2$ and $[\frac{me^2}{\hbar^2}] = 1/L$. It follows immediately that $\varepsilon \sim (\frac{me^2}{\hbar^2})^2$ and hence $E = (\frac{\hbar^2}{m})\varepsilon \sim \frac{me^4}{\hbar^2}$, the well known Bohr result.

Solutions for Chapter III.2

(2) After cleaning, the first order correction to the energy $\Delta\varepsilon$ is given by $\int_0^\infty dy\, y^4 (y^n + \cdots)^2 e^{-y^2}$ divided by wave function normalization. So $\Delta\varepsilon \propto g \int_0^\infty dy\, y^{2n+4} e^{-y^2} / \int_0^\infty dy\, y^{2n} e^{-y^2}$. The integral in the denominator is easily done by differentiating $\int_0^\infty dy\, e^{-ay^2}$ (which by dimensional analysis is equal to $a^{-\frac{1}{2}}$) with respect to a n times, thus giving* $(2n-1)(2n-3)\cdots 5\cdot 3\cdot 1$, and then setting a to 1. But we don't even need to do this. To obtain the numerator in the expression for $\Delta\varepsilon$, we have to differentiate with respect to a two more times. This brings down two more powers of n, and thus $\Delta\varepsilon \propto n^2$.

To obtain this result without messing with integrals and such, set $y^2 \sim \varepsilon \propto n$; the particle explores the potential out to $y \propto \sqrt{n}$. Hence the perturbing potential $y^4 \propto n^2$.

(4) Write $u = (\rho^n + a\rho^{n-1} + \cdots)e^{-\sqrt{\varepsilon}\rho}$. Then $\frac{d^2u}{d\rho^2} + \frac{1}{\rho}u = \{\varepsilon(\rho^n + a\rho^{n-1} + \cdots) - 2\sqrt{\varepsilon}(n\rho^{n-1} + \cdots) + (\cdots) + (\rho^{n-1} + \cdots)\}e^{-\sqrt{\varepsilon}\rho} = \varepsilon u = \varepsilon(\rho^n + a\rho^{n-1} + \cdots)e^{-\sqrt{\varepsilon}\rho}$. Note that there is no need to differentiate the polynomial twice. Also note that the coefficient of ρ^n was built in to match. Matching the coefficients of ρ^{n-1}, we obtain $\sqrt{\varepsilon} = \frac{1}{2n}$ and the exact answer $\varepsilon = \frac{1}{4n^2}$.

(6) For large x, write the differential equation $x^{\frac{1}{2}}\frac{d^2\zeta}{dx^2} = \zeta^{\frac{3}{2}}$ fly by night style as $\frac{1}{x^{\frac{3}{2}}} \sim \zeta^{\frac{1}{2}}$, so that $\zeta \sim 1/x^3$. (Check that this works.) For $x \to 0$, impose the boundary condition $\zeta(0) = 1$, expand in powers of $x^{\frac{1}{2}}$, and plug in to obtain $\zeta \to 1 - cx + c'x^{\frac{3}{2}}$ with some numerical constants c, c'.

Solutions for Chapter IV.3

(1) Since the only quantity with the dimension of energy is the magnitude of the acceleration, we obtain the stated result. Here is a fly by night argument.[†] Let our accelerated observer carry a detector designed to detect quantum fluctuations in, say, the electromagnetic field. The detector might consist of a quantum mechanical system with

---

*For the exact result, see *QFT Nut*, p. 523.
[†]I heard this argument from Bill Unruh (private communication).

energy levels E_i, $i = 0, 1, 2, \dots$. Every time the electromagnetic field causes a transition from some level i to some level j, the detector would beep. Now if the detector is being carried by a uniformly moving observer, we know that nothing would happen, since by Lorentz invariance, it might as well be sitting at rest. The reason is that a fluctuation that causes a transition from i to j would be quickly followed, in a time of order $\hbar / |E_i - E_j|$ (which we assume to be much shorter than the reaction time of the detector), by a counterfluctuation that would cause a transition from j back to i. But if the detector is being accelerated, then by the time the counterfluctuation comes along, it would be moving at a different velocity from before, that is, its rest frame would differ from what it was before. The electromagnetic field $\vec{E}(t, \vec{x})$ and $\vec{B}(t, \vec{x})$, when Lorentz transformed to the new frame, would be a tiny bit different, which would cause the transition from j back to i to occur at a slightly different rate. As a result of this mismatch, the detector would indicate the presence of a bath of radiation. (What? You're not convinced. Well, I did tell you that the argument was going to be handwaving.)

Solutions for Chapter VI.2

(4) Simply expand the Navier-Stokes equation to leading order as in the preceding exercise. Alternatively, note that viscosity has dimension $[\nu] = L^2 / T$. The only physical quantity we have around that we could use to get rid of the length unit, in order to obtain a rate, is the speed of sound $[c_s] = L/T$. Thus, $[\nu / c_s^2] = T$. But we know the attenuation rate Γ vanishes when ν vanishes. Thus we conclude that

$$\Gamma \sim \frac{\nu \omega^2}{c_s^2}$$

As the frequency ω vanishes, Γ vanishes like ω^2. Or equivalently, use the wavelength to get rid of the length unit L, and then convert to frequency.

Solutions for Chapter VI.3

(2) Let x denote the distance from the midpoint of the tunnel to the train and a the distance from that midpoint to the center of the earth. See the figure. By the Pythagorean theorem, the distance from the train to the center of the earth equals $r = (a^2 + x^2)^{\frac{1}{2}}$. Let M and R denote the mass and radius of the earth, respectively. Then the gravitational force acting on a train of mass m along the tunnel is given by

$$-\left(G\left\{M\left(\frac{r}{R}\right)^3\right\}m\right)\frac{1}{r^2}\left(\frac{x}{r}\right) = -\frac{GMm}{R^3}x$$

I have grouped the various factors for your convenience: the factor $(\frac{r}{R})^3$ follows from the two superb theorems, and the factor $\frac{x}{r}$ is due to decomposing the force along the x-direction.

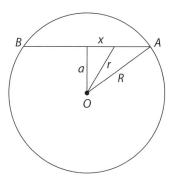

Calculating the transit time.

Newton's other law, $ma = F$, gives us

$$\ddot{x} = -\frac{GMx}{R^3}$$

Plugging in $x(t) = x_0 \cos \omega t$ gives us $\omega^2 = GM/R^3 = 4\pi G\rho/3$ and thus a travel time of

$$T = \frac{\pi}{\omega} = \pi\sqrt{\frac{R^3}{GM}} = \pi\sqrt{\frac{R}{g}} \simeq \pi\sqrt{\frac{6.4 \times 10^3 \text{ km}}{10 \text{ m/sec}^2}} \simeq 40 \text{ min}$$

Note that we have traded in M for g. Indeed, the distance a drops out. The dull function hypothesis holds resoundingly.

Solutions for Chapter VII.1

(2) Use $l = 1/n\sigma$. Recall the factoid that solar density is about the same as the density of water: $\rho \sim 1 \text{ g/cm}^3$, or compute $\rho = M_\odot/(4\pi R_\odot^3/3) \simeq 1.4 \text{ g/cm}^3$. Thus the number density of electrons $n \simeq \rho/m_p \sim (1.4/1.6) \times 10^{24}/\text{cm}^3$. Then plug in the Thomson cross section $\sigma_{\text{Thomson}} \simeq 6.6 \times 10^{-25} \text{ cm}^2$ from chapter II.3 to obtain $l \sim 1 \text{ cm}$. The actual mean free path is about an order of magnitude smaller, $\sim 1 \text{ mm}$, because the sun does not have uniform density.

Solutions for Chapter VII.3

(1) Plugging this equation of state into (5) to obtain $d(\rho a^3) + w\rho da^3 = 0$, we find $\rho \propto a^{-3(1+w)}$, and hence $a \propto t^{\frac{2}{3(1+w)}}$ by (3), thus recovering our previous results as special cases.

(3) Applying dimensional analysis to (6), we obtain $\frac{1}{t^2} \sim G\rho$, which is just what we found in chapter I.2.

Solutions for Chapter VII.4

(1) Use dimensional analysis: $[E] = ML^2/T^2$ and $[J] = M(L/T)L = [E]T$. Thus, we obtain the rate of angular momentum loss $\frac{dJ}{dt} \sim \frac{d\mathcal{E}}{dt} \times$ a characteristic time scale $\sim \frac{1}{\omega}\frac{d\mathcal{E}}{dt}$. If you want to push it further (which I think is a bit silly), you could of course invoke Kepler's law (see (7) and chapter I.2) to eliminate $\omega \sim \sqrt{Gm/r^3}$ and plug in $\frac{d\mathcal{E}}{dt} \sim \frac{G^4 m^5}{c^5 r^5}$ from (15) to obtain $\frac{dJ}{dt} \sim \frac{G^{\frac{7}{2}} m^{\frac{9}{2}}}{c^5 r^{\frac{7}{2}}}$. For your reference, the "exact" formula is

$$\frac{dJ}{dt} = \frac{32}{5}\frac{G^{\frac{7}{2}}}{c^5 r^{\frac{7}{2}}}(m_1 m_2)^2 (m_1 + m_2)^{\frac{1}{2}}$$

There are people walking around who have memorized this formula and marvel at the strange power $\frac{7}{2}$ without understanding that it just comes from taking the square root of Kepler's law. Better to fly by night first and then study the exact fly by day derivation. Just an opinion.

(2) Roughly, $\mathcal{E} \sim -Gm^2/r$ is the same order of magnitude as the potential energy. Plugging this into (15) and (being sloppy about signs when we can afford to be) we obtain $\frac{d\mathcal{E}}{dt} \sim (Gm^2/r^2)\frac{dr}{dt}$, and hence $\frac{dr}{dt} \sim \frac{G^3 m^3}{c^5 r^3}$. Solving the differential equation by canceling the ds (as we have been doing since chapter I.1), we find the characteristic time scale

$$t \sim \frac{c^5 r^4}{G^3 m^3}$$

Solutions for Chapter VIII.2

(1) The period of a tsunami is given by $2\pi/T = \omega = \sqrt{ghk} \sim (720$ km/hour$)$ $(2\pi/10^2$ km$)$ giving $T \sim 10$ minutes. Let's say the wave has amplitude ~ 1 m. A modern ocean-going ship going up and down through a meter or so in 10 minutes would be hardly something to write home about.

Solutions for Chapter VIII.5

(3) By Archimedes's law, $v_t \sim \left(\frac{\rho_b}{\rho_f} - 1\right)\left(\frac{ga^2}{v}\right)$. For plankton, this almost vanishes, but still they sink steadily. Wave action causing turbulent mixing could work both ways, bringing some plankton up while sweeping others down. Once a plankter falls beneath the zone of sunlight, it dies. Thus, according to Denny (*Air and Water*, pp. 121–122), it is imperative for a plankter to reproduce rapidly before eventual death.

(4) $(4\pi/3)/6\pi = 2/9$. No π! This is what I mean by having a sense of where the πs go.

Solutions for Chapter IX.1

(1) The process involves 4 photons and thus corresponds to $\mathcal{L} \sim KF^4$ with a constant K. Comparing this term with F^2, we see that $[K] = 1/M^4$. The process is proportional to e^4, since 4 QED vertices are involved. Therefore, $K \sim e^4/m^4$, with m the mass of the electron.

Appendix Del

(2) You can evaluate this integral directly. (For help, see *QFT Nut*, p. 31.) For a more elegant approach, consider

$$(-\nabla^2 + m^2)\left(\frac{e^{-mr}}{4\pi r}\right) = \int_{-\infty}^{\infty} \frac{d^3k}{(2\pi)^3} \frac{(-\vec{\nabla}^2 + m^2)e^{i\vec{k}\cdot\vec{x}}}{\vec{k}^2 + m^2} = \int_{-\infty}^{\infty} \frac{d^3k}{(2\pi)^3} e^{i\vec{k}\cdot\vec{x}} = \delta^3(\vec{x})$$

The identity in (4) is then a special case of this more general result.

Appendix L

(1) $u = \sin kr = (e^{ikr} - e^{-ikr})/2i$ represents an incoming wave bouncing off a brick wall with a 180° phase change. Similarly, ψ equals the superposition with a minus sign of an incoming and an outgoing spherical wave $e^{\pm ikr}/r$.

Suggested reading

I had some trouble compiling a bibliography for this book due to its very nature. For my textbook on gravity, for instance, I simply list some well known textbooks on Einstein gravity. But for fly by night physics, standard textbooks do not exist. I am writing from what I know, what I have read and learned about over the decades. No doubt I have absorbed stuff from many different sources. As mentioned in the preface, I did flip through some of the existing books on back of the envelope physics but confess that I am not ecstatic about all of them. I also tried not to read them in detail, lest they influence me unduly. In contrast, when I wrote about quantum field theory or Einstein gravity, of course I had been influenced (and will be influenced) by many of the standard texts.

Books by other authors

T. P. Cheng, *Einstein's Physics: Atoms, Quanta, and Relativity—Derived, Explained, and Appraised*, Oxford University Press, 2013.

D. Clayton, *Principles of Stellar Evolution and Nucleosynthesis*, University of Chicago Press, 1983.

E. Commins and P. Bucksbaum, *Weak Interactions of Leptons and Quarks*, Cambridge University Press, 1983.

M. Denny, *Air and Water*, Princeton University Press, 1995.

A. Garg, *Electromagnetism in a Nutshell*, Princeton University Press, 2012.

J. Jackson, *Classical Electrodynamics*, Wiley, 1962.

L. D. Landau and E. M. Lifshitz, *Fluid Mechanics*, Pergamon, 1959.

D. Maoz, *Astrophysics in a Nutshell*, Princeton University Press, 2007.

J. J. Sakurai and J. Napolitano, *Modern Quantum Mechanics*, Cambridge University Press, 2017.

J. Trefil, *Introduction to the Physics of Fluids and Solids*, Pergamon, 1975.

S. Weinberg, *Gravitation and Cosmology*, Wiley, 1972.

A. Zangwill, *Modern Electrodynamics*, Cambridge University Press, 2012.

Books by the author

In addition to the books listed above, I refer quite often, naturally, to the three textbooks and the three popular books I have written according to the following abbreviations: *QFT Nut, GNut, Group Nut, Fearful, Toy, G.*

Quantum Field Theory in a Nutshell, Princeton University Press, 2003, 2010.
Einstein Gravity in a Nutshell, Princeton University Press, 2013.
Group Theory in a Nutshell for Physicists, Princeton University Press, 2016.
Fearful Symmetry: The Search for Beauty in Modern Physics, Macmillan 1986; Princeton University Press, 2016.
An Old Man's Toy: Gravity at Work and Play in Einstein's Universe, Macmillan, 1989; retitled as *Einstein's Universe: Gravity at Work and Play*, Oxford University Press, 2001.
Unity of Forces in the Universe, World Scientific, 1982.
On Gravity: A Brief Tour of a Weighty Subject, Princeton University Press, 2018.

Index